# GERMAN RIGID AIRSHIPS

ADMIRALTY WAR STAFF
INTELLIGENCE DIVISION
FEBRUARY 1917

The Naval & Military Press Ltd

*Published by the*
The Naval & Military Press
*in association with the Royal Armouries*

© 2008

Unit 10 Ridgewood Industrial Park,
Uckfield, East Sussex, TN22 5QE
Tel: +44 (0) 1825 749494
Fax: +44 (0) 1825 765701

MILITARY HISTORY AT YOUR FINGERTIPS
www.naval-military-press.com

ONLINE GENEALOGY RESEARCH
www.military-genealogy.com

ONLINE MILITARY CARTOGRAPHY
www.militarymaproom.com

ROYAL ARMOURIES

The Library & Archives Department at the Royal Armouries Museum, Leeds, specialises in the history and development of armour and weapons from earliest times to the present day. Material relating to the development of artillery and modern fortifications is held at the Royal Armouries Museum, Fort Nelson.

For further information contact:
Royal Armouries Museum, Library, Armouries Drive,
Leeds, West Yorkshire LS10 1LT
Royal Armouries, Library, Fort Nelson, Down End Road, Fareham PO17 6AN

Or visit the Museum's website at
www.armouries.org.uk

*In reprinting in facsimile from the original, any imperfections are inevitably reproduced and the quality may fall short of modern type and cartographic standards.*

## PREFACE.

The object of this work is primarily to give the more general details of the construction and fittings of L 33 as being an example of the so-called "Super-Zeppelin." At the same time it was thought advisable to extend the scope of the book and to make comparisons, where needful, between the various types of Schütte-Lanz and Zeppelin practice; to include the available information on personnel and training of crews, and on any other matters closely relating to the handling and navigation of German rigid airships.

The work is the result of close co-operation between this Division and the Technical Departments of the Director of Air Services, Superintendent of Compasses, Director of Naval Construction, Engineer-in-Chief, Superintendent Electrical Engineer, and the Intelligence Section of G.H.Q., Home Forces, and is principally based on the actual examination of parts of the following airships: L 15, L 31, L 32, L 33, L 34, LZ 77, LZ 85, and SL 11.

The majority of the photographs have been taken by the Photographic Department of the R.N.A.S.

C.B. 1265.
**Plate 1.**
To face Page 3.

L 31 (sister ship to L 33) destroyed at Potters Bar, Oct. 1st, 1916.

Ordnance Survey, February, 1917.

# CONTENTS.

| | PAGE |
|---|---|
| CHAPTER I.—Development of German Rigid Airships | 4 |

Various Types; Their Characteristics; Details of Dimensions, Armament, &c.

CHAPTER II.—Training and Personnel - - - - - - - - - 6

German Naval and Military Airship Services before and since the Outbreak of War; Formation, Training, Number, and Disposition of Crews; Maintenance and Landing Parties.

CHAPTER III.—Construction - - - - - - - - - - 13

Type to which L 33 belongs; Capacity and Dimensions of L 33; General Arrangement of Hull, Keel, Cars, and Gasbags; Comparison with Previous Classes; Form; Wiring; Shear Ropes; Fins, Rudders, Elevators, and Controls; Keel Fittings; Water-Ballast Bag Arrangements; Bomb Stowage; Petrol Tanks and Piping; Ventilation; Access Hatches; Crew Space; Mooring Points; Suspension Strops; Gun Platforms; Tail Structure; Cars; Control Position; Wireless Telegraphy Cabin; Dimensions of Cars; Propeller Shaft Supports; Communications; Analyses of Metals; Tensile Tests; Observation Car used by German Rigids; Details of Schütte-Lanz Type Airships.

CHAPTER IV.—Machinery and Propellers - - - - - - - - 34

Motive Power of L 33; Details of Propellers; Description of Engines; Water Circulation; Connecting Rod Cooling Arrangements; Carburettors; Petrol Supply System and Storage; Lubricating System; Governor Control; Starting Arrangements; Magnetos; Transmission Gear; Engine Seatings; Petrol Filling Arrangements; Telegraphs.

CHAPTER V.—Fabric and Gas Valves - - - - - - - - - 46

Gasbags of L 33; Materials; Seams; Wire Supports and Nets; Filling; Gas; Temperature and Condensation of Moisture; Leakage of Gas and Ventilation; Repairs during Flight; Care of Gasbags; Details of Outer Covers of L 33 and how attached; Fabric of Outer Cover; Proofing and Dope; Lacing; Water Ballast Bags; Sundry Fabrics.

CHAPTER VI.—Bomb-dropping Gear and Electrical Installation - - - - 52

Bomb-dropping Positions of L 33; Releases; Electric Installation for Bomb Releases; Release Gear for Incendiary Bombs; Comparisons with Earlier Dropping Gear; Bomb-dropping Installation of LZ 77; Electric Lighting and Telephones of L 33.

CHAPTER VII.—Wireless Telegraphy Apparatus - - - - - - - 58

Forward Car Alternator in L 33; Results of Tests; Starboard, Amidship, and After Car Alternators; W/T Receiver; Aerial; Deck Insulator; Aerial Winding Gear; Wireless Cabin; Earth; General Arrangements.

CHAPTER VIII.—Bombs, Flares, Guns, and Ammunition - - - - - - 62

Details of Bombs carried by German Naval and Military Rigid Airships; High Explosive and Incendiary Bombs; Fuzes; Water Flares; Machine Guns, Rifles, and Ammunition.

CHAPTER IX.—Bomb and Machine Gun Sights - - - - - - - 72

Details of Bomb Sights; Variation in German Bomb Sights; Method of Use in Attack; Sights for Machine Guns.

CHAPTER X.—Navigation - - - - - - - - - - - 76

Central Position; Officers and Crew; Controls; Navigating Instruments generally; Calculating Ground Speed and Fixing Positions; Directional Wireless; Weather Reports; Courses of German Rigid Airships; Making a Landfall; Cruising and Full Speed; Heights attained; Airship Lights; Lightships; Position at Night; Course during Bomb Dropping; Course for Attack on Midlands; How German Rigids land by Day and Night; Mooring on Water; Sheds.

CHAPTER XI.—Instruments and Fittings - - - - - - - - 82

Compasses; Compass Position; Bearing Plate; Aneroid; Statascope; Electric Thermometers; Engine Evolution Indicator; Petrol Gauge of Storage Tanks; Fire Extinguisher; Life Belt.

CHAPTER XII.—Weather - - - - - - - - - - - 89

Summary of Weather on Nights of Raids to September 1915; Details of Weather during Raids to October 1916; Conclusions to be drawn.

APPENDIX A.—Production of Hydrogen in Germany - - - - - - 94

APPENDIX B.—Statistical Details of Machinery, &c. - - - - - - 95

APPENDIX C.—Further Statistical Details of Machinery, &c. - - - - 109

GENERAL INDEX - - - - - - - - - - - - 113

INDEX TO PLATES - - - - - - - - - - - - 117

# CHAPTER I.—DEVELOPMENT OF GERMAN RIGID AIRSHIPS.

**Zeppelin Types.**

Zeppelin airships can be generally classed under four types (*see* Plates 2, 3, 4, and 5).

There were 25 Zeppelin airships constructed between 1900, when the first experimental airship was completed, and the beginning of the war. Of this number 24 were of the 1st type and one was of the 2nd type. The 1st type of airship continued to be constructed after the outbreak of war up to the beginning of 1915, when the construction of a 3rd type commenced. This type of airship remained the standard type until the beginning of 1916 when the construction of L 30—the first of the 4th type —commenced. L 30 was completed in May 1916.

*1st Type.*—Includes all airships from 1900 up to the end of 1914 (with the exception under 2nd type). Characteristics:—

    (1) External keel gangway.
    (2) Two cars.
    (3) Four wing propellers—varying from two to four blades each.
    (4) Long parallel body with bluff bow and stern.
    (5) "Box" rudders and elevators.

*2nd Type.*—The naval airship L 2. (L 2 carried out its first flight on the 20th September 1913, and was destroyed in mid-air by fire when carrying out a trial flight over Berlin on the 17th October 1913. The cause of the fire was never ascertained.) Characteristics:—

    (1) Internal keel gangway.
    (2) Three cars—the foremost being for control only.
    (3) Four wing propellers—with four blades each.
    (4) Long parallel body; bluff bow; stern finer than first type.
    (5) "Box" rudders and elevators.

*3rd Type.*—The 1915 type.* Characteristics:—

    (1) Internal keel gangway.
    (2) Three cars—the foremost for control only. The two foremost cars close together and connected by a canvas joint, so as to look like a single car.
    (3) Four motors and four propellers of two blades each— one motor in the fore car driving a "pusher" propeller, three motors in the rear car driving two wing propellers and one "pusher" propeller.
    (4) Not quite so long a parallel body, bluff bow, and tapering stern.
    (5) Balanced "monoplane" rudders and elevators.

*4th Type.*—The 1916 type. This type, beginning with L 30, is the so-called super-Zeppelin. Characteristics:—

    (1) Internal keel gangway.
    (2) Five cars—two forward (combined as in third type), one aft, and two amidships abreast.
    (3) Six motors and six propellers of two blades each. The after one of the fore car and the side cars each have one motor driving direct a "pusher" propeller. The rear car has three motors, two of which, set side by side, drive the two wing propellers, while the third motor, set aft, drives direct a "pusher" propeller.
    (4) Short parallel body; long rounded bow; long tapering stern.
    (5) Type of girders much altered.
    (6) Balanced "monoplane" rudders and elevators.

---

\* Earlier airships of this type were built on slightly finer lines than the later ones.

Table showing the Development of Zeppelin Types.

(The figures given in columns 2, 3, 4, 8, and 11 must be regarded as approximate only.)

| Year of Completion. | Capacity in Cubic Feet. | Length in Feet. | Maximum Diameter in Feet. | No. of Cars. | Motors. | | | Propellers. | | Maximum Speed in Statute Miles per Hour. | Type. |
|---|---|---|---|---|---|---|---|---|---|---|---|
| | | | | | Type. | No. of. | H.P. of each Motor. | Wing. | "Pusher." | | |
| 1. | 2. | 3. | 4. | 5. | 6. | 7. | 8. | 9. | 10. | 11. | 12. |
| 1900 | 399,000 | 420 | 38·3 | 2 | Daimler | 2 | 16 | 4—3-bladed aluminium propellers. | — | 16·8 | 1st |
| 1905 | 399,000 | 420 | 38·3 | 2 | ,, | 2 | 85 | 4—3-bladed aluminium and steel propellers. | — | 24·6 | ,, |
| 1908 | 431,000 | 446 | 38·3 | 2 | ,, | 2 | 85 | ,, | — | 33·5 | ,, |
| 1908 | 530,000 | 446 | 42·6 | 2 | ,, | 2 | 105 | ,, | — | 28 | ,, |
| 1909 | 530,000 | 446 | 42·6 | 2 | ,, | 2 | 105 | ,, | — | 29 | ,, |
| 1910 | 565,000 | 472 | 42·6 | 2 | 2 Daimler. 1 Maybach. | 3 | 2 Daimler 115 each, 1 Maybach 140. | *2—2-bladed and 2—4-bladed aluminium and steel propellers. | — | 34 | ,, |
| 1910-11 | 680,000 | 485 | 45·9 | 2 | Daimler | 3 | 125 | ,, | — | 35·8 | ,, |
| 1911 | 628,000 | 460 | 45·9 | 2 | Maybach | 3 | 150 | ,, | — | 44 | ,, |
| 1912 | 680,000 | 485 | 45·9 | 2 | ,, | 3 | 150 | ,, | — | 47 | ,, |
| †1912 | 793,000 | 517 | 48·5 | 2 | ,, | 3 | 180 | ,, | — | 47 | ,, |
| 1913 | 734,000 | 489 | 48·5 | 2 | ,, | 3 | 180 | ,, | — | 47 | ,, |
| ‡1913 | 918,000 | 525 | 54 | 3 | ,, | 4 | 200 | 4—4-bladed wooden propellers. | — | 49 | 2nd |
| 1913-14 | 777,000 | 512 | 48·5 | 2 | ,, | 3 | 180 | 2—2-bladed and 2—4-bladed wooden propellers. | — | 47 | 1st |
| ¶1914 | 918,000 | 525 | 54 | 2 | ,, | 4 | 200 | 4—2-bladed wooden propellers. | — | 49 | 1st |
| 1915 | 1,059,500 | 525 | 60·7 | 3** | ,, | 4 | 3 of 180 each, 1 of 240. | 2—2-bladed wooden propellers. | 2—2-bladed wooden propellers. | 50-55‖ | 3rd |
| 1915 | 1,130,000 | 561 | 60·7 | 3** | ,, | 4 | 220 | ,, | ,, | 50-55‖ | ,, |
| 1916 | 2,000,000 | 643 | 78 | 5** | ,, | 6 | 240 | 2—2-bladed wooden propellers. | 4—2-bladed wooden propellers. | 60‖ | 4th |

\* The 2-bladed propellers were driven from the fore gondola and 4-bladed propellers from the rear gondola.
† The first naval airship L 1.    ‖ Probably slightly under-estimated.
‡ The second airship L 2.    \*\* The two foremost cars are merged into one.
¶ The third naval airship L 3.

### Armament.

Up to 1913 the armament against aircraft consisted of machine guns mounted on pivots fixed to the sides of the cars. For some time previous to 1913 there had been a climbing tube leading to the top of the envelope. This was originally used for enabling sights to be taken, and for observation purposes. Since 1913 it has been used for obtaining access to a duralumin-sheeted gun platform on the fore part of the envelope. On this platform are mounted two machine guns (see photograph of L 13, Plate 4). In the 1916 type Zeppelin ship an additional gun platform has been let into the tail abaft the vertical rudders. It is built into the framework so that the machine gun is clear of the envelope, while the floor of the platform is some 3 feet below the upper surface.

### Development on Schütte-Lanz Principles.

Until the end of 1914 Zeppelin airships had wing propellers only. In the 1915 and 1916 types direct driven " pusher " propellers, in accordance with the Schütte-Lanz principle, were introduced in conjunction with wing propellers. In addition, in the 1916 type two cars abreast were introduced, also in accordance with the Schütte-Lanz principle. Thus the 1916 type Zeppelin airship is in general appearance similar to the 1915 and 1916 type Schütte-Lanz airships, except that it is much larger, and has two wing propellers fitted to the rear car.

N.B.—(1) Compare the photograph of L 30 (see Plate 5) with that of SL 2 —1914 type—(see Plate 6). The only apparent difference between SL 2 and the 1915 and 1916 Schütte-Lanz types is that in the latter types the two foremost cars have been joined together.

(2) The following table shows the general dimensions of the Schütte-Lanz types (see Plates 6 and 7) for comparison with Zeppelin types.

The figures given in columns 2, 3, 4, 8, and 11 must be regarded as approximate only.

| Year of Completion. | Capacity in Cubic Feet. | Length in Feet. | Maximum Diameter in Feet. | No. of Cars. | Motors. | | | Wings. | Propellers. | Maximum Speed in Statute Miles per Hour. | Type. |
| | | | | | Type. | No. of. | H.P. of each Motor. | | "Pusher." | | |
| --- | --- | --- | --- | --- | --- | --- | --- | --- | --- | --- | --- |
| 1. | 2. | 3. | 4. | 5. | 6. | 7. | 8. | 9. | 10. | 11. | 12. |
| 1911* | 706,300 | 426 | 60·4 | 2 | Daimler | 2 | 250 | — | 2—3-bladed steel propellers. | 40 | 1st |
| 1914† | 812,000 | 472 | 60·4 | 5 | Maybach | 4 | 180 | — | 4—2-bladed wooden propellers. | 49·7 | 2nd |
| Early 1915 | 1,059,500 | 541 | 60·4 | 4 | Daimler | 4 | 240 | — | ,, | 49·7 | ,, |
| Late 1915 and 1916 | 1,275,000 | 600 | 66 | 4 | Maybach | 4 | 240 | — | ,, | 55 | 3rd |

\* SL 1—the first Schütte-Lanz airship constructed.
† SL 2—the second Schütte-Lanz airship constructed.

## CHAPTER II.—TRAINING AND PERSONNEL.

### I.—NAVAL AIRSHIP SERVICE.

#### A.—*Before the War.*

It was apparently not until 1910 that the German Naval Authorities began to take a very practical interest in aeronautics; but at the end of that year Kapitän z. S. Lübbert of the Imperial Navy Office was specially detailed to study aeronautical matters. In April 1912 the German Press announced that two naval officers and one naval engineer were undergoing instruction as airship pilots at Frankfurt-on-Main, and that in the summer a large number of engineers and warrant officers of the navigating branch ("Steuerleute") were to go there to become acquainted with the working of Zeppelin airships, though there is no evidence to show whether they were volunteers or men detailed for this service. On the 23rd April 1912 the contract for the first naval airship was made with the Zeppelin Company. Two months later the "three officers and about 12 men," who were to form her crew, were reported to be receiving instruction from the "Delag" Company ("Deutsche Luftschifffahrt Aktien Gesellschaft") at Frankfurt-on-Main.

At the end of June 1912 the passenger airship "Viktoria Luise," belonging to the "Delag" Company, made a very successful trip over the North Sea, with several naval officers, including Kapitän z. S. Lübbert, on board; and soon after this it was reported that the Naval Service would take her over; that she would be stationed at Hamburg, and that two or three officers and 11 petty officers would form her crew for instructional purposes. By the beginning of October 1912 the first naval airship, L 1, was ready to be commissioned. Her crew had completed their instruction at Hamburg-Fuhlsbüttel in the passenger ships "Viktoria Luise" and "Hansa," while a relief crew was about to commence training immediately at Berlin-Johannisthal. The crew of L 1 apparently consisted of a Kapitänleutnant (Lieutenant-Commander) in command, an Oberleutnant z. S. (Lieutenant), one engineer, one "Steuermann" (helmsman), four "Bootsmannsmaate" (boatswain's mates) or "Signalmaate" (yeomen of signals), and six "Maschinistenmaate" (engine room petty officers).

By the German Naval Bills of 1913–1914 the established strength of naval airships was to be 10 (eight in commission and two in reserve), in two squadrons of four each, and in June 1913 it was officially announced that it was proposed to divide the naval aeronautical personnel into two separate sections, (1) the Naval Airship Section, stationed for the moment at Berlin-Johannisthal, and (2) the Naval Flight Sections, stationed at Putzig. The autumn naval appointments for 1913 included the following officers for the Airship Section:—

    1 Korvettenkapitän in Command (Commander).
    4 Kapitänleutnants as Commanding Officers of Airships (Lieutenant Commanders).
    4 Oberleutnants zur See (Lieutenants).
    2 Oberingenieure (Engineer Lieutenants).
    3 Ingenieure (Engineer Sub-Lieutenants).

By September 1913 the second naval airship, L 2, was ready for her trials, and her crew, who had been trained in the "Viktoria Luise" and L 1, were prepared to take her over. She, however, was destroyed on October 17th, during one of her trials while carrying 28 persons, including the commanding officer, the naval engineer, and 13 of her regular crew.

There was now no naval airship in existence, for L 1 had also been wrecked a month earlier. Accordingly new arrangements had to be made to train crews for the succeeding airships. It was therefore decided that during November 1913 the crew of the future L 3 was to be trained in the "Viktoria Luise" at Frankfurt-on-Main, and that of the future L 4 in the "Sachsen" at Dresden. From the 1st December 1913, however, until L 3 was completed, the "Sachsen" was chartered by the navy, and was sent from Dresden to Hamburg-Fuhlsbüttel, where she and the whole of the Naval Airship Section were to be stationed.

The "Sachsen" arrived at Hamburg-Fuhlsbüttel on 8th December 1913, and was employed in training officers and men of the Naval Airship Section until 16th April 1914, during which time the following flights were reported:—

In December 1913, flights on 3 days.
In January 1914, flights on 8 days.
In February 1914, flights on 12 days.
In March 1914, flights on 8 days.
In April 1914, flights on 5 days.

On some of these days long trips were taken, including four to Heligoland and back; on others from one to six short flights were made, the crew being often changed between trips. On 24th February 1914 the "Sachsen" made a night trip to Heligoland; on the 27th a landing was effected after dark, and after a special trip on 2nd March, Korvettenkapitän Strasser obtained his pilot's certificate. In all during this period over 80 separate flights were reported. On 16th April the "Hansa" arrived at Hamburg-Fuhlsbüttel to replace the "Sachsen," which was sent to Potsdam to be overhauled.

On the 11th May 1914 L 3 made her first trial trip, and later in the month arrived at Berlin-Johannisthal. After staying there for a few days she went to Hamburg-Fuhlsbüttel. On 17th July 1914 she went for a 22 hours training trip over the North Sea, returning to Hamburg-Fuhlsbüttel, where she was stationed at the outbreak of war.

B.—*Since the Outbreak of War.**

### Formation of Crews.

At the outbreak of war, as there were probably not more than two trained naval crews in existence, volunteers were called for; but, early in 1915, this system was abandoned, and men are now transferred from the navy as required, those having some previous knowledge of internal combustion engines being primarily selected. The crews of the present naval airships, therefore, are almost entirely composed of naval ratings; thus, of the 21 prisoners taken from L 33, only one man is known to have done no naval service; of the 17 prisoners from L 15, all appear to be naval ratings; and of the seven prisoners rescued from L 7, six were men of the active German Navy, and one belonged to the second ban of the "Seewehr." The few exceptions to this rule appear to be mainly men employed before the war in the Delag Company (Deutsche Luftschifffahrt A-G), who have since been taken into the Naval Service.

The average ages of the crew of L 7, L 15, and L 33 are:—

|  | L 7. | L 15. | L 33. |
| --- | --- | --- | --- |
| Officers | — | 30 | 29 |
| Warrant Officers | 32 | 39 | 40½ |
| Petty Officers and men | 26 | 28 | 30 |

A crew is apparently formed at the beginning of training, and from the time it is so formed it is, in theory, always kept together; but changes are bound to occur, not only from sickness and other unavoidable wastage, but also from additional men being required on transfer to newer and larger ships, and it must inevitably happen that the original crew, as such, soon becomes greatly modified. As a rule, the period

---

* From information supplied by G.H.Q., Home Forces.

of training lasts from six to nine months, during which time the work is intensely specialised. For although every man learns to work the machine guns, and is required to have some slight acquaintance with the use of hydrogen in aeronautics, and sufficient knowledge of motors to enable him to take charge and do simple repairs in case of necessity, almost all his training is in his special job, and his special job only.

**Training.**

At the beginning of the war the preliminary training for the Naval Service used to be given at Leipzig, but in April 1915 the school was transferred to Dresden. The school is divided into two sections, navigational and technical. The length of the course at Dresden before transference to Hamburg depends on the aptitude of the pupils and the demand for crews, and in practice varies very considerably, although in theory it would appear to last about four months. There are said to be as many as 12 to 15 crews being trained at Dresden at the same time.

While at Dresden the crew is trained in practical flying in the old passenger ship "Viktoria Luise," and also, from time to time, in the "Sachsen," the flying consisting of short cruises in the vicinity of the airship shed.

There is a school of tailoring at Seifersdorf, in Silesia, and there is some ground for supposing that the sailmakers receive special instruction there in sewing and stitching.

The second phase of the training takes place at Hamburg–Fuhlsbüttel, where the older airships are used for this purpose, and instructional flights up to 80 miles out to sea are carried out. When a crew has gained sufficient practical experience in one of the older and smaller airships at Hamburg-Fuhlsbüttel (usually in about five months), it takes over one of the newer airships as a trained crew; thus the crew of L 7, which was shot down on May 4th, 1916, was expecting to go to a new ship in the middle of May.

This principle of transferring crews from the school ships to the older training ships and thence by successive steps to the newest ones, as they are completed, seems to be invariably adopted, as the following records show:—

    The crew of L 11 was turned over to L 30 in May 1916.
    The crew of L 13 was turned over to L 31 in July 1916.
    The crew of L 14 was turned over to L 33 in August 1916.
    The crew of L 16 was turned over to L 32 in July 1916.
    The crew of L 18 was turned over to L 21 in January 1916.
    The crew of L 21 was turned over to L 34 in August 1916.
    The crew of L 17 was turned over to L 35 in September 1916.

It is to be noted that the captain and crew when detailed to commission a new airship generally proceed to the building yard some six weeks before the airship is completed.

**Number and Disposition of Crews.**

The size of a crew carried by an airship is naturally influenced by the number of the motors. The complement required for all duties, except that of running the motors, is 10, and this seems to be a constant figure in all airships. This fixed total of 10 is made up of the following officers, warrant officers, petty officers, and men:—

    Commanding officer, usually "kapitänleutnant" (lieutenant-commander).
    Second in command, usually "oberleutnant z. S" or "leutnant z. S." (lieutenant).
    Obermaschinist (chief engine-room warrant officer).
    Obersteuermann (chief navigation warrant officer).
    *Two deck ratings for the elevators.
    *Two deck ratings for the rudders.
    Funkentelegraphistenmaat (wireless petty officer, 2nd class).
    Segelmacher (sailmaker).

While it would appear that this complement of 10 is the standard one for all airships in the naval service, it will be found to vary slightly in the different ranks in different ships.

The number of men carried in each airship, in addition to this fixed nucleus of 10, apparently depends directly on the number of the motors, there being always

---

* These ratings appear to be either "oberbootsmannsmaat" (chief boatswain's mate), "bootsmannsmaat" (boatswain's mate), "obersignalmaat" (chief yeoman of signals), or "signalmaat" (second yeoman of signals).

two men (one and a relief) for each motor. Thus for the airships of the 1914 type, fitted with three motors, the total crew consisted of 10 + 6 = 16 men; for L. 15 (1915 type with four motors) of 10 + 8 = 18 men; for L. 33 (1916 type with six motors) of 10 + 12 = 22 men.

As a matter of fact L 7 (1914 type with three motors) was actually carrying 18 men at the time she was shot down, but two of these were "maschinisten anwärter" (engine-room probationers), who were probably being carried for training purposes, as they would have been required to complete the number of the crew when they took over the new ship of the 1915 type, which was shortly expected to be ready for them.

The deck ratings and engine-room mechanics are specially trained in machine gun work. Those who are reliefs, when war stations are ordered, man the machine guns, and are referred to in the next paragraph as "machine gunners."

In the 1914 type of airship the "war stations" of the officers and men (16 in number) appear to be as follows:—

In the forward car:—

6 in No.
- Commanding officer.
- Chief navigational warrant officer.
- Deck rating in charge of the elevators.
- Deck rating in charge of the rudders.
- One engine-room mechanic.
- One machine gunner.

In the gangway:—

4 in No.
- *Second in command (responsible for bombs).
- Sailmaker (to attend to the gasbags and fabric generally).
- One engine-room mechanic (to attend petrol tanks).
- †Wireless operator.

In the after car:—

4 in No.
- Chief engine-room warrant officer (in charge).
- One engine-room mechanic.
- Two machine gunners.

On the upper platform:—

2 in No. } Two machine gunners.

In the 1915 type of airship, in which the number of officers and men is 18, the war stations appear to be the same as in the 1914 type, except that two additional machine gunners are stationed in the after car owing to the introduction of three motors in that car in the 1915 type.

In the 1916 type of airship, in which the number of officers and men is 22, the war stations appear to be the same as the 1915 type, except that one additional engine-room mechanic and one additional machine gunner are stationed in each wing car.

In all the above cases the war stations must be taken as giving only a general idea as to what the stations may be. From evidence received, it appears probable that much latitude in this matter is allowed to the commanding officers, who tell off their crew for war stations according to their own individual ideas.

It should be noted that no one is allowed to walk about in the airship without the commanding officer's permission, with the exception of the second in command, the sailmaker, and one other man who is detailed from the reliefs for attending the petrol tanks as necessary and for running messages.

**Maintenance Party.**

The maintenance party (Schiffspflege-gruppe) is drawn from the landing party and, in the 1916 type of ship, consists of 20 to 22 men, divided into port and starboard watches. Each watch appears to consist of two sailmakers, four or five mechanics, and four naval ratings. A maintenance party is attached to each airship, and is apparently under the orders of the officer commanding the airship.

---

* Is in the forward car when not required in the gangway.
† In the 1915 and 1916 types, the W.T. operator is in the forward car.

It takes over the airship on its return from any trip as soon as it has been safely housed in the shed by the crew and the "landing party" (see next paragraph). From that moment until the crew again takes it over for the next trip, the airship is in the care of the "maintenance party," who do any necessary repairs, cleaning, or adjustments, and are entirely responsible for it. From the maintenance party are drawn any spare numbers that may be required for the crew, either to replace men who are absent from sickness or other causes or to bring it up to the greater number required when taking over newer and larger ships. It is possible that crews just trained, but for whom no ship is at the moment available, are employed with the maintenance party.

### Landing Party.

The "landing party" consists of a detachment varying in number from 100 to 700 men, or even more, whose duty it is to assist in the landing and departures of airships. The "landing party" tows the airship into the shed; once safely housed it is taken in charge by the "maintenance party." The number of men required depends on the state of the weather and varies very considerably. It is uncertain from what branch of the service these men are drawn.

## II.—Army Airship Service.

### A.—*Before the War.*

The German military authorities became seriously interested in airships much earlier than the naval authorities, and from the time of the earliest experiments of Count Zeppelin and his rivals, their progress was eagerly watched and promoted. During the first years, up till 1906, the work was so tentative and experimental that the interest taken, though entirely benevolent, was not very active. At this time opinion in military circles had by no means crystallised on the question of rigid as against semi-rigid and non-rigid airships, and all over Germany experiments in each type were being carried out. In 1906 the army authorities were apparently most impressed by Count Zeppelin's designs, and in that year they ordered Z 1, the first army airship. In 1909 the results of experiments with this airship had been so favourable that they ordered a second one, Z 2, which was, however, wrecked in the same year. This misfortune, coupled with the fact that the M (Gross) airship had by now been produced by the military authorities themselves with fairly successful results, seems to have caused them to pause in their patronage of Zeppelin; and it was not till 1911, after three more passenger Zeppelin airships had been completed and had made successful flights, that Ersatz Z 2 was ordered. This ship and Z 3, which was taken over in 1912, were a complete success, and apparently convinced the military authorities that they would be well advised to confine their attentions to this type of airship. Up till this time the whole question had been in so experimental and indefinite a stage that there seems to have been no separate "airship service" or system of training crews. The crew of Z 1 was presumably trained by the Zeppelin Company at Friedrichshafen, while the crews of Z 2, Ersatz Z 2, and Z 3 were probably trained in Z 1. Late in 1912, as the military authorities seem definitely to have made up their minds to create a fleet of Zeppelins, an organisation of airship battalions was created as follows:—

*No. 1 Battalion*, at Berlin (Döberitz), with the school of instruction for aerial navigation attached to it.

*No. 2 Battalion*, No. 1 Company and headquarters at Berlin (Tegel) and No. 2 Company at Königsberg.

*No. 3 Battalion*, No. 1 Company and headquarters at Cologne, and No. 2 Company at Metz.

These three battalions were under the "Inspector of Air Troops and Mechanical Transport," who in his turn was under the command of the "General Inspector of Military Communications."

Each airship battalion had a large technical staff of civilians attached to it, and probably controlled two airships.

It is believed that the crews of the five Zeppelin airships delivered to the army in 1913 received their final training in the already existing military airships at the headquarters of the airship battalions, although this is not definitely known.

Prelimininary training in flying was carried out in the old M airships, M 1 and M 4.

By the new Army Bill of 1913 this organisation was considerably increased. Prussia was to form five airship battalions, to which Würtemberg and Saxony were each to contribute one company, the total complement consisting of:—

|  | Officers. | N.C.O.'s | Men. |
|---|---|---|---|
| Prussia | 81 | 275 | 1,995 |
| Würtemberg | 4 | 20 | 149 |
| Saxony | 4 | 20 | 149 |

Prussia, Würtemberg, and Saxony maintained joint jurisdiction over airship affairs. Bavaria appears to have furnished one airship battalion in addition to her aeroplane service and maintained a separate organisation under the Engineer Corps at Munich.

The airship battalions were stationed as follows:—

*Airship Battalion No. 1:—*
    Staff and 1st Company
    2nd Company
    Airship Instructional Establishment
    Transport Detachment
    } At Berlin (Tegel).

*Airship Battalion No. 2:—*
    Staff and 1st Company and yards - At Berlin (Tegel).
    2nd Company - At Hanover (Königsberg temporarily).
    3rd Company (Saxon) - At Dresden.

*Airship Battalion No. 3:—*
    Staff and 1st Company - At Cologne.
    2nd Company - At Düsseldorf.
    3rd Company - At Darmstadt (Metz temporarily).

*Airship Battalion No. 4:—*
    Staff and 1st Company - At Mannheim.
    2nd Company - At Metz.
    3rd Company - At Lahr (Gotha temporarily).
    4th Company (Würtemberg) - At Friedrichshafen.

*Airship Battalion No. 5:—*
    1st Company - At Königsberg (Allenstein temporarily).
    Staff and 2nd Company - At Graudenz (Liegnitz temporarily).
    3rd Company - At Schneidemühl (Posen temporarily).

### B.—*Since the Outbreak of War.*[*]

**Formation of Crews.**

It is not known whether the men are volunteers or are detailed for this duty. They are sent in the first instance to one of the airship battalions which existed before the war. These battalions are now performing the functions of "depôts" for collecting untrained men for the airship service. From there the men are drafted, as required, to a "Schulkommando," where they receive instruction, obtain their certificate, and are finally detailed for an airship. When they no longer form part of the crew of an airship they return to the "Schulkommando" in order to await a new ship and, if necessary, to qualify for a higher rating.

**Training.**

The training of the officers and crews of army airships is carried out at "Schulkommandos" which have been formed since the outbreak of war.

The chief "Schulkommando" was at Reinickendorf, near Berlin, where preliminary instruction in flying was carried out in the old M airships, M 1 and M 4. About January or February 1915, the "Schulkommando" moved to Johannisthal, near Berlin, where instruction was carried out in the old Delag airship, "Hansa." Meanwhile, the central training school, which was being constructed at Niedersgörsdorf, about 40 miles S.S.W. of Berlin, was nearing completion.

---
[*] From information supplied by G.H.Q., Home Forces.

On its completion, about March or April 1915, the "Schulkommando" was moved there. The army airships, Z 4 and LZ 72, which were to be the training ships, were sent there and housed in the double airship shed which had been constructed for them. About January 1917, it appears that a new "Schulkommando" was started at Spich, where LZ 72 was sent. Whether this is to be the only training centre in future, or whether it is to be additional to Niedersgörsdorf is uncertain.

The course of instruction is not supposed to last more than six weeks, but, in practice, its duration is governed by the number of new airships to be commissioned.

Instructional trips at Niedersgörsdorf are usually of two or three hours' duration. They are mainly intended to practice landings—both by day and by night. As many trips as possible are carried out each day.

All ratings are instructed in the use of the machine gun, with special reference to its use on airships. Bomb-dropping practice, however, is not carried out at Nieders Görsdorf.

Although the principal training is carried out at Nieders Görsdorf, there is evidence to show that men have also been sent to the Schütte Lanz works at Mannheim to undergo special instruction, while some have gone to Potsdam to follow the construction of gasbags, and officers have gone to Dresden for training in bomb-dropping.

The W/T operators under a three months' course at the "Kriegstelegraphenschule" at Spandau, near Berlin.

On completion of courses the men cease to be simple "Luftschiffer," and become specialists, and are distinguished by the title of the branch in which they have specialised.

The average age of the crew of LZ 85 was as follows:—

| Officers. | Men. |
|---|---|
| 32 | 26 |

### Number and Disposition of Crews.

Military airships carry four officers:—
(1) Commanding officer.
(2) Wireless and armament officer.
(3) Navigation officer.
(4) Engineer officer (in charge of the motors).

The crew is divided into two sections—the first and second crew—numbering (exclusive of officers) 8 to 10 men each. When it is desired to add to the first crew, the second crew is drawn on. The men are allotted to the first and second crews, according to seniority and merit, but there is naturally a correct proportion of each rating in either crew.

It is probable that, as in the case of naval crews, the total number of men carried in military airships is dependent on the number of motors, and that the war stations are analogous to those of naval airships. It is, however, probable that wide latitude in this direction is given to the commanding officer.

### Maintenance Party.

Every military airship seems to have attached to it a party of men, in number about 20, whose duties are analogous to those of the naval "maintenance party." They are also called "Schiffspflege-gruppe," and wear the same uniform as the "Luftschifftrupp" (see next paragraph), namely, "Feldgrau" (field-grey) uniform, black shoulder straps with "L" in red, and a special badge representing a propeller, on the arm.

### Landing Party.

In addition to the "maintenance party," but under the command of an independent officer, there are the "Luftschifftrupp" or landing party, attached to each airship station. They come under the heading of "Garnisontruppe" (garrison troops), and are drilled as infantry. Their duties with regard to the airships consist in inflating the gasbags, and assisting in landings and ascents. These companies appear to be recruited from the classes of auxiliary services named "Garnisondienstfähige" (men fit for garrison duty only), and not from the Landsturm. The "Luftschifftrupp" usually consists of about 140 men; but this number would naturally be increased for large airship bases.

## CHAPTER III.—CONSTRUCTION.

L 33 belonged to the class of which L 30, completed about end of May 1916, was the first ship. She was laid down by the Luftschiffbau Zeppelin Aktien Gesellschaft, at Friedrichshafen, in the early part of July and completed on August 30th, 1916, and was a sister ship to L 31 (completed in the middle of July and destroyed at Potters Bar, October 1st, 1916); to L 32 (completed in the middle of August and destroyed at Billericay, September 24th, 1916); and to L 34 (completed in the beginning of September and destroyed near Hartlepool, November 27th, 1916).

L 33 was damaged by gunfire over London on September 24th, being hit in the after gasbags, and after making an attempt to return to Germany, she was forced by loss of buoyancy to land at Little Wigborough, in Essex, where her crew gave themselves up after the ship was on fire. When examined next morning, the wreck was found to be fairly complete, except as regards the fabrics, of which only very small portions remained; the control car, of which little but the bottom plating was left; the keel structure over the control car and over the after power car; and the forward part of the after power car (see Plate 8). Her general particulars are as follows:—

| | |
|---|---|
| Length overall | 196 metres (643 feet). |
| Extreme diameter | 24 metres (78·7 feet). |
| Extreme width, wing propeller blades horizontal | 28·8 metres (94·4 feet). |
| Extreme height | 28 metres (91·8 feet). |
| Gasbag capacity, 95 per cent. full | 54,000 cubic metres (= 1,907,000 cubic feet). |
| Total lift, 95 per cent. full | 57·8 tons. |
| Disposable lift, 95 per cent. full | About 22 tons. |
| Maximum elevation, feet | 13,000 to 15,000. |
| Machinery | Six Maybach engines, 240 b.h.p. each. |
| Total h.p. | 1,440. |
| Speed | About 60 m.p.h. |

With regard to the disposable lift and maximum elevation, the crew state that the ship could rise statically to 15,000 feet, but it is not clear what weight, in addition to her crew, the ship could take to this height. After a raid on London, with 2,500 kilograms (5,511 lbs.) of petrol and 200 kilograms (441 lbs.) of oil on board, and presumably all bombs discharged, heights of 3,800 metres (12,500 feet) have been logged—but this may have been obtained partly by the use of her elevators. The crew also state that the maximum lift that could be obtained through using the elevators was 1,320 lbs. per motor, and the additional height, 3,000 to 4,000 feet, obtainable in 3 or 4 minutes.

The following figures show approximately the weights which would be carried on a raid to England, and may be considered as fairly correct:—

| | Kg. |
|---|---|
| (a) Bombs | 3,240 (7,142 lbs.). |
| (b) Ballast | 10,300 (22,707 lbs.). |
| (c) Petrol | 5,000 (11,023 lbs.). |
| (d) Oil | 550 (1,212 lbs.). |
| (e) Crew, 22 at 80 kilograms each | 1,760 (3,880 lbs.). |
| (f) Machine guns and ammunition (eight at 70 kilograms) | 560 (1,234 lbs.). |
| (g) W/T | 100 (220 lbs.). |
| | 21,510 |

The total gasbag capacity has been calculated to be 57,000 cubic metres (2,013,000 cubic feet), corresponding to 62,700 kilograms (138,229 lbs.) total lift, reckoning the net lift of hydrogen at 1·1 kilogram per cubic metre. At 95 per cent. inflation, this would give 54,200 cubic metres (1,914,158 cubic feet) capacity and 59,400 kilograms (130,954 lbs.) lift respectively. The nominal gasbag capacity is understood to be 54,000 cubic metres (1,907,095 cubic feet), whilst 55,000 cubic metres (1,942,411 cubic feet) of gas are said to be allowed in filling, but the latter figures may be presumed to include gas lost during inflation.

It appears, therefore, that the Germans adopt 95 per cent. inflation as the basis for calculating the nominal gasbag capacity of Zeppelin airships.

*General Description of L* 33.—Broadly speaking, L 33 may be said to have been essentially a bomb-dropping airship. All things considered her speed was very moderate, her armament was clearly intended only for repelling attacks by aeroplanes, and her bomb stowage was very large. Her form is exceedingly fine, but the diameter is unusually large in relation to length.

The gasbags, 19 in number, are generally 10 metres (32·8 feet) in length, *i.e.*, only about 40 per cent. of the diameter amidships. It is clear that experience has taught the Germans that, in airships which may be exposed to shell fire, it is an advantage to keep down the relative capacity of individual gasbags, especially towards the ends of the ship.

The arrangement of cars is as follows :—

Forward at about 22 metres (72 feet) from the bow, a control car is hung under the hull, on the middle line. This car contains the controls for rudders, elevators, gasbag-valves, water ballast (including the emergency water ballast); electric bomb release switchboard and bomb-sights; engine telegraphs; trail and handling-rope release controls; voice-pipe and bells to forward gun platform, crew space and bomb positions, and telephone to after gun platform. It also contains the W.T. cabinet and the commander's quarters.

Immediately abaft the control car is the forward power car containing a single engine driving, through a spur reduction gear, a two-bladed propeller at the after end of the car.

The after end of the control car is nearly the full width and depth of the car, and is formed by a flat vertical plate. The forward end of the power car is of the same size and shape and is placed only 2 or 3 inches abaft it. The gap between the two is faired externally by a fabric strip laced to each, and the two cars are shaped so that, when in position, they present the external appearance of a single long car. Their combined air resistance is accordingly low. At the same time the vibration of the engine is not transmitted to the control car containing the W.T. gear.

Amidships, about 83½ metres (279 feet) from the bow, there are two wing cars abreast, one on each side of the ship, each containing a single engine driving, through spur reduction gear, a propeller at the after end of the car. These cars are each 6·78 metres (22·1 feet) from the middle line, measuring to the centres of the propeller shafts, and access from the keel is obtained by means of a passage across the lower part of the ship, leading to hatches over the cars. It is interesting to note that the release catches to these hatches are operated by wires leading to the control car, so that the mechanic cannot leave his work without the commander's knowledge.

Further aft, about 121 metres (396 feet) from the bow, there is a large car on the middle line, containing three engines; two of these are abreast near the forward end of the car and each drives one wing propeller through bevel gearing and long inclined propeller shafts; the third is placed near the after end of the car and drives a propeller at the after end through spur reduction gear.

The after car also contains an auxiliary position for steering and elevator control at its forward end, but none of the other controls can be worked from this position.

Access to each car is obtained through a hatch in the hull above it.

All the cars are attached to the hull by means of struts and suspension wires.

The ship is fitted with an internal triangular keel extending from just forward of Frame 5 right aft to Frame 39 abreast the rudders and elevators.

This keel is the main corridor of the ship, and is fitted with a footway down the centre extending its full length. It contains the main mooring point (at its forward end), the water ballast, petrol tanks, bomb-stowage and dropping gear, handling rope buckets, and crew space. The various control wires, petrol pipes, and electric leads run along its lower part.

The fins, rudders, and elevators are arranged in the manner first seen in LZ 77 at Revigny, and followed in all subsequent Zeppelins. This is said to have been copied from the Schütte-Lanz ships. The essential features are that the after ends of the fins are some distance forward of the tail of the ship—10 metres (32·8 feet) in L 33— and that there is a single balanced rudder or elevator at the after edge of each of the fins.

The armament consists entirely of machine guns distributed as follows:—

> In the FORE CONTROL CAR a gun position on each side with probably one gun only, which could be shifted to either position as required.
>
> In the WING CARS, one gun position in each, placed on the outboard side. It is believed, however, that guns were not generally carried in these cars.
>
> On the FORE GUN PLATFORM, on top of the ship, two machine guns on tripod mountings are carried. Traces were also found at the wreck, of what is believed to have been a mounting for a third gun on this platform.
>
> On the AFTER GUN PLATFORM, a single gun can be arranged so as to have a clear range of fire abaft the beam on both sides. The maximum downward depression of this gun is 21° right aft, and about 43° on each beam.

It is doubtful whether or not any guns were carried in the after car.

It is to be noted that none of the guns in the cars can fire right aft on account of the propellers.

The bomb armament actually carried would vary according to circumstances. Bomb-dropping hooks are provided for eight bombs of not more than 300 kilograms (661 lbs.), 40 of not more than 100 kilograms (220 lbs.), and 30 incendiary bombs of 10 kilograms (22 lbs.). Spaces are also provided for frames carrying an additional 90 incendiary bombs.

The crew numbered two officers and 20 men.

*Comparison with Previous Classes.*—So far as can be ascertained, L 33 is about 69 per cent. greater in gasbag capacity than the preceding class, of which LZ 85 (destroyed at Salonika, May 5th, 1916) is believed to have been representative, and about 28 per cent. greater in extreme diameter; her form is considerably finer, the length of parallel body being only 45 metres (147·6 ft.), *i.e.*, about 23 per cent. of her length. The arrangement of her control car, forward power car, and after power car is generally similar to that of the corresponding features of LZ 85 and L 20 (wrecked at Stavanger, Norway, May 3rd, 1916), but L 33 is, in addition, fitted with two wing cars amidships, as mentioned on page 14, to provide the additional power consequent upon her increased size.

The keel, both in structure and arrangement, is very similar in L 33 to that in LZ 85, but the petrol tanks are fitted in such a manner as to facilitate dropping them in case the ship should be short of lift, and the bomb positions are more extensive. (*See* Plate 18, Fig. 73.)

As regards armament, the principal innovation in L 33 is the gun platform on the stern.

*Form.*—The unusual fineness of her form is shown by the following figures:—

> Length forward of parallel body - - 56 metres (183·7 feet).
> Length of parallel body - - - - 45 metres (147·6 feet).
> Length abaft parallel body - - - 95 metres (311·6 feet).
> Ratio $\dfrac{\text{total length}}{\text{extreme diameter}}$ - - - - $\dfrac{196}{24} = 8\cdot17$.

So far as can be ascertained, the ratio $\dfrac{\text{total length}}{\text{extreme diameter}}$ has been rather over 9 in previous recent Zeppelins, and in the earliest types it was over 10. In recent Schütte-Lanz ships the ratio is about 9·2, but, apart from this, it may be said that the general characteristics of the form of L 33 approximate very nearly to those of the Schütte-Lanz types.

In transverse section the external form, except at the stern, is a 25-sided polygon, of which all the corners lie on a circle, and of which all the sides are equal except the bottom one, which is about 25 per cent. longer than the others. At about the forward end of the fins, the number of sides is reduced to 13, and abaft the rudders a further reduction to six is made.

As usual, there is a longitudinal girder at each corner of the polygon, and the large number of the sides, as compared with previous ships, is accounted for by the increased girth of the ship, and by the desire to avoid an increase in the spacing of the longitudinals. This will be explained further in dealing with the transverse frames.

*Structure.*—Z 4, completed in 1913, is the earliest Zeppelin of which any detailed information is available. In this ship the main transverse frames were generally 8·5 metres apart, and were built of girders, equilateral in section, with 160-mm. side. The bracing pieces used in the girders were nearly the same in shape and arrangement as those used in all subsequent Zeppelins down to and including classes built in 1915. The longitudinal girders in Z 4 are said to have been square in section with 130-mm. side.

The intermediate transverse frames extended round the lower half of the hull only. There were no diagonal wire bracings in the panels formed by the longitudinal and transverse frame girders. The keel was external, and built of rectangular girders, but nothing is known as to its wiring.

In L 3, built in June 1914, the main transverse frames and the longitudinals appear to have been generally triangular. Girders of the W type (*see* Plate 9) were used, but the positions are not known. The intermediate transverse frames extended right round the ship. Diagonal bracing wires were probably used in the panels formed by the longitudinal and transverse frame girders.

LZ 85, completed in September 1915, was brought down at Salonica in May 1916, and the following information was obtained from her wreck.

The girders, like those of the preceding classes, were characterised by the considerable width and thickness of the bracing pieces, and by the general use of bracings alternately diagonal and square (*i.e.*, at right angles to the girder edges) on the sides of the girders. In cases where additional girder stiffness was required double diagonal bracing was used, as shown in Plate 9, by adding diagonal bracings sloped in the reverse direction. It should be noted that in the latter case, the diagonal bracings were not connected together at their centres, where they cross each other.

The continuous members of the girders were as follows:—

*Triangular Girders.*—A channel at the apex and an angle at each side of the base.

*Rectangular Girders.*—Four angles.

*W-type Girders.*—Three channels and two angles.

The edges of the flanges of all the angle and channel sections were slightly turned in, to give greater stiffness.

All the triangular girders of the old type which have been seen have an apex angle of 40° and base angles of 70° each.

The main transverse frames were built of W-type girders, the form of the whole frame being a simple polygon without trusses. An intermediate frame of triangular section was fitted midway between each pair of main transverse frames. In LZ 85 the distance between main transverse frames was about 10 metres.

The longitudinals were triangular girders except the top and the two bottom ones, which were of the W type.

All the girders of the main hull framework were approximately of the same depth, necessitating rather heavy joints at their points of intersection in order to obtain the required continuity in the girders.

The keel was internal and was very similar to that found in L 33 and described hereafter.

In L 33 the structure is chiefly notable for—

(*a*) the use of a new type of girder throughout.

(*b*) the novel arrangement of the Main Transverse Frame.

In the new type of girder, the bracings are narrower, thinner, and more deeply fluted than in the old types, and diagonal bracings only are used—the "square" bracings have disappeared.

The diagonal bracings are invariably doubled, each pair forming a cross, the two parts of which are firmly riveted together at the centre. The bracings thus form a series of rigid X forms along each face of the girder (*see* Plates 10–12). The upper and lower faces of the W-type girders are an exception to the above (*see* Fig. 44).

The continuous members of the girders are arranged as in the old types, with the important exception that the triangular girders have channel sections for all three members, a feature which greatly increases their strength in compression. All continuous members have turned-in edges as before.

The girders are triangular in the ordinary main and intermediate longitudinals, in all transverse frames, and in other parts where convenient; rectangular in many struts; W-type in the top ("A") longitudinal and in the two bottom ("G") longitudinals.

The equilateral form of triangular girder is now used freely for girders whose principal loads are compressional, such as the transverse frames.

Girders, such as the longitudinals, which are exposed to lateral loading are however still made of isosceles triangular form, with the longest axis in the plane of the load.

Throughout the ship the girder bracings are very light, compared to the loads to which the girders are exposed, and it has accordingly been considered necessary to introduce additional bracings, often of increased thickness, in the sides of the longitudinals close to the main joints, and in other girders at points where heavy lateral loads are applied to them (see Plates 10 and 33).

In Plate 34, transverse sections of a number of the principal girders in L. 33 are shown.

The new type of girder has been found to be stronger, weight for weight, than the old type, and the principles which it embodies accord generally with the results of the various experiments which have been made for the Admiralty during the past year. At the same time, the extremely thin bracings employed render the girders more liable to failure under loads applied in an abnormal manner, such as might occur in the event of any damage from gunfire or accident.

Regarding (b) the arrangement is shown in Plate 33 and Plate 12, Fig. 47.

It will be seen that structurally each main frame has only 13 sides, of which the bottom one, forming the bottom of the keel triangle, is five-eighths of the length of the others. The remaining 12 sides are each stiffened by a simple truss with a kingpost which is extended outwards to carry the intermediate longitudinal. The girders forming the 12 sides are of equilateral triangular section, with one reverse channel, and weigh only 0·8 lb. per foot. Tested in compression as a strut, a length of 17 feet has been found to stand a load of over 10,000 lbs. By this system each of the sides is rigidly held against deflection at the middle, and its strength against compression is thereby increased very greatly. The kingposts and trusses also serve the purpose of transmitting the load on the intermediate longitudinals (see (3) below), due to gas pressure, away to the main joints without throwing a lateral load upon the main transverse girder. At the ends, namely, forward of frame 5 and aft of frame 35, these kingposts are unnecessary owing to the reduced diameter and are accordingly omitted.

The intermediate transverse frames are of the simple type without trusses. (See Plate 33.)

The longitudinals are of three types:—

(1) The A and (two) G girders are of "W" section, as noted above. These girders are, by their positions, the most heavily loaded of the longitudinals, and they are therefore the strongest.
(2) The other 10 of the longitudinals which pass through the corners of the main transverse frames are of a very deep and wide triangular section. These girders, in conjunction with those at (1), provide the main longitudinal strength of the ship.
(3) The longitudinals which pass through the ends of the kingposts are also triangular, but of smaller size. They contribute but little to the longitudinal strength of the ship as a whole, and extend only as far aft as frame 35.

The main longitudinals are of unusually large dimensions and of great stiffness and have clearly been designed to withstand the considerable lateral loads due to gasbag pressures, to which they are exposed in a ship of this large diameter. The "B" and "C" main longitudinals, port and starboard, which are exposed to the pressures near the tops of the gasbags, have bracings and channels of greater thickness than those in the other longitudinals.

*Wiring.*—The wiring of the ship presents many interesting features.

(a) Transverse frame wiring (see Plates 33, 35, and 36).

Generally, there are in each main transverse frame six doubled 3-millimetre diametral wires, three of which are clamped together by a special fitting at the point where they cross. This fitting (see Plate 37) is connected by steel wire ropes with the similar fittings in the adjoining main tranverse frames, and there is thus a continuous chain of wire ropes and fittings running along the axis of the ship between frames 3 and 37. The wire ropes pass through the gasbags by means of a special joint described in the chapter on Fabric. Forward of the fitting in frame 3 and abaft that in frame 37, the wire rope divides into four ropes whose further ends are attached at frames 1 and 39 to the top, bottom, and two side joints of the frame (see Plate A, in packet). By this means the tension in the wire rope is distributed to the structure of the ship.

A similar longitudinal wire rope was found in the Schütte-Lanz airship SL 11, brought down at Cuffley on September 3rd, 1916, but it does not appear that LZ 85 or previous Zeppelin or Schütte-Lanz airships were so fitted. Its function is to support the diametral wires in the transverse frames against the fore and aft pressure to which they are subjected by the gasbags when the ship has a large longitudinal inclination, or when there is a difference of pressure between adjoining gasbags.

Returning now to the transverse frame wiring : in addition to the diametral wires there are 3-millimetre wires led from the joints of G girders, and from the top of the keel, to the joints at C and D longitudinals.

Also there is an elaborate system of $2\frac{1}{2}$-millimetre wires, which contribute to the strength of the frame, and also serve the important function of filling in the large gaps between the diametral wires. They provide valuable support to the ends of the gasbags when a difference of pressure occurs between adjoining gasbags.

In frames 1 and 39 the tranverse wiring is designed chiefly to support the end pressure of the adjoining gasbag, and is composed of wires $2\frac{1}{2}$ millimetres in diameter.

In certain frames where special loading occurs small variations in the wiring are made, but the general arrangement of the wires is unaltered. (See Plates 35 and 36.)

The arrangements of the diagonal bracing wires between the longitudinal and transverse girders is clearly shown in Plates 13, 33, B, and C, and no description is required. It should, however, be noted that none of the wires is attached to the intermediate longitudinals, and the latter do not, therefore, take a primary part in resisting the varying bending moments of the ship. The wires, at the points where they cross over these longitudinals, are held in position only by small plate clips which fit loosely over them and allow them to slide freely.

The size of the diagonal bracing wires is greatest, viz., 3 millimetres, at the sides of the ship, where the shear forces are greatest, showing that the main structure of the ship is designed to take its share in supporting these shear forces.

The inner diagonal wires (see Plate C) are stretched rather loosely between the inner edges of the longitudinals, and their purpose is to support the gasbags. In diameter they are graduated amidships from $2\frac{1}{2}$ millimetres at the top of the ship to 2 millimetres at the bottom.

From frame 3 to the bow an additional system of wires is fitted on the outside of the structure, forming a closely spaced network, designed to support the outer cover against the external air pressure to which it is exposed when the ship is proceeding at high speed (see Plates B and 13).

The ends of the main diagonal bracing wires are very strongly made by binding and soldering as shown in Plate B (in packet).

The ends of the inner diagonal wires are roughly turned three or four times round the wire and not soldered. Both types of wire end are covered with rubbered tape to prevent chafing. Turnbuckles are not used in any of the standing wires of the structure. All wires are of carbon steel, cold galvanised. Test results are given at the end of this chapter.

*Shear Ropes.*—The name "shear rope" has been given to a series of wire ropes placed diagonally in the ship on the middle line (see Plate A). Each of them is attached at its upper end to the top joint of a main transverse frame, and at its lower end to the top of the keel at the next main transverse frame, and it passes through the gasbag by means of special fittings described in the chapter on Fabric. These shear ropes, of which there are 15, are used to distribute directly to the adjacent main frames heavy loads applied to the keel, and in general they may be said to "take the peak" of the shear force curves. Shear ropes were first seen in LZ 85 at Salonika. No traces of them were found in the Schütte-Lanz ship at Cuffley.

*Fins, Rudders, and Elevators.*—The general arrangement of these is shown in Plates A and 14. The fin construction follows the usual Zeppelin practice, and is seen in Plate 14. Each fin is bounded at its inner edge by one of the ship's longitudinal girders and at its outer edge by a strong triangular girder, the apex channel of which forms the extreme edge of the fin. Connecting these two girders, and rigidly attached to both, is a number of rectangular booms. The boom forming the after edge of the fin is, however, triangular, with its apex channel aft, so as to reduce air resistance. Diagonal bracing wires are fitted in the fins as usual, and these wires are in duplicate, *i.e.*, one set at each face of the fin, so as to support the fabric covering.

The outer edge girder of each fin is extended beyond the after edge of the fin far enough to carry the outer bearing of the rudder or elevator.

The structure of the rudder and elevator planes differs considerably from that seen in previous Zeppelins, notably in the omission of the steel tube formerly fitted along the axis of each plane (*see* Plate 15).

In the present instance a neatly designed duralumin "W" girder is fitted at the axis of each plane, and the construction is such as to give a fairly uniform strength over the whole area of the plane. The air pressure is finally transferred to the operating sector by the kingpost wires. The operating sector, as usual, is very rigidly constructed. The pintles are in every case fitted with ball bearings and ball thrust bearings, which, in the case of the inboard pintles, are carried in a spherical seating, giving self-alignment.

The principal dimensions of the fins, &c., are as follows:—

*Rudders and Elevators.*

|  |  | Upper Rudder. | Lower Rudder. | Elevator. |
| --- | --- | --- | --- | --- |
| Maximum width (fore and aft) | Metres | 4·2 | 4·2 | 4·2 |
| Length at axis | ,, | 6·3 | 4·08 | 6·3 |
| Thickness | Millimetres | 208 | 208 | 208 |
| Area | Square metres | 26·9 | 16·8 | 26·9 |
| Weight | Lbs. | — | 180 | — |
| Diameter of operating sector |  | All 2·04 metres. | | |

*Fins.*

|  |  | Horizontal and Upper Vertical. | Lower Vertical. |
| --- | --- | --- | --- |
| Extreme length | Metres | 23·28 | 18·3 |
| Depth at after end | ,, | 6·48 | 4·26 |
| Thickness | Millimetres | 208 | 208 |
| Area | Square metres | 89·4 | 41·2 |

It is interesting to note that the prisoners stated the ship to be less handy than her predecessors, though she steered fairly well except in bad weather.

*Rudder and Elevator Controls.*—The arrangement of controls at the stern is shown in Plate 38. A wire rope is attached to each side of each of the operating sectors, and is led over guide sheaves to the lower part of the keel corridor, where it connects to a solid drawn wire running forward to the control positions in the cars.

The excess weight of the elevators abaft their axes is balanced by means of springs attached to the control wire ropes, and to the structure at the lower part of frame 39. Preventer wires are also fitted to the elevators to prevent the latter from taking an excessive angle in the event of the control wire rope breaking, and to hold the elevators in the horizontal position when desired.

The control wire ropes are of remarkably flexible construction and are 6·1 millimetres (·24 inches) in diameter, with 11 strands each of 19 wires. There is a hemp core. A sample of this wire rope has been tested, and broke under a load of 3,750 lbs.

A notable point is the completeness with which the guide sheaves are guarded, to prevent the ropes from slipping off (*see* Plate 16).

The prisoners from this ship stated that the elevators were very stiff to work whilst the rudders were comparatively easy, and this would appear to indicate either that the balance springs did not function well or possibly that they would have been better placed at the after edges of the horizontal fins.

*Keel.—*

For arrangement of keel in the ship (shown in dotted lines) *see* Plate A (in packet).

For detailed arrangement of keel *see* Plates D to G (in packet).

For sectional views of keel at various parts of the ship *see* Plates 33, 35, and 36.

For views of keel at various parts *see* Plate 17.

Structurally the keel constitutes a triangular girder extending from frame 5 to frame 39, having four continuous girders throughout its length, viz., the two " G " girders, the apex girder, and the walking-way girder on the middle line between the " G " longitudinals.

In addition, between frames 7 and 31 there are two rectangular girders situated about 80 centimetres (amidships) below the top of the keel, whose primary function is the support of bombs, tanks, water-ballast bags, &c. These girders are all converted into a single complex girder by side struts and base struts of about $2\frac{1}{2}$ metres pitch in combination with diagonal wire bracings in all the sides.

The extreme depth of the keel is $3\frac{1}{4}$ metres (10·8 feet) amidships, reduced at the forward end to 2·6 metres (8 feet) and at the after end to 2 metres ($6\frac{1}{2}$ feet).

The clear head-room is 6 feet 3 inches amidships, reducing gradually towards the ends and reaching a minimum of 4 feet 6 inches at frame 31.

The width of gangway is least—namely, 15 inches—abreast the petrol tanks.

At each main transverse frame, struts are fitted between the apex girder of the keel and the " F " joints on each side in order to stiffen the lower part of the transverse frame and to steady the keel.

Considering the keel as a part of the whole elastic structure of the hull, it is clear that all the longitudinal members of the keel share with the ship's longitudinals the stresses arising from the bending moments to which the ship is subjected. The longitudinal members of the keel, therefore, contribute to the strength of the ship.

From this point of view, however, the material of the transverse struts, &c. of the keel contributes nothing and that of the longitudinal members would be better placed if all of it were put into the ship's main longitudinals where it would be at the greatest distance from the ship's neutral axis. Similarly, the diagonal bracings of the keel share with the ship's bracings and shear wires the shear forces of the hull, but their effect is not important.

The keel may also contribute to the strength of the hull against bending by virtue of its moment of inertia about its own neutral axis. As this is, however, less than 1 per cent. of that of the transverse section of the hull, the relief which it affords in reduction of hull stresses under this heading is proportionately small.

It is therefore clear that the keel structure does not materially assist the hull in resisting the general bending moments and shear forces, by virtue of its own girder strength.

It is, accordingly, to be expected that its function is the provision of local strength at the points where concentrated loads occur. Such points are the main car strut attachments, where the car weights are partly taken and the landing shocks wholly taken; the main mooring point forward and the various handling rope attachments; the supports of water ballast, petrol tanks, bombs, &c. All of these points are ranged along the keel, and, at all of them, concentrated forces of relatively large magnitude are applied. In a ship with so large a disposable weight these points cannot without inconvenience all be located at the main transverse frames, and a girder keel is therefore introduced to distribute the loads to the main frames.

It follows from the above that not only is the keel not the backbone of the ship, in the sense in which the term is sometimes used, but its strength is, in fact, only adequate to deal with the local loads which are applied to it.

At frame 18 a transverse " keel " structure with walking-way is fitted across the ship between longitudinals F, port and starboard, to provide access to the hatches over the wing cars. The arrangement of the structure is shown in Plates 19 and 39.

*Walking-Way.—*The walking-way is 30 centimetres (about 1 foot) wide, with no guards at the sides, such as existed in LZ 85, to prevent the feet from slipping off.

The wood of the walking-way is 3-ply, the centre ply being ·129-inch thick, of a white wood, probably aspen, with the grain running longitudinally, and the two outer plies of birch, each ·032-inch thick, with the grain running transversely.

Both the top and bottom of the walking-way is covered with cotton fabric similar to that used generally in the ship, weighing 130 grams per square metre.

*Water-ballast Bag Arrangement.*—(*See* Chapter V. for detailed description of water-ballast bags, and also Plate H, in packet.)

As usual the ballast bags are arranged so as to facilitate the substitution of ballast for other weights, such as petrol, bombs, &c. When a certain ballast bag is not required, it is entirely removed, together with its valve and pipes, as far as the flange coupling in the horizontal part of the pipe (Plates D to G). In emergencies, when it would be desired to lighten the ship as rapidly as possible, the fabric of the bag is cut away just above the valve, and the bag only is thrown overboard.

The valve and its attachment to the bag are shown in Plate 21, Figs. 82–3.

The manner of attaching the bag to the girder above it is shown in Plate 21, Fig. 84.

The emergency ballast is situated at frames 5 and 33 (*see* Plates D to G). At both frames four bags, each holding 80 kilograms (176 lbs.) of water, are suspended from two tubular frames slung from the upper part of the frame by wire ropes. Over each bag a slip hook operated from the control car is fitted on the tubular frame. Under each bag is fixed an aluminium bucket having a flap bottom 9·35 inches in diameter and closed by a light spring.

The bag is of the shape shown in Plate 75, Fig. 193.

Normally the end of the fabric discharge pipe is held up to the slip hook. When it is desired to lighten the ship rapidly at either end, as might be desirable in landing, the end of the fabric pipe is slipped from the hook and discharges water into the bucket below, whence the water escapes freely through the bottom, which is opened against its spring by the weight of the water.

It is understood that the emergency water-ballast is very rarely used.

*Bomb Stowage.*—The general arrangement of the bomb stowage is shown in Plates E and F. Under each bomb position there is an opening in the outer cover above which is placed a horizontal shutter, 3 feet 4 inches by 7 feet 3 inches, composed of an aluminium tubular frame with fabric covering. At the ends of the shutter, rollers are fitted running in grooves on the frames, enabling the shutter to slide freely in a transverse direction. One of these shutter frames is clearly shown in Plate 17, Fig. 67, and Plate 40.

Before dropping the bombs the shutter is moved outboard into a position clear of the opening in the outer cover, the operating wires for this purpose being worked from the control car, or locally, as desired.

The bomb release hooks for the 300-kilogram bombs are attached to the main rectangular longitudinals of the keel, but those for bombs of weight not exceeding 100 kilograms (on which, however, bombs of 50–60 kilograms were commonly carried), are attached to short girders introduced for this purpose at a lower level. The incendiary bomb frames (Plates 40 and E) are slung about 2 feet 3 inches above the shutters.

It is clear that importance was attached to suspending all bombs as close down to the shutters as possible, and, as a further precaution against fouling, guide plates are fitted at the lower parts of the keel uprights in way of the bomb positions. These are clearly seen in Plate 17, Fig. 65.

*Petrol Tanks and Piping.*—The arrangement of petrol tanks is shown in Plates D to G. Positions are provided for 54 tanks, but it appears that the number of tanks carried in a raid to England was usually about 30. Each tank carried 200 kilograms (440 lbs.) of petrol, its total capacity being, however, 74 gallons (*see* Chapter IV.; its diameter is 2 feet $1\frac{1}{4}$ inches, overall length 3 feet $7\frac{1}{2}$ inches, and weight, when empty, but with fittings in place, $27\frac{1}{2}$ lbs.

Each tank is fitted with a skeleton casing consisting of two circular bands of aluminium strip fitting round the tank near the top and bottom respectively, and a similar strip which fits under the bottom and passes up the forward and after sides of the tank, ending just above the top of the tank in two eyelets through which the whole weight of the tank is transmitted to the suspension strops or links. This strip is connected to both the circular bands at the points where it crosses them, and at each point two projecting lugs are fitted to it, arranged so as to fit easily against the two edges of a vertical girder attached to the structure. The arrangement is seen in Plates 22, 70, and 72.

About two-thirds of the tanks are supported by means of wire strops passing through the two eyelets and attached to a slip release bolt on the girder over the centre of each tank. The bolt is operated locally only. The arrangement is shown in Plates 72–3. On withdrawing the bolt the tank is free to drop, and is guided in its descent by the projecting lugs sliding down the vertical slide girders at each side.

Presumably the outer cover is arranged so as not to obstruct the fall of the tanks, but no details of this are available.

The remaining petrol tanks are fixed, *i.e.*, they cannot readily be dropped, the means of support being short links bolted to the tops of the slide girders and to the eyelets mentioned above.

The pipe system is described in Chapter IV.

In the keel above each engine there is a radiator water tank and an oil tank, both resting horizontally on crutches. Plate 23, Figs. 90 and 91 show the arrangement of these over a wing car.

*Ventilation.*—The automatic gas valves being near the bottoms of the bags, some of the hydrogen escaping from them must pass into the keel space. To enable this to exhaust into the atmosphere five ventilator shafts are fitted to the top of the keel. The position of these is shown in Plate A (in packet).

The three shafts in the midships part of the ship are elliptical in section, built of channel rings about 21 inches apart, with four longitudinal channels, the outside dimensions being about 30 inches by 16 inches. They are covered with nets only —not with fabric.

The shafts are attached to the diametral wire system for support and their upper ends are attached to the hull girders at B longitudinal as shown in Plate 23. On the outside of the ship a light wooden hood with an opening on the after side only is provided over the end of each shaft to keep out the weather and to induce a draught when the ship is in motion (*see* Fig. 93). As these shafts have no fabric covering, any gas in the upper part of the ring space can escape freely through the hoods. The other two ventilators are made of aluminium tube ·6 inch diameter wound spirally, with cane spacing-pieces running longitudinally. These shafts are covered with fabric on the outside, and their external diameter is $10\frac{1}{2}$ inches. Fig. 92 shows the remains of one of these. Their upper ends are at D longitudinal and open into the ring space between the gasbags and the outer cover.

*Access Hatches.*—A sliding access hatch $2\frac{1}{2}$ feet square is fitted in the bottom of the keel over each middle-line car and in the hull over each wing car. In addition, a similar hatch is fitted between frames 14 and $14\frac{1}{2}$, presumably for access to the keel when on the ground. The hatch covers are of aluminium plate.

*Crew Space.*—Between frames 12 and 13, a plywood flooring 6 feet 8 inches (2 metres) wide is fitted in the keel, and it appears that this is the crew space. The plywood is supported by a girder system very similar to that used for supporting the plywood fitted in way of the wing propellers (*see* Plate 50).

At the forward end of the crew space is a small cleat and fairlead intended for use in connection with a boat to be slung under the hull at that part (*see* Frontispiece). It is stated by prisoners that this boat was considered useless and was not carried; they state also that experiments were made for carrying a second boat on top of the hull of other ships.

*Rudder and Elevator Control.*—As stated in describing the rudders and elevators, each of these is controlled by two wires running along the lower part of the keel from aft to the control positions. The wires are solid drawn steel 3 millimetres diameter, galvanised, breaking strength about 2,450 lbs., and two pairs are led along each side of the keel. The pairs for the lower rudder and port elevator are at the port side, the other two pairs being on the starboard side.

The control wires are painted at intervals as follows:—

| | |
|---|---|
| Upper rudder | Blue. |
| Lower rudder | Black. |
| Starboard elevator | Green. |
| Port elevator | Red. |

The wires are supported at each transverse frame (*i.e.*, every 5 metres (16·4 feet)) on fairleads of types shown in Plate 16, type A being used when there is a change of direction in the vertical plane, and type B where there is a straight run.

Over the forward end of the after car an arrangement of two transverse shafts with sprocket wheels, clutches, and chains, is introduced; these two transverse shafts are connected, respectively, to the auxiliary steering and elevator handwheels in the after car by means of sprocket wheels and driving chains.

The arrangement is shown in Plate 41. It will be seen that one of the two shafts is connected to the wires from the elevators, and the other to those from the rudders. Each of them carries a sprocket wheel, through which the shaft can be driven by means of a 1½-centimetre pitch chain from the steering or elevator pedestal in the car below.

The two sprocket wheels at each end of each shaft are loose upon the shaft and are fitted with clutches so arranged that the rudders and elevators can be worked from the after car or from the control car. In the former case only the sprocket wheels for the chains and wires running forward run loose upon the shafts. In the latter the sprocket wheels are coupled together, but both run loose upon the shaft, so that the drive will come from the forward car without rotating the shaft.

At the control car forward the wires are led directly down to the steering and elevator pedestals which contain the clutches for disconnecting either rudder or either elevator.

The rudder and elevator pedestals in the cars consist simply of a casing, bolted to the car structure and supporting a bronze shaft with fixed sprocket wheel and handwheel; the shaft also carries a disc, with spiral groove on one face, for operating a helm indicator pointer on top of the casing. The diameter of the handwheel is about 20 inches.

*Minor Controls.*—The minor control wires are all 1 millimetre diameter solid drawn steel, and are led along the sides of the keel as shown in Plates D to G (in packet).

In every case the arrangement at the objective end of the control wire is as shown in Fig. I., the arrangement being such that the short length A of the wire is

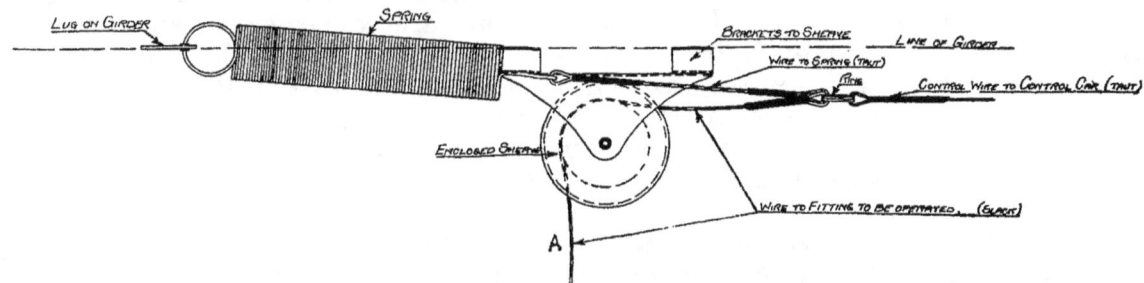

FIG. I.—TYPICAL ARRANGEMENT OF CONTROL WIRE, L 33 (END NEAR FITTING TO BE OPERATED).

normally slack, the remainder of the wire being held taut by the spring. The operator in the control car requires to pull the wire about 2 inches (extending the spiral spring while doing so) before A becomes taut and begins to do its work.

In Plate 16 are shown the types of fairleads used for the control wires in the keel. Where a considerable change of direction occurs at a fairlead, fine stranded steel wire is used instead of solid-drawn wire.

In the keel over the control car the wires are separated into groups according to their functions, and are led down to the control car over nests of sheaves.

The arrangements in the control car for operating these wires are known only in the case of the manœuvring gas valves and the water-ballast valves. Both these sets of controls are led to a fitting, shown in Plate 42, by means of which it is possible to operate the water ballast valve controls in different combinations as may be desired. The manœuvring gas valves are operated simply by pulling the wooden handles on the ends of the wires below the casing.

The control wires for the manœuvring gas valves are led from the keel straight up to the valve levers, those parts of the wires which pass between the gasbags being cased in aluminium tubing. Branch wires, with springs, as shown in Fig. I., are attached to the control wire close to the valve lever, and also in the keel near the point where the control wire turns up towards the valve.

Throughout the ship, sheaves for water-ballast control wires are painted black, whilst those for gas control wires are painted red.

*Mooring Point, &c.*—The mooring point in L 33 was much damaged by fire, but enough was left to show that the arrangement was very similar to those found in LZ 85 and L 15. Plates 24 and 25 refer to the arrangement in LZ 85, but the following remarks apply to L 33.

As shown in Plate 26, two large tubs for the trail-ropes are situated on the floor of the keel, forward of frame 5. These tubs are 2 feet 7½ inches in diameter and about 2 feet 5 inches high, and, like the handling rope tubs described below, they are made of aluminium sheeting ·025 inch thick. They have open tops, and flat bottoms fitted with a release catch worked through control wires from the control cars.

The forward trail-rope, which is understood to have been 340 feet long, was attached to a short length of steel wire rope passing out through a notch in the flat bottom, and connecting to a double wire strop of 0·785 inch diameter rope. This strop has a circular iron thimble at each end—one thimble being held by the slip bolt (shown in Plate 24) and the other being attached to the following systems of wire:—

(a) To two wires ·629 inch diameter, six-strand rope, which are in turn connected to the upper joints of frame 5.

(b) To steel wire ropes running forward and aft, as shown in Plate 24, by means of which the longitudinal pull is distributed to the joints of the "G" girders with frames 3, 7, 9, and 11.

The slip bolt is forged steel 2⅜ inches in diameter in the body; one end being reduced to 1 inch in diameter and screwed to fit a nut carrying a sprocket wheel. The latter is rotated by a chain drive from a crank handle on a low pedestal immediately above, and the nut being prevented from moving laterally, the rotation withdraws the bolt, releases the thimble of the double strop, and so slips the trailing rope.

It was stated by the crew that this slipping device is hardly ever used.

The handling rope tubs are six in number, 1 foot 4 inches in diameter and 1 foot 9 inches high, and they are placed as shown in Plates D to G (in packet).

The arrangement of the handling ropes is described in Chapter X., page 81. Those attached to points on the G longitudinals are connected to steel wire rope strops passed round castings in the joints of these longitudinals with the main transverse frames. It is to be noted that these castings are exactly similar to those which take the upper ends of the struts for the three middle line cars.

At frames 7, 17, and 31, strops, with wire rope pendants 21 feet to 23 feet long, are attached to the F joints on both sides, but no castings of the type mentioned above are fitted at these joints.

The tubs described above are fitted for the handling ropes which are carried in the ship, but strops and castings are also fitted at the G longitudinal joints at every main frame between 5 and 31, except No. 23, for use with ropes carried by the landing parties.

The trail ropes of L 33 were 3·31 inches in circumference; weight, 1·43 lbs. per fath.; breaking load, 6,550 lbs. The handling ropes were 1·5 inches in circumference; weight, 0·27 lb. per fath.; breaking load, 1,455 lbs. They were fitted with wooden toggles to facilitate handling.

*Suspension Strops.*—Strops for suspending the ship in the shed are fixed in A girder at frames 5, 7, 9, 11, 13, 15, 17, 21, 25, 27, 29, and 31, and in both the B girders at frames 7, 9, 11, 13, 21, 23, 25 and 27. These strops are simply passed round the radial wire castings in the main joints.

*Forward Gun Platform.*—This was found in fairly complete condition in L. 33, and is illustrated in Plates 27 and 43.

It will be seen that the structure of the platform is exceedingly strong and deep, and at the sides additional support is provided by diagonal struts, which distribute a part of the load to frames 6 and 7 and to C longitudinal (*see* Plate A, in packet).

It would appear that this platform is capable of carrying heavier guns than the two or three machine guns actually mounted upon it. Presumably the heavier guns would be mounted if hostile rigid airships were known to exist.

The platform is covered with corrugated duralumin sheeting ·032 inch thick. It is not known whether any means were provided for preventing water from finding its way through the platform. The gun pedestals can be housed down and a cover laced over the platform, so as to present a fair external surface.

The gun tripods are described in the armament chapter. The structure immediately below the tripods is shown in Plate 43, the top plate being of duralumin ⅛ inch thick.

Access from the keel to the platform is provided by an access shaft about 2 feet 10 inches by 2 feet 4 inches in cross-section on the starboard side just aft of the platform. It is constructed of channel rings about 16 inches apart, with four longitudinal channels, and is fitted with a hatch-cover at top and bottom.

Communication to the platform from the control car is made through a "voice-pipe," 1⅝ inches diameter, with separate whistle tube of aluminium, and by two gongs worked by wires. The voice-pipe, whistle tube, and wires pass up the access tube.

In addition to the two tripod mountings and guns actually found on the platform, traces were found which were believed to indicate that a third gun was carried on the A longitudinal girder which passed over the middle of the platform.

*After Gun Platform.*—The after gun platform is shown in Plate 28. It is made of three-plywood, ·2 inch thick, covered with light cotton fabric on top. A small aluminium scupper pipe is fitted at each after corner of this platform. The gun is mounted on the A girder just abaft the platform so as to command a wide range. A notice upon the platform states that not more than two persons are to be upon it at one time.

Access to the platform is obtained by means of a sloping ladder consisting of a triangular girder with crossbars of channel on the upper face.

The foot of this ladder rests upon a small plywood platform just abaft frame 39, and level with the keel walking way, which, as stated previously, terminates at that frame.

Communication from the control car is by telephone, which, however, is stated to have been disconnected owing to fear that sparking might occur and produce a fire.

A screen of outer cover fabric was fitted in frame 39 to prevent gas in the ring space from passing aft, inside the outer cover, to the after gun platform.

*Tail Structure. See* Plates 18 and 28.—At the extreme after end a cap is fitted carrying a horizontal flagstaff.

*Cars.*—The number and arrangement of the cars in L. 33 have been described in dealing with the general arrangement of the ship. Generally speaking, the structure of the cars follows closely the practice followed in previous Zeppelin ships.

The cars are all of good stream-line form, with rounded bows and fine sterns, permitting a good flow of air to the propellers placed at their after ends. In this connection the control and forward engine cars are considered in combination.

All the cars are covered in, the lower portions being plated, while the upper parts are covered with canvas stretched on a light framing, a part of which is made portable to facilitate removal of the machinery. The car structures are entirely built of duralumin, the bottom plating and the lower parts of the sides being each ·048 inch thick; while the upper strakes of plating are corrugated sheeting ·029 inch thick. The transverse framing consists generally of narrow girders formed of two opposed channels connected together by bracings (*see* Plate 46). Stronger frames of rectangular or triangular section are introduced as necessary.

There are two deep longitudinal girders under each engine extending to the after end of the car so as effectively to support the propeller gear box, &c., and to preserve the alignment of the shafting.

In the wing cars these girders are built up of channels as shown in Plates 46-7, in the forward and after engine cars they have lightened plate webs of 3½ mm. duralumin plate, with channel stiffeners and stays, and fitted along the top edge with two channels and top plate similar to those shown in Plate 47. The longitudinal girders under the two forward engines of the after car are arranged so as to connect to the girders of the after engine.

The upper parts of the cars are framed transversely in a similar manner to the lower parts, with lattice girders running longitudinally in the roof of the cars.

In the tops of the wing cars there is a central rectangular girder running the whole length of the cars, *see* Plate 46.

In the control car a triangular girder runs completely round the edge of the top (*see* Plate 48).

Light wooden battens are fitted between the frames in the curved parts, so as to give the correct shape to the canvas covering.

Cellon windows are fitted as necessary for lighting the engine cars.

All of the cars are fitted with handrails of wood or steel tubing low down on each side, for use by the landing party when the ship is being handled on the ground.

Drain valves are fitted in the bottoms of all the cars.

Landing buffers are fitted under all except the wing cars as shown in Plate A. Each of these consists of a "coracle" about 7 feet 6 inches long by 3 feet 6 inches wide and 1 foot 6 inches deep, of shape somewhat like a small dinghy. The framework is of strong cane bound together with cords and covered externally with strong canvas, protected underneath by a number of short lengths of rope sewn on to the canvas. The "coracle" is secured firmly to the bottom plating of the car by leather straps threaded through castings riveted to the plating. Inside the "coracle" are two bags of single-ply rubbered fabric inflated with air.

Access to the cars from the hull is obtained in each case through a hatch, 2 feet 6 inches by 2 feet 6 inches, in the top of the car. A similar hatch is fitted in the hull cover, and a wooden ladder completes the arrangement.

The control car is also fitted with a door on the starboard side, for use when the ship is on the ground, and a door in the after end for access to the forward engine car. A corresponding door is fitted in the forward end of the latter car. The after car was not fitted with a door, but steps were fitted up one side, leading to an opening in the canvas covering.

The suspensions, as indicated in Plate A, are of the usual Zeppelin type, the cars being connected to the hull rigidly by struts and wire ropes.

The struts are in some cases steel tubes 5·8 centimetres in diameter with streamline casing of ply wood; the remaining struts are of pine wood shaped to a streamline section externally and hollowed out for lightness. In all cases the struts have end socket castings of aluminium alloy, attaching to universal joints in the hull girders and on the cars. In former ships a ball and socket joint was used for this purpose, but the neat universal joint now used is superior.

The wire ropes for car suspensions are very carefully faired, as shown in Plate 49, to reduce resistance.

*The Control Car* has the shape and dimensions shown in Plate 48.

An important feature—noticeable also in photographs of L 20—is the manner in which the upper part of the forward end is built up from the edge of a projecting shelf 16½ inches wide. The bomb sights are fitted over two windows in this shelf, and can by this means be used over a considerable range extending to a line vertically downwards.

The bomb switchboards are placed at each side of the car close to the sighting positions. The bomb shutters under the bomb positions in the keel and the safety catches of the bomb-dropping hooks are probably controlled from the starboard side of the car.

Inside the forward end of the car a transverse shelf of duralumin, 2 feet in greatest width, carries the compass and the steering wheel and gear box.

The elevator control wheel and gear box are on the port side, and close to them the controls for gasbag valves and water ballast. Opposite to these, on the starboard side, are the engine telegraphs, and the remaining controls (to handling rope tubs, &c.)

The W/T apparatus and the Commander's quarters are in the after part of this car, which also contains a storage battery of 30 ampère-hours capacity for lighting and for driving a motor alternator for W/T purposes.

*The Forward Engine Car*, as previously stated, is shaped so as to form a continuation of the form of the control car, and is separated from the latter by a gap of 2 or 3 inches, which is covered in externally by a canvas strip laced to eyelets on both cars and is filled in with packing to deaden the vibration. The machinery arrangements of all the engine cars are described in the chapter on machinery.

*The Wing Cars* (Plates 29, 44, and 45) are chiefly remarkable for their neat and compact form. They are each fitted with a gun position on the outboard side, but it appears that these guns were not usually carried.

*The After Car* is the principal machinery compartment, and must have weighed nearly 2½ tons complete. The side plating is carried higher than in the other cars, no doubt with a view to imparting the greater strength required. An auxiliary control position is arranged at its forward end. An internal shelf carrying a steering wheel, compass, &c., is fitted as in the control car, and an elevator control wheel is also fitted on the port side, but no other controls can be worked from this position.

The dimensions of the respective cars are as follows:—

|  | Ft. | ins. |
|---|---|---|
| Control car:— | | |
| Length, including projection at fore end | 31 | 4½ |
| Beam, excluding projection at fore end | 6 | 9 |
| Height | 7 | 3 |
| Forward engine car:— | | |
| Length | 18 | 9 |
| Wing cars:— | | |
| Length | 19 | 4 |
| Beam | 5 | 2½ |
| Height | 7 | 4 |

After car:—

|   | Ft. | ins. |
|---|---|---|
| Length | 39 | 6 |
| Beam | 8 | 2 |
| Height | 7 | 3 |

*Propeller Shaft Supports and A Brackets for Wing Propellers.*—The arrangement of these is shown in Plates 20, 36, and 49, which also shows the manner in which the main frame, No. 25, has been strengthened locally by the introduction of strut girders between the tops of the kingposts. In order to support efficiently the long propeller bevel-gear box two A brackets are fitted, one attached to frame 25 and one attached to a partial transverse frame 1 metre abaft 25 and extending up to D′ longitudinal.

Each A bracket is built of duralumin tubes, with cast aluminium alloy junction pieces. The attachments of the tube ends to the frames are universal joints of the type shown in Plate 20. The tubes are all fitted with stream-line fairing of plywood shaped as shown in Plate 49. The thrust is transmitted to an aluminium tube which is connected to the forward end of the gear box by a ball and socket joint, and which branches, abreast frame 24, into two similar tubes, the forward ends of which are connected to frame 23 at E and F longitudinals, respectively, by ball and socket joints.

Each of these three tubes is built out of two pieces of aluminium sheeting with soldered joints, each piece of sheeting forming one-half the length of the tube. Two of the tubes when tested under compression failed at a mean load of 3,000 lbs.

Additional tubes and wire rope stays are introduced to support the A brackets as shown.

The propeller shafts have three intermediate supporting brackets, one of which is suspended from the A brackets and two from the hull girders. The two latter have universal joints at their attachments to the girders. These brackets are made of elliptical tubes to reduce air resistance.

A minor point worth notice is that all wire ropes which, if they broke, might foul the propellers are fitted with preventer wires. These consist of short lengths of light stranded wire attached at one end to the structure close to the ends of the wire ropes, and at the other end to the wire ropes themselves clear of the fittings, *e.g.*, turnbuckles, so that if the latter should fail the wire ropes would be prevented from falling out of position.

*Plywood Panels in way of Propellers.*—These are fitted to protect the gasbags, &c., from the fragments of ice which are thrown off by the propellers under certain conditions. They are shown in Plates 50 and A (in packet).

**General.**

The workmanship of L. 33 is generally good. The rivet holes in the girders are, however, not very regular, and the English practice of using jigs in drilling them has clearly not been followed. No washers are used under the rivets. The rivet points appear to have been knocked up first with a flat-pointed punch and then rounded down with a snap: probably machine riveters were used. Rivets 3 millimetres diameter are used throughout, except in special places, where 4 millimetre and 5 millimetre rivets are used, and the proportions of the heads and points are approximately as in Fig. 2.

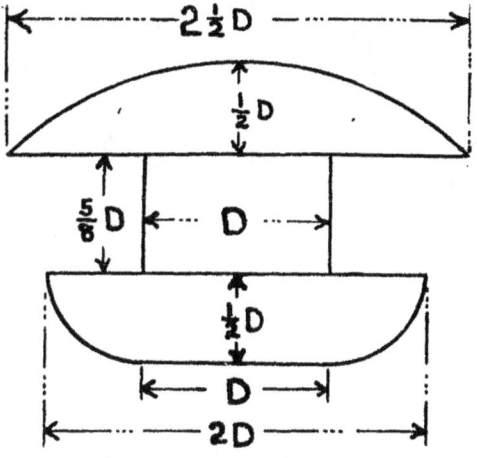

Fig. 2.—Enlarged View of Rivet.

No protective varnish or paint is applied to the duralumin structure.

L 33.—*Chemical Analysis of Materials.*

| | Longitudinal Channels of Main Girders. | Cross Bracing of Main Girders. | Rivets. | Heavy Sheet (Control Gear). | Corrugated Sheeting (Car Covering). | Girder (Petrol Tank Support). | Casting of Lever Mounting (Clutch Gear). | Main Casting from Clutch. | Propeller Strut (Oval Tube). | Pulley Housing. | Hull Casting. | Thimble Casting. | Fitting for Diametral Wires in Transverse Frames. |
|---|---|---|---|---|---|---|---|---|---|---|---|---|---|
| Silicon (as silica) | 0·54 | 0·57 | 0·45 | 0·52 | 0·52 | 0·55 | 0·61 | 0·62 | 0·51 | 0·53 | 0·48 | 0·46 | 0·49 |
| Silicon | — | — | — | — | — | — | — | Tr. | — | 0·11 | — | — | — |
| Lead | — | — | — | — | — | — | — | — | — | — | 0·16 | 0·11 | — |
| Copper | 3·96 | 4·05 | 0·93 | 4·20 | 4·08 | 4·07 | 0·52 | 0·66 | Tr. | Tr. | 2·10 | 3·80 | 4·30 |
| Iron | 0·80 | 0·73 | 0·51 | 0·58 | 0·47 | 0·41 | 0·48 | 0·57 | 0·44 | 0·53 | 0·30 | 0·64 | 0·45 |
| Zinc | Tr. | Tr. | — | Tr. | Tr. | Tr. | — | 12·18 | — | — | 13·29 | 10·10 | — |
| Manganese | 0·50 | 0·52 | 0·41 | 0·54 | 0·51 | 0·52 | 0·05 | Tr. | Tr. | Tr. | Tr. | Tr. | 0·52 |
| Magnesium | 0·51 | 0·51 | 0·15 | 0·55 | 0·51 | 0·51 | Tr. | Tr. | Tr. | Tr. | — | — | — |
| Aluminium (diff.) | 93·69 | 93·62 | 97·55 | 93·61 | 93·91 | 93·94 | 98·34 | 85·97 | 99·05 | 98·83 | 83·67 | 84·89 | 94·24 |
| | 100·00 | 100·00 | 100·00 | 100·00 | 100·00 | 100·00 | 100·00 | 100·00 | 100·00 | 100·00 | 100·00 | 100·00 | 100·00 |

### Communications.

It is understood that in the daytime the general scheme of communications was by means of flags in L 33, as appears to be the case in other German rigids. Some of the methods of signalling orders was exhibited on coloured cards in each car. A copy of one of these from L 32 is shown in Plate 30.

Ship's telegraphs are fitted between the control car and the various engine cars, and were operated by means of 1 millimetre steel wires. These telegraphs were stated to be satisfactory.

Loud-speaking telephones are fitted between the control car and the after machine gun platform and keel gangway, as has already been remarked, but their use was discouraged owing to danger from sparking.

Speaking tubes are fitted from the control car to the crew space, bomb positions, and forward gun platform. They were reported as unsatisfactory, owing to noise and vibration.

It appears that in L 33 orders were frequently transmitted by a messenger.

### Remarks on Analysis of Materials.

*Main Structure.*—It will be seen from the table on page 28 that the structural material generally consists of a typical duralumin alloy containing about 4 per cent. of copper and ½ per cent. each of manganese and magnesium. The impurities are not particularly heavy, except that in the main girders the proportion of iron is higher than previously found. This may be an intentional addition, but probably arises from accidental contamination (due to iron or steel tools) in manufacture. The high copper content places the duralumin among the hardest varieties of that alloy, and the comparatively low ductility found in mechanical tests agrees with this.

*Rivets.*—In all previous Zeppelin structures examined the rivets were practically "pure" aluminium, but in the present case there is nearly 1 per cent. copper with a little manganese and magnesium, suggesting a "dilute" duralumin. The amount of magnesium is sufficient to produce a small amount of hardening, and the rivets are probably considerably stronger than those of pure aluminium.

It is noteworthy that the pulley housings and similar spinnings were made of an impure aluminium without any attempt at hardening.

### Tensile Tests.

*(a) Rolled and Cast Duralumin and Aluminium Alloys.*

| Item. | Mean Yield Stress. Tons Inches. | Mean Ultimate Stress. Tons Inches. | Percentage Elongation on 2 Inches. |
|---|---|---|---|
| Main girders, channel bar ·048 inch thick | 14·47 | 24·30 | 16·0 |
| Plate from gondola ·026 inch thick | 14·09 | 25·62 | 11·0 |
| Plate from engine girder, ·144 inch thick | 18·1 | 28·85 | 11·0 |
| Tube (2·3 inches by ·06 inch) | 22·6 | 26·3 | 16·0 |
| Hull casting | — | 8·7–11·0 | — |
| Diametral wire fitting in main transverse frame | — | 16·6 | 17·0 |

*(b) Drawn Steel Wires.*

| | Mean Yield Stress. Tons Inches. | Mean Ultimate Stress. Tons Inches. | Percentage Elongation on 2 Inches. |
|---|---|---|---|
| 3-millimetre rudder control wire | 93·5 | 99·2 | — |
| 3-millimetre major diagonal wire | — | 98·1 | — |
| 2-millimetre minor diagonal wire | — | 97·0 | — |

*(c) Stranded Steel Wires.*

| Wire. | Diameter. | Construction. | Breaking Load. |
|---|---|---|---|
| | Millimetres. | | Lbs. |
| Car suspension | 6·2 | 1 by 75 | 10,000 |
| "      " | 6·2 | 6 by 19, hemp core | 5,000 |
| Central wire through gasbags | 6·2 | 1 by 75 | 8,750 |
| Flexible rudder control | 6·1 | 11 by 19 hemp core | 3,750 |

**Appendix to Construction Chapter.**

The following information with reference to L 32 has been obtained from various sources.

*Speeds of L 32.*                                                                 Kilometres per Hour.

    With all motors, maximum power - - - - - - 100
    With all motors, normal - - - - - - - - 95·2
    Without port after motor - - - - - - - 90
    With three after motors - - - - - - - - 75·6
    With three forward motors - - - - - - - 68·5
    With both side motors - - - - - - - 63
    With forward and after motors - - - - - - 61½

*Consumption of Petrol in L 32.*—A table is attached showing consumptions and also distribution for two typical long journeys.

An analysis of the petrol consumption at varying heights and speeds shows the following figures:—

| Height in Metres. | Temperature in Degrees C. | Time of Flight. Hours. | Average Speed. Km/Hour. | Petrol in Kg/Hour. | Weather. |
|---|---|---|---|---|---|
| 300 | 19° | 4 | 68·2 | 245 | Misty. |
| 350–2,000 | 19°–5° | 3¼ | 63·3 | 263 | ,, |
| 2,000–3,800 | 0° | 4¼ | 65·5 | 230 | ,, |
| 3,800–400 | 5° | 3½ | 93·5 | 259 | Raining. |
| 2,000 | 0° | 5 | 69 | 250 | Clear. |
| 3,800 | 5° | 3 | 68½ | 235 | ,, |
| 0–800–0 | — | 1½ | 75 | 310 | Cloudy. |
| 1,000 | 5° | 2½ | 68 | 245 | Clear. |

| Tank No. | Trip No. 13. | | | | | Trip No. 8. | | | | | |
|---|---|---|---|---|---|---|---|---|---|---|---|
| | 2.30 p.m. | 6 p.m. | 8 p.m. | 10 p.m. | 12 p.m. | 2 p.m. | 4 p.m. | 6 p.m. | 8 p.m. | 11 p.m. | 2 a.m. |
| 2 | — | — | — | — | — | — | — | — | — | — | — |
| 3 | — | — | — | — | — | — | — | — | — | — | — |
| 4 | — | — | — | — | — | 190 | 190 | 190 | 190 | 190 | 190 |
| 5 | 190 | 190 | 120 | — | — | 190 | 110 | — | — | — | — |
| 24 | — | — | — | — | — | — | — | — | — | — | — |
| 25 | 200 | 140 | 120 | — | — | 200 | 200 | 200 | 150 | 150 | 50 |
| a2 | 200 | 200 | 200 | 200 | 250 | 200 | 200 | 200 | 200 | 100 | 85 |
| a3 | 200 | 100 | 100 | 100 | | 200 | 200 | 200 | 135 | 110 | 100 |
| b1 | 200 | 200 | 200 | 200 | — | 200 | 200 | 200 | 200 | 100 | 100 |
| b2 | 200 | 105 | 105 | 105 | — | 200 | 195 | 195 | 130 | 100 | 75 |
| b3 | 200 | 200 | 200 | 200 | — | 200 | 200 | 200 | 200 | 100 | 75 |
| b4 | 200 | 115 | 115 | 115 | — | 200 | 185 | 185 | 125 | 100 | 100 |
| 26 | 200 | 200 | 180 | 180 | 180 | — | — | — | — | — | — |
| 27 | — | — | — | — | — | 190 | 190 | 190 | 145 | 140 | 60 |
| 6 | 190 | 190 | 190 | 190 | 190 | 190 | 110 | — | — | — | — |
| 7 | — | — | — | — | — | — | — | — | — | — | — |
| 8 | 190 | 190 | 150 | 150 | 150 | 190 | 110 | — | — | — | — |
| 9 | 200 | 30 | — | — | — | 190 | 190 | 190 | 140 | 140 | 100 |
| c2 | 200 | 140 | 140 | 140 | 140 | 200 | 195 | 195 | 140 | 100 | 100 |
| c3 | 200 | 200 | 200 | 200 | 100 | 200 | 200 | 200 | 200 | 100 | 100 |
| c6 | 200 | 100 | 100 | 100 | — | 200 | 200 | 200 | 180 | 120 | 80 |
| c7 | 200 | 200 | 200 | 200 | — | 200 | 200 | 200 | 200 | 130 | 80 |
| 10 | 190 | 85 | 80 | 30 | — | 150 | — | — | — | — | — |
| 11 | — | — | — | — | — | — | — | — | — | — | — |
| 12 | 190 | 190 | 80 | — | — | 190 | 190 | 90 | 40 | 40 | — |
| 13 | — | — | — | — | — | 195 | 195 | 195 | 195 | 190 | 190 |
| 28 | — | — | — | — | — | — | — | — | — | — | — |
| 29 | 190 | 80 | 80 | 30 | — | 155 | 135 | 80 | 40 | 40 | — |
| 14 | 190 | 190 | 190 | 190 | — | 195 | 195 | 195 | 125 | 125 | — |
| 15 | 190 | 190 | 190 | 190 | — | — | — | — | — | — | — |
| 16 | 190 | 190 | 190 | 190 | — | 190 | 190 | 190 | 190 | 190 | 190 |
| 17 | 190 | 190 | 190 | 190 | — | — | — | — | — | — | — |
| 18 | — | — | — | — | — | — | — | — | — | — | — |
| 19 | 190 | 190 | 190 | 190 | — | 190 | 190 | 190 | 190 | 190 | 190 |

| Tank No. | Trip No. 13. | | | | | Trip No. 8. | | | | | |
|---|---|---|---|---|---|---|---|---|---|---|---|
| | 2.30 p.m. | 6 p.m. | 8 p.m. | 10 p.m. | 12 p.m. | 2 p.m. | 4 p.m. | 6 p.m. | 8 p.m. | 11 p.m. | 2 a.m. |
| 30 | 190 | 190 | 190 | 190 | — | 195 | 195 | 195 | 195 | 190 | 190 |
| 31 | — | — | — | — | — | — | — | — | — | — | — |
| 20 | — | — | — | — | — | — | — | — | — | — | — |
| 21 | 190 | 190 | 140 | 110 | — | 195 | 195 | 195 | 150 | 150 | 65 |
| 22 | 190 | 140 | 130 | 90 | — | 195 | 85 | — | — | — | — |
| 23 | — | — | — | — | — | — | — | — | — | — | — |
| 32 | — | — | — | — | — | — | — | — | — | — | — |
| 33 | 190 | 190 | 190 | 190 | — | 190 | 190 | 190 | 190 | 190 | 190 |
| d1 | 200 | 200 | 200 | 200 | — | 200 | 200 | 200 | 200 | 120 | 120 |
| d2 | 200 | 125 | 125 | 125 | — | 200 | 195 | 195 | 140 | 100 | 80 |

## Comparison of Petrol Tank Numbers.—L 32 & L 33.

| L 32. | L 33. | L 32. | L 33. |
|---|---|---|---|
| 2 | 46 | 10 | 105 |
| 3 | 47 | 11 | 106 |
| 4 | 48 | 12 | 107 |
| 5 | 49 | 13 | 108 |
| 24 | 51 | 28 | 109a |
| 25 | a1 | 29 | 109b |
| a2 | a2 | 14 | 121 |
| a3 | a3 | 15 | 123 |
| b1 | b1 | 16 | 125 |
| b2 | b2 | 17 | 126 |
| b3 | b3 | 18 | 127 |
| b4 | b4 | 19 | 128 |
| 26 | 91a | 30 | 129a |
| 27 | 91b | 31 | 129b |
| 6 | 92 | 20 | 136 |
| 7 | 93 | 21 | 137 |
| 8 | 94 | 22 | 138 |
| 9 | 95 | 23 | 139 |
| c2 | c2 | 32 | 141 |
| c3 | c3 | 33 | 143 |
| c6 | c6 | d1 | d1 |
| c7 | c7 | d2 | d2 |

**Observation Car used by a German Rigid Airship.**

The following is a description of an observation car dropped in one of the home counties by a German military rigid airship taking part in the raid of the 2nd–3rd September 1916 (*see* Plate 31):—

    Length overall - - - - - - - 13 feet 6 inches.
    Maximum depth - - - - - - - 4 „ 3 „
    Maximum width across forward part of body - - 2 „ 10 „
    Width across fins (of which there were two vertical
      and two horizontal) - - - - - - 4 „
    Weight - - - - - - - - - 122 lbs. approximately.

The body is built of thin aluminium sheeting, stiffened with channel frames ($\frac{3}{4}$ inch by $\frac{3}{4}$ inch by $\frac{3}{4}$ inch) about 15 inches apart. In the top of the body is a sliding access hatch about 2 feet wide and 2 feet 7 inches long. At the forward end are six Cellon windows — two on each side and two larger ones (22 inches long and from 9 to 18 inches wide) in the under part of the bow. In the lower part of the body is a light wooden platform, on which the observer rested face downwards in a convenient position for making observations through the two large windows in the lower part of the bow. A hinged chart table is fitted in the bow, above the large windows. A telephone is placed at the port side of this table.

The car was slung by four wire slings, attached to eyeplates riveted on the side. In order to distribute the tensions, taut wires were passed under the body of the car connecting the eyeplates on opposite sides of the car. It may be assumed that the upper ends of the four wire slings were attached to a steel ring, to which also the suspension rope from the airship above would be secured. The suspension wire rope is 5½ millimetres diameter, galvanised steel wire rope, seven strands of 14 wires each, with an insulated telephone cable as core. Its weight is 6·5 ozs. per fathom and its breaking strain 5,170 lbs. This suspension wire rope was 3,600 feet long, and was attached at its upper end to a winch.

The winch is constructed for the most part of aluminium or aluminium alloy parts, the exceptions being the gear wheels, the shafts, and the brake details. The winding drum is 8 inches in diameter and 12½ inches long. The drum is operated through spur gears from a secondary shaft, with a centrifugal brake, and also a hand-operated brake. The secondary shaft can be worked by hand or by means of a chain driving a sprocket upon it; the means by which this chain was driven are not known. A band brake is fitted upon the main winding drum (*see* Plate 32, Figs. 113 and 114).

The dropping of the above car would appear to have been the result of an accident. The German airship stopped her engines and lowered the car, and the wire cable to which it was attached broke. It is probable that the pawls of the winch becoming disengaged allowed the cable to run out suddenly. Marks on the teeth of the gear wheels of the winch, which later was dropped some distance away, were possibly caused by a bar having been thrust between the teeth to check the descent of the car.

According to information obtained an observation car is provided for many of the German rigids, but is not usually carried. Its original form was a species of basket in which an observer was lowered when the ship was only drifting; such was the type used at the bombardment of Antwerp. A more elaborate type was then developed into which the observer was strapped in a sitting posture, the land beneath being visible between his feet. This again appears only to have been used when the airship was drifting, and the maximum depth to which it was lowered was stated to be about 2,250 feet. The type described above was said to be initiated by the Schütte-Lanz firm, who were experimenting with it at their Leipzig works. These cars are intended for use in foggy or cloudy weather, and if used when the ship is under power it is considered essential that the ship's course be not suddenly altered.

**Schütte-Lanz Airships.**

Schütte-Lanz airships are mentioned at various parts of this chapter in comparison with Zeppelin practice; but they are of sufficient importance to merit a brief separate description.

It is shown in Chapter I. that Schütte-Lanz airships have closely followed Zeppelin development in regard to their general dimensions, horse-power, and speed. The later types appear to have been entirely successful, and it is understood that the principal reason for the relatively small numbers completed is the slow rate of their construction, due, in part, to bad workshop management.

The following details refer more particularly to SL Nos. 7–9.

*Form.*—The form is characterised by very fine lines, the parallel body being short and the tail long and tapering to a fine point. In transverse section the form is a polygon with 20 sides, the top and bottom sides being 1 metre shorter than the others.

*Structure.*—The structure is arranged in the same manner as in Zeppelin ships, as will be seen from Plate 51. The girders are, however, built entirely of wood, the use of metals being confined to such parts as the steel connection plates at the main joints, steel wires, wire ropes, and fittings, steel frames to rudders and elevators, &c.

Typical girders are shown in Plates 52–53. The ingenious use made of 3-ply wood in these girders is very notable and the strength obtained is remarkable, being comparable, weight for weight, with that obtained from duralumin girders. The connections are all made with a special casein cement, not only in building the girders, but also in connecting them together at the main joints.

The ply-wood is made up of three layers of aspen.

The triangular girders are used for main transverse frames and for the ordinary longitudinals, the apex of the triangle being turned inwards in the former and outwards in the latter. The rectangular form is used for the two top longitudinals and for the keel walking way. The longitudinals at the main joints ride on the outside of the transverse frames instead of intersecting them as, for example, in Zeppelin LZ 85. The intermediate transverse frames, however, are level with the longitudinals; i.e., they intersect the latter in the usual way.

*Keel.*—The keel is internal and has no girder strength beyond that required for supporting the bombs, &c.

*Hull Wiring.*—The wiring in the main transverse frames is mainly radial, with the addition of four chord wires ranged round the frame so that each subtends an angle of about 90° at the centre.

In SL 11 (Cuffley) a longitudinal wire rope was stretched along the axis of the ship as in L 33, and was connected to the centre ring of each main transverse frame.

The diagonal wiring on the outside of the hull is very light, consisting of steel wires 1 mm. diameter doubled on each diagonal of each panel.

There are no inner diagonal wires, as in L 33, for support of the gas bags, but, as stated in the chapter on fabric, their place is taken by circumferential wires about 15 inches apart, ·8 mm. in diameter, cotton covered, which pass completely round the ship on the inside of the longitudinals, and are attached to each of the latter.

*Fins, Rudders, and Elevators.*—The fins are shaped as in recent Zeppelin practice, but the girders are wooden, and in each fin there are longitudinal girders spaced about 1 metre apart, in addition to transverse girders or booms 2 to $2\frac{1}{2}$ metres apart.

The rudders and elevator frames are said to have been made of light steel tubing with acetylene welded joints. The tube at the axis of each plane is extended inwards to connect by a flange joint to the axis tube of the opposite plane. The two rudders and the two elevators are thus rigidly connected together in pairs, and each pair is operated by a single sector on the part of the axis tube within the hull.

*Cars.*—There are five cars—a control and a separate power car forward; two wing power cars amidships; and a power car aft. Each of the power cars contains a single engine of about 240 horse-power driving a pusher propeller through spur reduction gear. This type of propeller drive was first introduced by Schütte-Lanz designers and was borrowed from them by Zeppelin. The wing cars are placed as far as possible from the middle line and well up the ship's sides; their propellers are fitted with reversing gear. The wing cars are held in position by suspension wires and by horizontal transverse struts fitted between them and the hull girders. The forward and after power cars have no reversing gear. All three middle line cars are suspended from the hull by wire ropes only, without the suspension struts fitted to Zeppelin cars. In consequence, these cars are less liable to damage in landing, and it is not found necessary to fit them with landing buffers.

In case the ship should come to ground with a slightly excessive downward speed, the ship is relieved of the weight of the cars as soon as the latter touch the ground, and the hull of the ship is therefore rapidly checked by its own excess of buoyancy. It would appear that in a Zeppelin ship, under similar conditions, the hull would be relieved of the weight of the cars when these touched the ground, and that the hull would then be checked, not only by its own excess of buoyancy, but also by the expenditure of energy required to break the car suspension struts.

The Schütte-Lanz cars have light wooden hulls and the engine cars have also strong internal longitudinal girders built largely of steel tubes and sections under the engine and transverse gear.

*Armament.*—This is understood to be similarly placed to that of the 1915 type of Zeppelin airship.

## CHAPTER IV.—MACHINERY AND PROPELLERS.

The motive power of L. 33 was provided by six independent Maybach motors of 240 h.p. each, normal revolutions 1,400, developing together an estimated total of 1,440 h.p. Each motor drove a two-bladed propeller. The motors were situated as follows:—

| | |
|---|---|
| Forward car | 1 motor. |
| Starboard car | 1 motor. |
| Port car | 1 motor. |
| Aft car | 3 motors. |
| Total | 6 motors. |

The power is transmitted to the propellers through Hele Shaw type friction clutches. A dog clutch is also provided between the plate clutch and the propeller to allow the motors to be run on the ground, as the friction clutches do not "free" completely, and also admits of the alternator being run with the propeller declutched.

Propellers which are driven through plain reduction gearing are geared down about 2·6 to 1. The two wing propellers are driven through a reversing bevel gear box, the gear ratios being 1·35 to 1 up to the wing shaft, and 3·65 to 1 down in the wing box; thus the wing shaft rotates at 1,885 r.p.m., with engines turning at 1,400. All propellers are fitted with band brakes, so that when the ship is landing the non-reversing propellers can be held horizontally, otherwise there is considerable risk of their being broken in case the cars touch the ground, and to facilitate operating the reverse gear in the case of the reversible propellers.

### Propellers.

*Diameters and Pitch of Propellers.*—The propellers examined vary somewhat in diameter. In L. 31 the diameter is 5·1 metres (16·7 feet); in L. 32,[*] 5·0 metres (16·4 feet); in L. 33, 5·3 metres (17·3 feet); and in L. 34, 4·7 metres (15·4 feet). The available propeller remains were not sufficient to tell definitely whether there was any variation of diameter between propellers on different parts of any one ship. The largest diameter, 5·3 metres, is on the wing drive of L. 33 (*see* Plates 54 and 59). The propellers are all two-bladed.

In the Schütte-Lanz ship (S.L. 11) destroyed at Cuffley the diameter of the one whole propeller found was 5·3 metres (17·3 feet), with a pitch of 3·6 metres (11·8 feet).

These large diameters are necessitated by the slow rate of revolution, *i.e.*, 516 r.p.m. on the wing propellers, and 540 r.p.m. on the stern propellers.

It is the English practice to run propellers at about 1,000 r.p.m. and to use smaller diameters. Whilst this practice is sound for aeroplanes—and especially for seaplanes, which work over a range of from 40 to 80 knots—for airships, which work at approximately a constant speed, this slow-running large-diameter propeller has a decided advantage. The blades of one of the propellers examined are between 16 and 17 inches wide, and it has, therefore, been possible to make a ratio of thickness to chord as small as ·08 and ·09 on the working parts of the blade, thus further increasing the efficiency. It is considered that the latter may be as high as 80 per cent.—not including losses in gearing.

The plan form is between a Chauvière and an Admiralty shape, the trailing edge being straight and lying in the plane of revolution.

The constant incidence method of design has not been followed. All the propellers appear to be a constant 4·0 metres (13·1 feet) pitch.

---

[*] Notes have been found which state "Pusher, 5·1 long (16·7 feet), pitch, 4; side (this may mean either wing propellers or side cars), 5·2 (17 feet).

*Hub-fitting.*—The propellers fitted abreast are inward turning, while the middle line propellers are right-handed. They are fixed on to the hub as shown in Fig. 142, the annulus of this hub being filled with plaster of paris. The long taper makes it difficult to remove the hub from the shaft.

*Weight of Propellers.*—The weight of an undamaged propeller from L. 33, including the steel bearing through the centre of the hub, is 148 lbs. A similar one from L. 32, slightly broken at the end of one blade, and with no bearing in hub, weighs 121 lbs.

*Details of Leading and Trailing Edges.*—A brass beading covers the leading edge for its entire length and passes down approximately one-third of the trailing edge. The method of attachment is interesting, and is shown in following sketch (*see* Fig. 3):—

FIG. 3.    FIG. 4.

A copper wire is carried up the trailing edge of both blades of the propeller, making the connection at one end with the brass beading and at the other end with the framework of the engine (*see* Fig. 4). It is considered that the purpose of this wire may be to prevent any considerable accumulation of electric charge taking place at any part of the propeller blade. The effects of a large accumulation of charge would be:—

(*a*) A possible sparking.

(*b*) An electro-magnetic field would be set up when the propeller is rotated. A big charge rotated at a great speed might produce a field strong enough to affect a compass in the vicinity.

It is also suggested that the object is to bring the framework of the airship to the same electric potential as the atmosphere through which the airship might be passing.

The timber used has been examined by Professor Groom, of the Imperial College of Science, and the following details are from his report.

*Veneer.*—All the blades examined are externally clothed with veneer. The latter is mainly of walnut, though on one blade of one of the Billericay propellers the inmost veneer is composed of a worthless white wood. The outermost veneer had its grain parallel to the length and the two others parallel to the breadth of the blade. Each layer of the veneer commences at the tip of the blade.

The innermost layer is carried to within seven feet, the intermediate layer to within five feet, and the outer layer to three feet from the centre of the propeller. Thus the three layers of walnut do not constitute a true three-ply. All veneers are knife-cut. The wood lying within is plane-scratched to increase adhesion, and the external veneer is, in places, folded over the margins of the blade, so that its joints are along the broad faces of the latter. But in the case of one blade, composed entirely of walnut, the veneer ceases near the thick margin and a lamination is exposed.

*Functions of the Walnut Veneer.*—Fully seasoned walnut is unsurpassed for "standing well" when it has been finished and polished, so that it is excellently adapted for protecting wood lying within it from warping when exposed to drying or wetting influence. This is probably the main significance of the walnut veneer, which thus allows the use within it of woods which, unprotected, would warp considerably (*e.g.*, elm used in S.L. 11, the Cuffley airship). One of the defects of an opaque covering concealing the laminations is that it provides facilities for the dishonest manufacturer to use defective wood. In one propeller, for example, which belonged to the Cuffley airship a piece of ash which was quite rotten in parts was found—also a short piece of ash about one foot in length was employed to fill up a deficiency.

*Woods composing the Laminations.*—The following kinds of woods were observed in the propeller blades:—

(1) West African Mahogany: Sapele.

(2) West African Mahogany: a kind among several exported from the Cameroons under the name of Bonamba.

(3) West African Mahogany: a third more indeterminate type, probably one of the kinds imported under the name of "brown cherry mahogany."

(4) Central American Mahogany: Honduras (with tiered rays).

(5) Central American Mahogany: Honduras (without tiered rays).

(6) West African "acacia": (satin-wood, "yellow wood").

(7) American walnut.

(8) European walnut.

(9) European elm.

(10) European ash.

(11) A porous red, probably tropical, wood.

(12) A tawny unrecognised wood recalling mahogany.

Except in the case of blades composed wholly of American walnut, the propellers showed the contiguous laminations composed of different kinds of woods, though the same kind of wood tended to be repeated in alternate laminations. It cannot be safely assumed that the variety of woods used in German rigid airship propellers is used merely on account of shortage of appropriate wood. The alternating combination of different kinds of woods (all arranged with their main grain generally parallel) as a means of checking warping is familiar, as, for instance, in billiard cues. Moreover, the alternation of harder, stiffer, and softer, more flexible woods in the blades provides a means of regulating bending and resilience. The composition of the propeller blades of these airships proves that, with an external coating of walnut veneer, cheaper woods may be used to replace Honduras mahogany or walnut.

In English practice the purpose is served by fabric-covering the propeller.

One important fact in reference to the woods used in the German propellers is that a number of them (especially West African mahoganies) are truly cross-grained (with "interlocking fibre"), so that these will not split so readily along the grain as, for example, does Honduras mahogany. They may, however, if unprotected, tend to twist more when exposed to draught or moisture.

*Orientation of the Laminations.*—The cross sections of the laminations show that these boards are not accurately cut along the radii ("quarter sawn"), nor perpendicular to this, but are mainly intermediate between these as regards cut. As seen in cross section of the propeller blade the contiguous laminations are so arranged that their rays (radii) impinge at angles varying from nearly 90° to nearly 180°. Thus there appear to be no special arrangements made, as regards "cut" and orientation of the laminations, to secure symmetrical shrinkage, bending, or rigidity.

The laminations are, in the majority of blades, 22 mm. thick, though in Billericay blades I., II., and V. they were 16 mms. thick.

Many laminations are composed of two pieces glued side by side, and, in a few cases, scarf joints are found in the length of the lamination.

The following table refers to Plate 60, Figs. 167 to 171. The first column gives the name of the place in which the particular airship fell. The second column gives the distinguishing number of the blade. The third gives the number of the laminations.

## SPECIAL DESCRIPTIONS.

| Airship. | Blade. | Laminations. | Wood. | Comments. |
|---|---|---|---|---|
| Cuffley | I. | 1 and 2 | Honduras mahogany | Merely a marginal fragment of a blade. |
|  | II. | 1<br>2<br>3<br>4 | Walnut, European elm, European walnut, American ash, European. | Only a fragment of a blade. |
| Little Wigborough | I.<br>(See Fig. 167.) | 1 | West African mahogany. | A dark variety. |
|  |  | 2, 4, 8 | Central American mahogany. | No. 2 probably being Honduras. |
|  |  | 3, 5, 7 | West African mahogany. | Wood being similar to, but not identical with, Sapele mahogany, and being one of the kinds exported from Cameroons under the name of "Bonamba" mahogany. |
|  |  | 6 | West African mahogany. | Similar to another Cameroon mahogany exported as Bonamba mahogany. |
|  | II.<br>(See Fig. 168.) | 1, 2, 6 | West African mahogany. | Bonamba-like, agreeing with 6 of blade I. |
|  |  | 3, 5, 7 | West African mahogany. | Sapele type. |
|  |  | 4 | Unrecognised | A porous red wood, probably tropical. |
|  | III.<br>(See Fig. 169.) | All | American walnut. |  |
| Potters Bar | I.,<br>a fragment.<br>(See Fig. 171.) | 1 | West African mahogany. | Wood agreeing with lamination 1 of blade I. from Little Wigborough. |
|  |  | 2 and 4 | West African mahogany. | Bonamba type, agreeing with laminations 1, 2, 6 of blade II., and lamination 6 of blade I., from Little Wigborough. |
|  |  | 3 | Unrecognised | Resembling a yellowish mahogany. |
|  |  | 5 | West African satinwood. | Wood also known as West African "acacia," West African "yellow-wood," and even misnamed "white mahogany." |
|  | II. | All | American walnut. |  |
| Billericay | I. | All | American walnut. |  |
|  | II. | All | American walnut. |  |
|  | III.<br>(See Fig. 170.) | 1, 6, 8 | West African mahogany. | Sapele type. |
|  |  | 2 and 4 | Central American mahogany. | Probably not Honduras. |
|  |  | 3 and 7 | West African mahogany. | Of the more ordinary cross-grained type, agreeing with one of the varieties exported from the Cameroons. |
|  |  | 5 | West African mahogany. |  |
|  | IV. | All | American walnut. |  |
|  | V. | All | American walnut. |  |

**Machinery of L. 33.**

The engines of L. 33 are 6-cylinder vertical water-cooled Maybach engines, 150 by 180-mm. bore and stroke (*see* Plate 55). The compression ratio is 5·68 to 1. Each cylinder has five valves working vertically by means of an overhead rocking gear, which is lubricated by pipes from main oil system. There are three exhaust and two inlet valves in each cylinder (*see* Plate 56, Figs. 151 and 152). The cylinders are separate, and consist of a steel barrel threaded at the top end, and screwed into the cast-iron cylinder head and water jacket combined; the compression joint is metal to metal, the top face of cylinder barrel jamming hard up against a shoulder turned in cylinder head.

The cylinders are bolted down to the crank case by four claws and bolts, which run through the crank case and secure the main bearing caps, thereby permitting of the use of a light crank chamber, as the explosion stresses are taken by these bolts and not by the crank case itself.

The bottom edge of the water jacket beds down on to a rubber ring which is fitted into a groove turned in the cylinder barrel. Any one cylinder can be removed from the engine without disturbing the remainder, and it is said that this operation occupies about one hour.

The water joint between the cylinders, which, when bolted in place, stand about $\frac{5}{16}$ inch apart, consists of an aluminium ring C, over which is a thick rubber ring B, and over all a brass strap A, which is tensioned up by means of a screw. As the brass strap is tightened up the rubber bulges out, and is forced tightly against the water jacket faces D, so preventing water leaks (Fig. 5). This form of joint gives trouble should the jacket temperature exceed 80° C., owing to the rubber softening:—

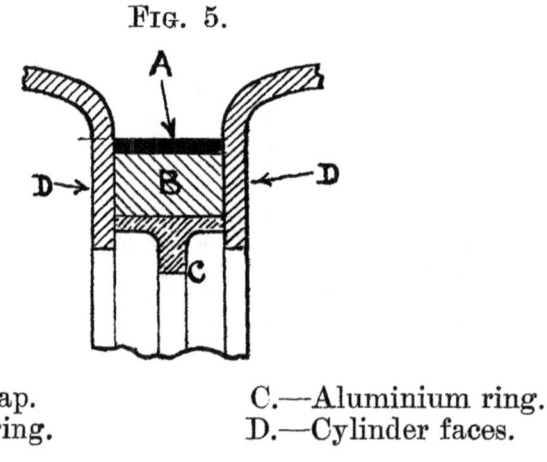

Fig. 5.

A.—Brass strap.　　　　C.—Aluminium ring.
B.—Rubber ring.　　　　D.—Cylinder faces.

*Water-cooling System* (*see* Plate 61).—The water-cooling system is Thermo syphon, with an accelerator of the propeller type driven at twice crank speed by means of bevel gears. This accelerator is at the forward end of the engine, and is in a vertical position. A bad feature in the arrangement of this pump is that the spindle gland is situated inside the timing gear case; thus, in case of any leakage, the circulating water finds its way direct into crank case. There is a large screw-down greaser fitted to this stuffing box to help to prevent leaks.

The water circulation is as follows (*see* Plate 61, Fig. 173):—It is forced by the accelerator through the forward carburettor, thence through cylinder jackets through aft carburettor, through the jacket round the exhaust branch to an aluminium tank mounted in the framework above the engine and holding about 14 gallons. The radiator is about 4 feet square, and is kept full of water by means of this tank, which is situated above it. The outlet from the radiator is led to the accelerator inlet. The radiator weighs about 220 lbs. empty and holds about 5 gallons of water. The radiator tubes are $\frac{1}{4}$ inch diameter by ·005 inch thick, $5\frac{7}{8}$ inches long, and expanded at the ends to ·28 inch diameter. The radiator is provided with a blind to regulate the amount of air passing through it.

*Connecting Rod Cooling Arrangements.*—Owing to trouble which occurs with the big ends of connecting rods it has been found necessary to cool them by means of air currents. The cooling arrangement is as follows:—Opposite each crank pin on each side of crank case is an opening about 2 inches diameter. On one side of the crank case these are fitted with gauze-covered cowls, while those on the other side are connected to an aluminium manifold which is coupled to a large cowl fixed over the side of car and facing aft (*see* Fig. 6, also Plate 56, Fig. 154). The speed of the ship through the air induces a draught through the crank case, thus cooling the bearings and incidentally ventilating the crank chamber and the car. The tube leading from the manifold to cowl is fitted with a butterfly valve to control the draught through crank case. From the appearance of the inside of this cowl it would appear that a considerable quantity of oil is lost overboard by this cooling arrangement.

Fig. 6.

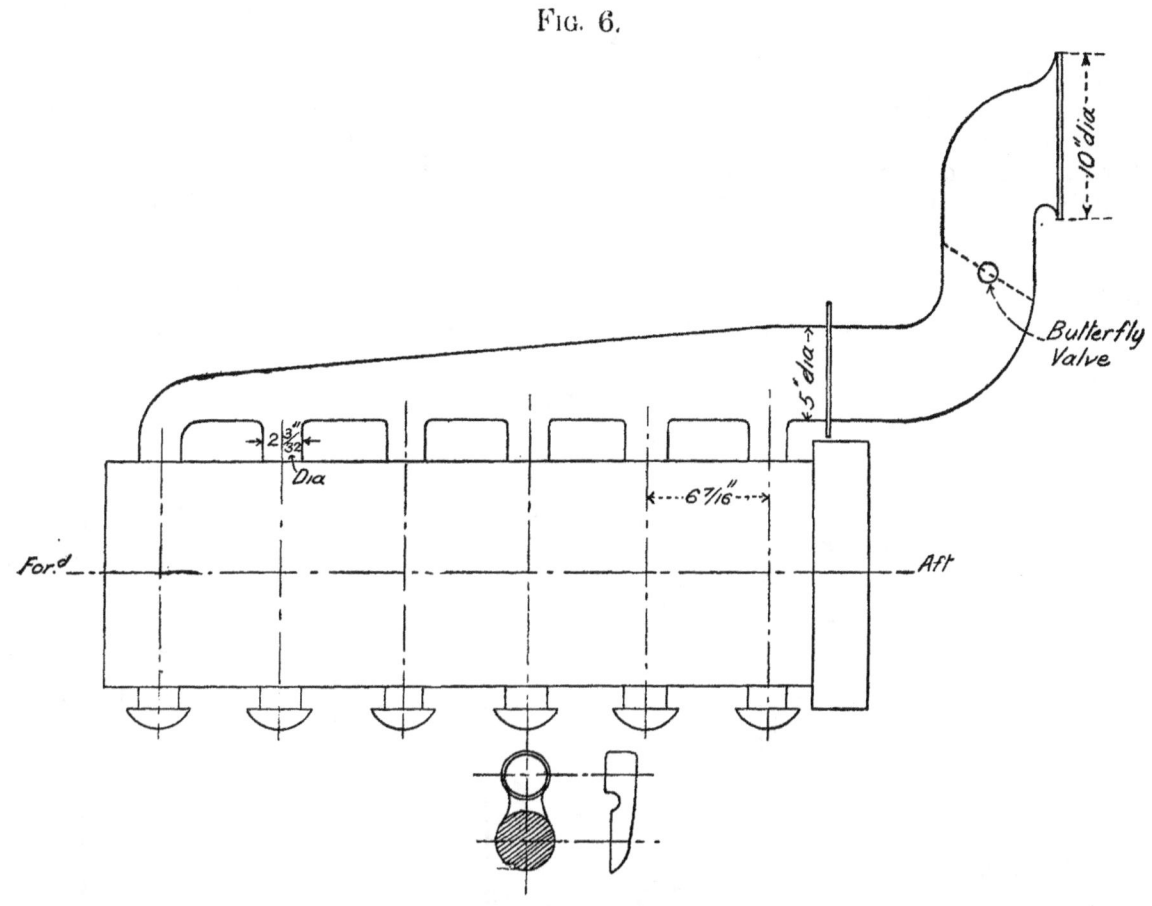

Crank Case Vent.

*Carburettors.*—There are two carburettors, situated one at each end of the engine bolted on to the cylinders.

The jet, choke, and throttle all being interconnected by levers, open and close together.

The jet consists of a 2·5-mm. hole in a plate, and over this is mounted eccentrically another plate with a countersunk hole in it. This latter plate rotates over the jet by means of levers coupled up to the throttle lever, and in this manner the jet orifice is varied to suit different throttle openings much in the same way as in the White and Poppe carburettor.

The choke is really a shutter which rises and falls over the main air orifice. The throttle is of barrel type, made so as to uncover an extra air port to the mixing chamber, allowing air to enter without passing over the jet.

The carburettors have no float chamber. The petrol is pumped into a small rectangular box holding about one-eighth of a pint situated above the jet, and an overflow from this tank leads back to a float chamber from which the pump sucks its petrol. This box maintains a constant head of $2\frac{1}{4}$ inches of petrol over another box arranged with a cylindrical overflow weir which maintains a constant level slightly below that of the jet itself. This double overflow system is probably necessary in order to obtain sufficient steadiness on the level actually supplying the jet itself. There is also a needle valve in the carburettor which is screwed into the jet, thereby rendering the strength of the mixture easily adjustable when required. The main function of this needle valve, however, is to shut off petrol from the carburettors in case of oil pressure failure or of the engine racing. This will be explained in more detail later on. There is no adjustment provided for altitude.

*Petrol Supply System* (*see* Fig. 7).—The petrol is supplied to the carburettors by means of a small plunger pump driven off the crank shaft by means of spiral gears. The bore and stroke of the pump are 40 mm. and 20 mm. respectively, and it is single acting. This pump draws its petrol from a float chamber situated beneath the engine, and supplies it to the small reservoirs placed above the jets, whence it flows by gravity to the space surrounding the jet. The surplus from this space overflows and joins the overflow from the reservoir, and is led back to the float chamber.

The intermittent nature of the discharge from the slow-running single-acting petrol pump is not entirely eliminated by the small air vessel fitted. The small reservoir provides a means of securing a steady flow.

The float chamber is merely a brass chamber containing a brass float and valve arranged to cut off the petrol supply from the main tanks in the keel at a certain level. Should the float stick for any reason with the valves open, petrol is lost overboard through an overflow pipe. This is stated to have occurred on occasion.

A hand pump is fitted to prime the carburettors when starting up.

An interesting point in connection with the petrol pump is the fact that no petrol enters the cylinder; air only enters the cylinder. The air pulsates up and down the pipe, the valves being arranged in a chamber fitted to the bottom of crank case.

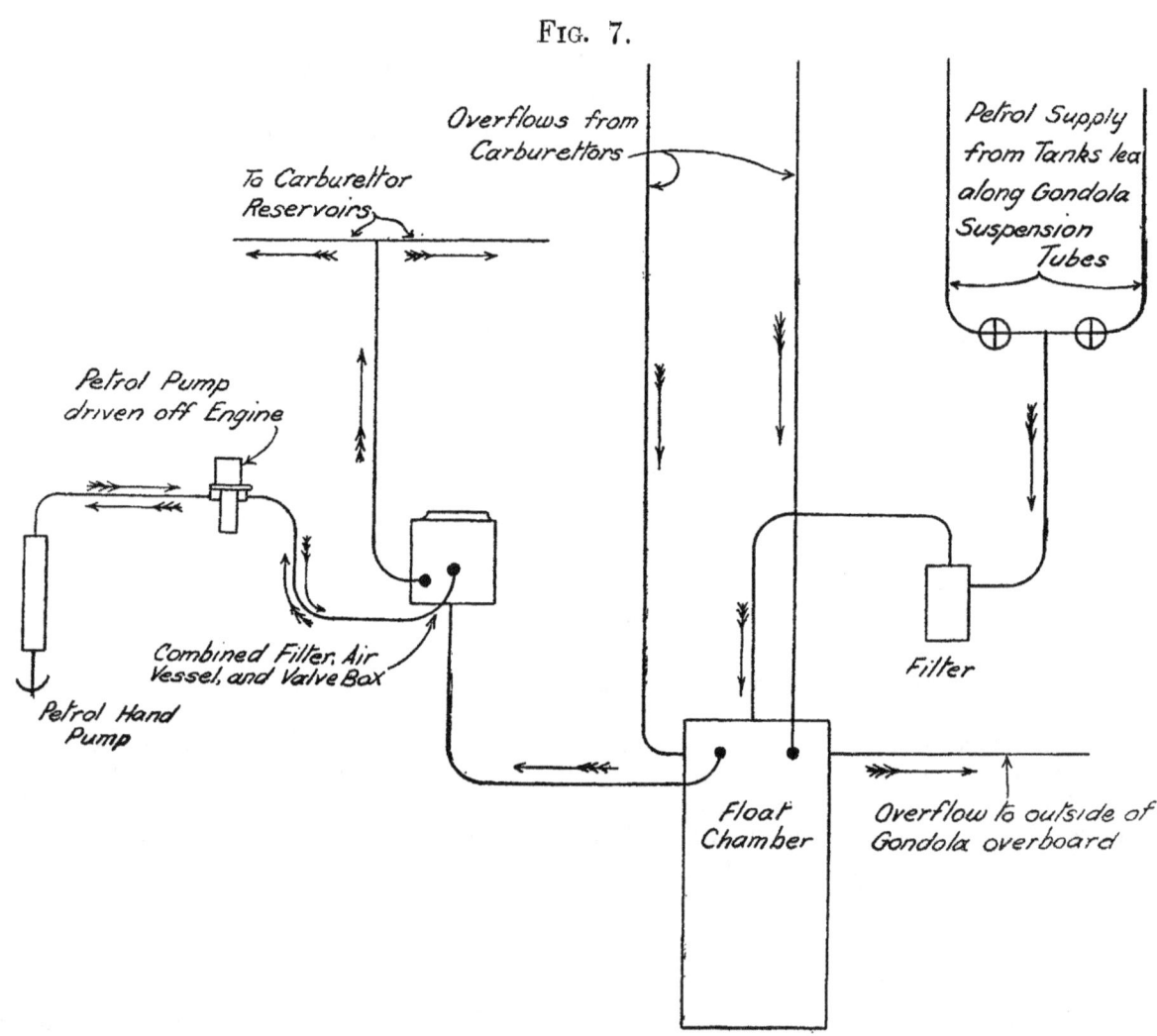

PETROL SYSTEM.—DIAGRAMMATIC ARRANGEMENT.

*Lubricating System* (*see* Fig. 8).—The crank case is kept empty, the oil running by gravity from the base to an oil tank fixed under the engine (*see* Plate 57, Fig. 157, and Plate 62, Fig. 175). There is a reciprocating pump P, driven off the same crank pin as the petrol pump. This pump sucks its oil from the sump beneath the engine and discharges to two filters FF arranged in parallel; thence to an air vessel A, thence to a honeycomb radiator R fitted outside the gondola. The outlet from the radiator is taken to another filter fixed to the bottom of crank case, and thence to the main bearings, each of which has a small filter of its own. A pressure gauge is fitted to the discharge, reading to 5 kg. per square cm.

(70 lbs. per square inch). A relief valve is also fitted to the discharge of the pump, and is led to an open funnel on the sump in full view of the mechanic in charge. The radiator can be cut off from the system at will by means of a cock, so that the oil can be warmed up quickly on starting up.

An aluminium reserve tank holding 28 gallons is fitted to each power unit, and each tank has a float indicator working on a spiral.

The outlet from the tank is led through a cock to the suction side of pump, so that when fresh oil is needed the pump may suck clean oil and discharge it direct to the filter under the crank case and so to the bearings.

A hand pump is provided for priming the system, and also two other hand pumps for pumping dirty oil from the bottom of sump.

The oil-cooling radiator is 1 foot by 1 foot by 2 inches deep, with 12-mm. diameter circular tubes belled out to a hexagonal form at each end to facilitate soldering up.

Lubrication of all main bearings is forced, but the big ends and gudgeons are lubricated centrifugally by means of triangular banjos fixed to each side of crank webs to catch the oil that flows from the cheeks of the main brasses.

Baffles are fitted to each cylinder to prevent an excess of oil from reaching the piston.

Fig. 8.

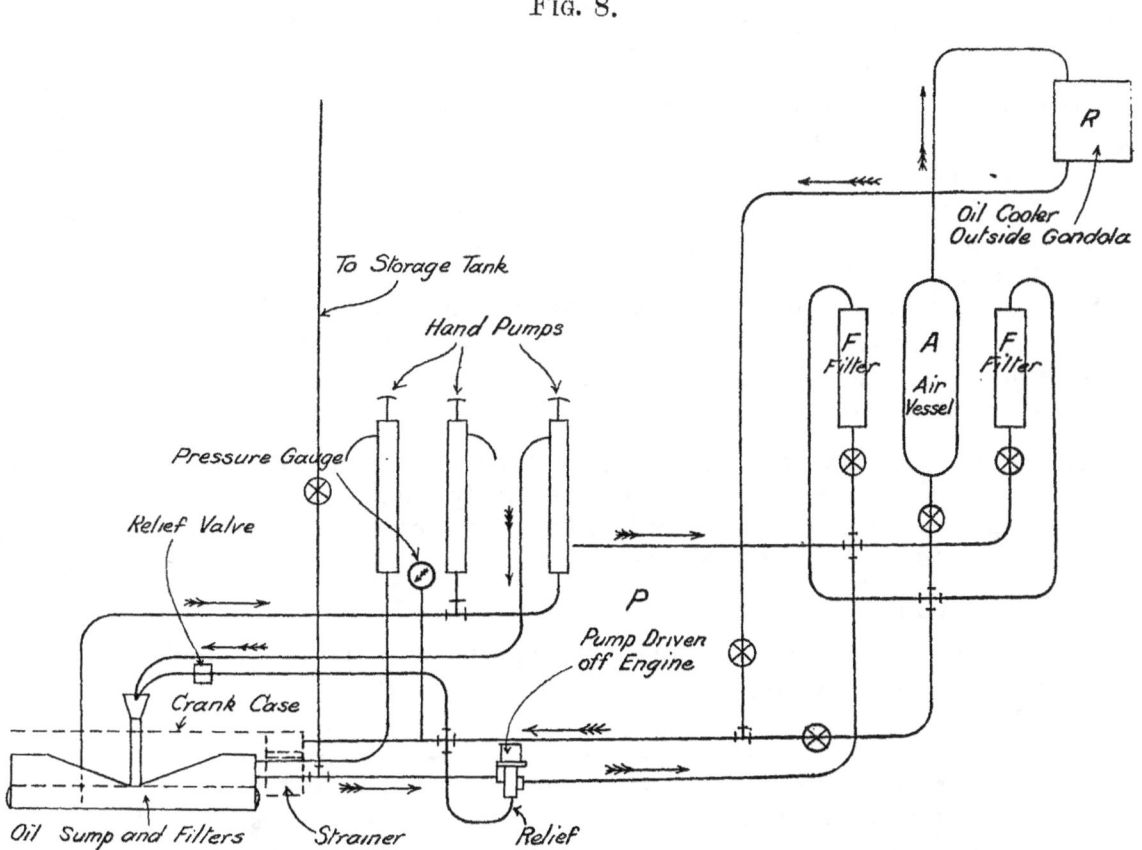

LUBRICATING OIL SYSTEM.—DIAGRAMMATIC ARRANGEMENT.

*Governor Control.*—There is a small centrifugal governor driven by the pumps' shaft which is arranged to stop the engine in case of either excessive speed or a failure of oil pressure (*see* Plate 63). The stoppage of the engine is effected in either case by shutting off the petrol supply by means of needle valves which are arranged to close up the jet orifices. The operation is as follows :—

The needle valves are connected by levers to a spring loaded plunger Q passing into a cylinder at the forward end of crank case. The spring is put in compression by the end of the plunger resting against the periphery of a spring loaded piston P, arranged at right angles to it. This latter piston is arranged to cover up, when oil pressure is sufficient, the hole through which the end of the plunger would project. It will thus be seen that when the oil pressure on the spring loaded piston is normal the hole through which the spring plunger is attempting to force its spindle is blanked off by the piston, and it remains so until the oil pressure is removed from the piston by the action of the governor, in the following manner:—

The governor actuates a spindle R with three collars turned on it, as if it were a small piston valve. This valve works in a brass cylinder fixed in crank case, and the cylinder has three ports cast in it. These connect with the main oil pressure system, the top of the spring loaded piston described above, and with the crank case. In the normal running position the valve admits oil to the above piston. Should the engine race, the governor forces the valve along its cylinder so that the oil pressure is bye-passed to the crank case, thereby relieving the pressure on piston. The piston rises and allows the plunger spring to force the plunger through the hole the piston has now cleared, thus shutting off the petrol from jets. In this way the engine is stopped in cases of low oil pressure or of excessive speed.

When starting, the plunger has to be held against its spring until sufficient oil pressure is obtained to blank off the hole through which it would otherwise slide.

To stop the engine a quick threaded screw is provided to remove the load from governor spring, allowing the governor to act as though the engine were racing.

*Starting Arrangements.*—A shaft runs along each side of the engine at the top of the tappet plunger guides, and is fitted with small cams which engage lugs machined on the tappet inlet and exhaust plungers. When turned through 30° these shafts lift every valve in the engine. At the same time, by means of levers, a cock in the main exhaust manifold closes the passage to the silencer, but opens it to a large exhauster hand pump. The carburettors are then primed by the hand petrol pump referred to in connection with the petrol system, and the exhauster pump is vigorously worked. This draws mixture into the cylinders through the carburettors and inlet valves.

The hand lever is now released; the valves close, the exhaust pipe opens and the charge of gas in the cylinder which happens to be on the firing stroke is ignited by means of a Bosch hand magneto. The governor plunger has, of course, to be held in the running position until the oil pressure has become normal.

There is also fitted to the above valve lifting gear a small link which closes an aluminium flap on each carburettor air inlet, thus increasing the suction on the jet and causing a rich mixture.

*Magnetos.*—The magnetos are Z.H. 6 type, and there are two fitted to each engine, two sparking plugs being fitted in each cylinder. The magnetos are driven through spur gearing from cam shafts at the forward end of the engine, the main timing gears being at the flywheel end.

*General Remarks.*

[Further data as to weights and analyses will be found in Appendix B.]

| | |
|---|---|
| Number of cylinders | 6. |
| Bore | 150 mm. |
| Stroke | 180 mm. |
| Compression ratio | 1 to 5·68. |
| Length of connecting rod between centres | $12\frac{1}{8}$ inches. |
| Length of piston | $5\frac{5}{8}$ inches. |
| Thickness of piston head | $\frac{7}{16}$ inch. |
| Shape ,, ,, | Flat. |
| Diameter of crank shaft journals | $2\frac{5}{8}$ inches. |
| Length | $2\frac{3}{4}$ inches. |
| Type of bearing | White metal in brass housing. |
| Diameter of crank pin | $2\frac{5}{8}$ inches. |
| Length | $2\frac{3}{4}$ inches. |
| Section of crank web | $3\frac{3}{4}$ inches by $1\frac{5}{16}$ inch. |
| Diameter of flywheel | $19\frac{1}{4}$ inches. |
| Diameter of inlet valves | $1\frac{7}{8}$ inches, lift $\frac{5}{16}$ inch. |
| ,, ,, exhaust valves | $1\frac{3}{8}$ inches, lift $\frac{5}{16}$ inch. |

*Clutch.*—The clutch is not rigidly attached to the inside of the flywheel, but the drive is transmitted through eight driving blocks fitted partly in a flanged rim bolted to the flywheel and partly in the clutch casing. The driving blocks consist of a number of strips of leather glued together and secured with brass binding pieces, which latter form the driving surfaces of the blocks. This provides a very good flexible drive.

The clutch is of the Hele Shaw type, having 11 bronze driving plates and 11 steel-driven plates.

The plates are 16 w.g. thick, the mean diameter of the "V" grooves is 165·5 mm., and the grooves 16 mm. wide by 15·5 mm. deep.

The load on the plates is supplied by a single spring, giving 1,300 lbs. when compressed to its running condition. The clutch is operated by a lever working two eccentrics, one over and one below the spring, which further compress the spring and remove the load from the plates.

*Transmission Gear.*—Plate 64, Fig. 177, shows a general arrangement of the transmission gear. A ball bearing and flexible coupling are fitted aft of the clutch and the spur drive casing is fitted at the after end of car in a rigid casing of cast aluminium.

The inboard end of the propeller shaft is supported in the bottom of an intermediate casting which also supports a spherical type flexible coupling on the engine shaft, and a clutch formed by the splined end of the shaft engaging with the coupling sleeve. Between this clutch and the engine clutch a pulley is fitted for driving a W/T alternator by means of a belt. A jockey pulley is fitted for tightening the belt, which is arranged to run on either of two pulleys giving alternative speeds for the alternator shaft for the same speed of engine. The after clutch enables the alternator to be run with the propeller disconnected.

A propeller brake is fitted in the engine shaft abaft the dog clutch.

The spur drive has wheels with straight teeth $2\frac{1}{4}$ inches wide ($7\frac{1}{2}$ module) and 9-inch centres, giving 17 teeth on the pinion and 44 in the propeller shaft, thus reducing the engine speed of 1,400 r.p.m. to 540 r.p.m. at the propeller.

The engine clutch and the propeller brake are operated from the starting end of the engine.

Two engines are fitted abreast in the forward end of the after car, with their transmission gears forward of the engines, so that the space between these engines and the after engine is the control position for all three (*see* Plate 65). The clutches are similar to those for the other engines and connect through a coupling to a bevel gear reversing box (*see* Plate 58, Fig. 163).

The three bevel wheels are always in gear and either of the two on the engine shaft are clutched to the shaft according to the direction of rotation of propeller desired (*see* Plate 58, Fig. 165).

The wheels on the engine shaft have 35 teeth and on the athwartship transmission shaft 26 teeth (7 module), giving a speed of rotation to athwartship shaft of 1,885 r.p.m. at an engine speed of 1,400 r.p.m.

The athwartship shaft is $2\frac{13}{32}$ inches diameter and ·132 inch thick and 40 feet long between centres of reverse gear and propeller boxes. The shaft is in two main lengths and one short length bolted together by flanged couplings. No flexible couplings are fitted. The short length next the propeller box is of larger diameter than the remainder, *i.e.*, $2\frac{3}{4}$ inches tapering at its ends to $2\frac{3}{8}$ inches, this being apparently to assist in taking the bending stress due to the propeller torque.

The maximum unsupported length is 15 feet 6 inches, and it has a first whirling speed at about 440 r.p.m. The engine speed for this speed of shaft is 325 r.p.m., and the engines are stated to have a critical speed at 700 to 900 r.p.m.

It is a matter of interest to note that such a long shaft, with a low whirling speed and supported on a structure of necessarily light design, should be fitted without any flexible couplings. In this connection, the fact that the shaft broke away before the ship landed is significant. The pinion wheel in this shaft has 17 teeth and gears with a crown wheel of 62 teeth (7 module) on the propeller shaft, giving a speed of propeller of 516 r.p.m. (*see* Plate 58, Fig. 164).

The propeller boxes (*see* Plate 58, Figs. 160–2) are supported by aluminium tubes and the hull is stiffened in way of this gear, as described in Chapter III. on Structure.

The propeller thrust is taken by tubes made from aluminium sheet, and taper from 2 inches diameter at the ends to 5 inches at the centres. The plate from which the tube is formed is bent in one piece, and the edges acetylene welded.

*Engine Seatings.*—The type of engine seatings is shown on Plates 44 and 66–9 and consists of two longitudinal vertical plates in way of the engine bearer brackets and horizontal plates connecting these to the sides of the car. Transverse vertical plates are fitted at intervals between the longitudinal plates and the sides of the car. The vertical transverse and longitudinal plates have large lightening holes which give access for cleaning and examination. The tops of the vertical longitudinal plates are fitted with "U" channel bars, on which the engine and forward intermediate bearing are supported (Plates 66 and 69). "U" and angle diagonal bracing bars are provided between the vertical longitudinals and between the longitudinals and the car secured to gusset pieces worked into the structure.

The after intermediate bearing frame and the main gear box are supported between the vertical longitudinals.

The propeller thrust bearing, which is in the after gear box, transmits the thrust to the longitudinal plates and so to the car.

The engine casing is supported on two longitudinal wooden girders built up in layers as shown on Plate 69, Fig. 183. The wooden girders have steel channels fitted to the top and bottom edges with side brackets for connection by pin joints to the brackets fitted on top of the aluminium channel bars over the vertical plates.

Each wooden girder is bolted to, and rests on, two additional brackets P secured rigidly to the inside of the main longitudinal girders. This would appear to indicate that supporting the engine on pin joints was not successful.

Except where otherwise mentioned the cars and seatings are of duralumin.

*Petrol Storage Arrangements.*—The petrol is carried in aluminium tanks each of about 74 gallons total capacity. Plate 70 shows a tank with its aluminium slings in which the tank is supported with the axis vertical. Each tank is fitted with :—

(1) A running-down valve fitted to the projecting portion of the bottom, which forms a sump for the collection of water or sediment. A short internal pipe from the valve connection to the top of the sump prevents any water collected from passing into the petrol system. A plug enables the sump to be completely drained when the petrol has been used to the level of the internal pipe (*see* Plate 71, Fig. 186).

(2) A connection with two separate passages to the tank, one for a short ¼-inch air pipe and the other for connecting the filling pipe used when transferring petrol from other tanks (*see* Plate 71, Fig. 187).

(3) The main filling connection, 1½ inch diameter, with filter formed of a strip of thin aluminium with one edge crinkled wound spirally to form a tube, the end of which is closed. The plain edge of the strip overlapping the crinkled edge forms a large number of small passages for the petrol (*see* Plate 71, Fig. 185).

(4) A depth gauge, which consists of a brass float 2 inches diameter attached to a silk cord and guided in a perforated tube.

The silk cord is led over a wheel in a gauge screwed in the tank top which operates a pointer through suitable gearing showing the weight of petrol in kgs. remaining in the tank. The figures on the gauge and the pointer are covered with luminous paint.

The tanks are suspended in the keel as shown in Plate 72, either on links as fixed tanks or on wires capable of being released, and thus allowing the tank to slide on the guides and through the bottom of the keel.

The fitting for releasing the tanks is shown in Plate 73, and consists of an aluminium casting with a sliding tube operated by a lever. A safety peg which enters a hole in the side of the tube when in position for holding the sling wires prevents the tank being inadvertently slipped or released by vibration.

Provision is made for carrying 54 tanks at various positions along each side of the keel, but the actual number carried is evidently varied to meet the requirements of each particular voyage as regards anticipated length of journey, trim, and the weight of bombs carried. A diagrammatic arrangement of tanks and pipe arrangement is attached (Plate 74).

In the positions where tanks are shown on one side of the keel only ballast bags can generally be carried on the opposite side.

The tanks are numbered as shown, and a key-board is fitted in the control car to indicate the disposition of the tanks on board, on starting and during a voyage.

The groups of tanks over or in way of the cars are fixed tanks, and are numbered separately from the remaining tanks and lettered A, B, C, or D. The remaining tanks are fitted with the slipping device.

These tanks would generally be kept supplied from the slipping tanks and supply the cars by gravity through the running-down valves. The position of these tanks in relation to the cars give ample head in any condition of trim of the ship.

*Pipe Arrangement.*—The running-down pipes are of copper, and, except where otherwise stated, are ⅜ inch external diameter and 18 w.g. thick. A main lead runs along under the keel footway, secured to the structure by aluminium clips, with fabric wound round the pipes in way of the clips, and extends from the forward to the after car.

Each group of fixed tanks on either side of the keel is connected to the main lead with a shut-off valve in each common pipe from the group.

On the tank side of the shut-off valves pipes are led to the cars down the side of suspension tubes for the forward and after cars, and down the side struts of the wing cars. The connections led along the forward struts of wing cars are taken off the main lead with shut-off valves in the main lead to enable these connections to take their supplies from the fixed tanks forward or aft as desired.

Except for these connections, the supplies to the cars can be taken direct from the adjacent fixed tanks as well as from the main run.

The branch pipes to the wings and forward cars are ½ inch external diameter and 18 w.g. thick.

The ends of the supply pipes in the keel over the forward and after cars are fitted with hose connections from, which it appears that all or part of the piping from these connections to the car is flexible, and provides for slight relative movement of these cars and the hull when landing.

Near each group of slipping tanks (except aft of frame 28) valves, each with two branch connections, are fitted to the main run of piping.

One of the branches has a plain end with circumferential grooves suitable for pushing the end of a rubbered hose over, and connecting to a similar connection on the outlet valves of the slipping tanks. The other branch is fitted with a screwed cap, and may be used either for taking the suction pipe of the semi-rotary hand pump (which will be referred to later), or for attaching a hose for running petrol overboard. It is observed that no means are provided for rapidly discharging petrol overboard except by dropping tanks. Sectional valves are fitted in the main lead at frames 11a, 18, and 21.

*Filling Arrangements.*—The 1½-inch connection in the tops of the tanks, each fitted with a strainer, would be used for filling when the ship is in its shed. Arrangements are also provided for transferring petrol from the slipping to the fixed tanks. A main lead of ¾ inch diameter brass piping 24 w.g. thick runs under the footway of the keel, extending from the forward to the after cars, as in the case of the running-down pipe. Branches controlled by valves are fitted to the brass pipe, and led to the tops of one of each pair of fixed tanks, and could be used for filling both by allowing the fuel to level up in the other through the bottom valves. Branches, with valve and screw cap, are also fitted near each group of slipping tanks.

Semi-rotary hand pumps are fitted at frames 11, 21, and 30 to the longitudinal girder which carries the petrol tanks. The suctions to the hand pumps may be taken from the branches on the running-down pipe and the discharges to the connections on the brass filling pipe. Alternatively, the hand pump suction may be taken from a portable pipe, which can be placed in the top of the tank on removing the filter. Thus, the petrol may be transferred from any tank to any other as desired.

It will be noted that, although the petrol tanks can be carried aft of the after car, the main leads of the running-down and filling connections do not extend aft of this car.

It is probable that these tanks are not fitted with running-down valves, and are similar in this respect to the tanks dropped in Essex from L. 15.

The hand pump fitted near these tanks would take its suction from a portable pipe in the top of the tank and discharge to the nearest connection on the brass filling pipe.

*Telegraphs.*—Plate 62, Fig. 174, shows the orders on the telegraph dial in a wing car.

Separate pointers are used for the engine orders and those relating to the machine-gun and lighting of the car.

The transmitters are fitted in the control car forward, but the number fitted could not be ascertained, owing to the wrecked condition of the car.

It is probable that the numbers of transmitters are:—
- 1 for each wing car.
- 1 for forward pair of engines of after car.
- 1 for aft engine of aft car.
- 1 on bulkhead of control car for the forward engine.

The telegraphs are operated by stranded steel wire over the wheels, each wheel having two turns of wire secured to the wheel at the top of the loop to allow equal lengths of wire to run off the wheel either way, and to prevent relative movement of the wire and wheel.

Piano wire is used for the straight lengths along the floor of the keel. Fairleads are fitted at the ship's frames, and each lead runs through a separate space in the group of fairleads.

The system adopted appears to be that of an endless lead which runs over the transmitter wheel in the control car and its corresponding wheel on the receiver in the machinery compartment, each lead passing over a barrel operating a gong for each order given.

The transmitters and receiver for the forward pair of engines in the after car would require orders for reversing the direction of rotation of their propellers.

## CHAPTER V.—FABRIC AND GAS VALVES.

**Gas Bags.**

*General Arrangement of Gas Bags.*—Except at the extreme ends, the gas bags are made to fill as completely as possible the space inside the hull (*see* Plate A (in packet)). Each gas bag extends the full length between consecutive main transverse frames. In L 33 this distance was 10 metres (32·8 feet) generally. The gas bags are made to fit over the triangular keel space, and are laced to the top girder of the keel. There is said to be no other attachment of the gas bags to the structure, except by the sleeves of the manœuvring valves, where fitted, at the top. There were 19 gas bags in L 33, which occupied the whole length of the ship with the exception of 1 metre (3·28 feet) at the bow and 10 metres (32·8 feet) at the tail.

In Schütte-Lanz ships the arrangement of the gas bags is similar to that described above.

*Gas Bag Materials.*—The gas bags are usually very completely destroyed in case of fire, and as all of the airship wrecks to which British experts have obtained access, except L 15, have either been burned in the air or have been set on fire by their crews after landing, the information available is somewhat slender. In the case of L 15, samples of gas-bag fabric were found to be constructed of a single-ply cotton which, as regards weight and strength, is comparable to Admiralty "D" quality. It is stated to come from the Cellenstoffabrik G.m.b.H. Templehoff, Berlin, and also from a firm in Augsburg.

The cloth is lined on the inside with two layers of goldbeaters' skin. Adhesion is obtained by means of gelatine or glue, of which there is a considerable quantity present. It would appear that the cloth is well impregnated with glue before the skins are applied.

The proportions by weight of the various components of the fabric are roughly:—

| | |
|---|---|
| Weight of cotton | 60 Gm/m². |
| Weight of skin | 55 Gm/m². |
| Weight of glue and varnish | 45 Gm/m². |
| Total | 160 Gm/m². |

The approximate strength is 880 kg/metre warp.
" " " 556 kg/metre weft.

*Seams.*—The seams examined are of two types:—
(a) Overlap joints 20–30 mm. wide, by means of which the separate pieces of cloth are joined side by side, and the ends of pieces joined. These seams are not stitched or taped, but merely stuck—apparently with glue. The joints are made before the application of the skins, which are laid over them.

(b) Overlap joints 50–60 mm. wide. These are made after lining with skin, and the skin is included in the joint. These seams appear to be used to build the separate sections, which have been made up, using type (a) of seam, into complete gas bags. In some places a layer of skin is found over the outside of the joint, evidently to stop leaks.

*Other Gas-Bag Fabric examined.*—The only other specimens of Zeppelin gas-bag material examined consist of a few small pieces of unburnt gas bag and some unburnt portions of a roll of material intended apparently for use in repairing gas bags, obtained in both cases from the wreck of L 33.

The former of these two materials was found to be identical with that obtained from L 15—a notable fact when the difference in diameter (and consequently in the gas pressure) of the two ships is considered.

In recent Schütte-Lanz ships the gas-bag fabric has been the same as in Zeppelins, and is stated to be obtained from the same fabric works. It is, however, reported that, until the time when the rubber shortage in Germany became important, Schütte-Lanz gas bags were made of rubbered single-ply cotton, having a total weight of about 250 g/r$^2$.

*Gas Bag Supports and Protection.*—In connection with this matter, the special measures adopted by Zeppelin designers to support and protect the gas bags may be described. They are as follows:—

(i) WIRE SUPPORTS.—In the chapter on Structure mention has been made of the special wiring between the longitudinals for supporting the circumferential surfaces of the gas bags, and of the additional wires at the main frames, whose function it is to fill in the larger gaps between the diametrical wires (*see* Plate 33, Chapter III.), and so to support the ends of the gas bags.

(ii) NETS.—Between the longitudinal girders nets are fitted right round the ship, on the inner side of the special wiring, in way of the gas bags. These nets support the gas bags in the spaces between the special wiring. They are not fitted across the ends of the bags. The nets are composed of cotton cord, having a breaking strength of about 160 lbs., and weighing 4·3 grammes per metre. There is no jackstay round the edges of the individual nets, and the latter are simply tied to the longitudinal and transverse girders by cord at intervals of about 50 cms. (19·7 inches). The mesh of the nets is about 35 cms. (13·8 inches).

There is no evidence that nets were fitted along the sides of the keel, but at the sides of the structure forming the gangway across the ship to the wing cars, nets were found in position having a mesh of only 13 cms. (5·1 inches), the cord being the same as for the larger nets.

It does not appear that covering strips were fitted along the longitudinal or intermediate transverse frames to protect the gas bags, but these girders were designed so as to avoid any exposed edges which might do damage. Cover pieces of cotton fabric were fitted at the sides of the kingpost and truss girders of the main frames which projected in between the gas bags.

All loose wire-ends were carefully covered over with rubbered tape.

In the Schütte-Lanz ships it is stated that nets are not used, and that the gas bags are supported entirely by steel wires ·8 mm. in diam., cotton covered, which extend completely round the ship inside the framework. These wires are spaced about 15 inches apart, measuring fore and aft, and are secured at each longitudinal. This method of support is simpler and lighter than that adopted by Zeppelin designers.

*Filling.*—According to prisoners' statements the gas bags are filled simultaneously, to avoid undue straining of the structure of the ship. The gas is carried to the various bags by fabric hose about 10 inches in diameter from the hydrogen main in the shed. It is said that the attachments to the keel are cast off during the process of filling. An automatic pressure alarm is fitted to give notice when the pressure is excessive. But this is primarily intended for use when the ship is rising rapidly. The signal is given by the gas bag pressing on a plate kept in position by a spring, thereby closing an electric bell circuit to the foremost car.

*Gas.*—The hydrogen is generally prepared by the well-known "water-gas" system of blowing steam over red-hot iron. As the prisoners state that the gangway of the ship usually "smells of gas," it would appear that no very special precautions are taken to purify the gas (*see* Appendix A on Hydrogen).

*Temperature and Condensation of Moisture.*—A thermometer is suspended in the gas bag for finding the temperature of the gas. It is stated that the electrical thermometers tried in L 33 were found satisfactory (*see* Chapter XI.). Exhaust gas is not used to heat the hydrogen. Considerable condensation of moisture is stated to occur even when the ship is not flying but is stationary in the sheds.

*Leakage of Gas and Ventilation.*—It is further stated that considerable quantities of gas are usually present in the ring space and the gangway of the keel, so much so that telephones and other electrical connections likely to give rise to sparks are looked upon with suspicion by some commanders. Ventilation of the gangway is effected by shafts leading from the top of the gangway to the upper part of the ring space and to exhaust cowls on the top of the ship. Details of these will be found in Chapter III.

*Repairs during Flight.*—According to the evidence of the sailmaker of L 33, rents less than 4 feet long and triangular tears and punctures up to 18 inches or 2 feet can be repaired during flight. The repair is carried out simply by gathering up the slack of the gas bag round the hole and tying it off. Small punctures appear to be repaired by sticking on a patch of rubbered fabric by means of rubber solution. The larger holes, after being tied off, can be patched in a similar way.

*Care of Gas Bags.*—The sailmaker of L 33 further asserted that considerable vigilance must be exercised during flight in order to adjust the gas-bag attachments to suit varying conditions, whether the variation be due to surging of the gas when the ship rolls, to the rapid rise of the ship, or to variation in the amount of gas in individual bags. The fabric needs to be looked after carefully to keep it in good condition, but no trouble had been experienced owing to the growth of organisms in the skins.

*Special Fittings in Gas Bags for Shear Wires and Central Wire Ropes.*—It is stated by prisoners that these fittings are jointed to the gas bags after the latter have been delivered at the shed. The type of fitting shown in Plate 78 is used for shear wires, and for the four ends of the axial wire-rope system which attach to frames 1 and 39. In every specimen examined the fabric had been burned away, except between the metal discs, and it is impossible to determine precisely the relevance of the numerous layers of fabric and canvas which in every case were found.

In all cases, however, a disc of rubber was found next to each of the metal discs. Starting from the outer disc, the sequence was somewhat as follows:—Metal disc, rubber disc, cotton fabric, fine canvas, cotton fabric, cotton fabric, fine canvas, cotton fabric, &c., skinned cotton fabric (one or two layers), cotton fabric, fine canvas, rubber, metal disc (inner). The number of layers varied considerably.

The type of fitting shown in Plate 37, was used generally for the axial wire rope. The innermost layer of the joint was formed by a sleeve of rubber tubing. The gas-bag sleeve was, naturally, made of rubbered fabric, and canvas was bound round it to protect it from being damaged by the binding cord.

In S.L. 11, the Schütte-Lanz ship which fell at Cuffley, an axial wire rope was found, but there were no indications of any fittings suitable for making a joint with the gas-bag fabric. Possibly the wire rope passed through a sleeve extending from end to end of the gas bag.

### Outer Covers.

*Arrangement on Ship.*—A careful examination of the wreck of L 33 gives the following information (*see* Plate 79):—

Over the length of the ship, between frames 1 and 39, the outer cover sheets are about 5 metres (16·4 feet) wide in a fore and aft direction, *i.e.*, they each extend the distance between consecutive transverse frames. In the circumference of the ship there are three sheets, viz., one sheet on each side of the ship extending from A to G longitudinal, while the third or bottom sheet is stretched between the two G longitudinals. The side sheets are attached to A longitudinal by lacing hooks hooked over the flanges of the girder angles. In all the other edges lacing eyes are fitted. The vertical edges, *i.e.*, those at the transverse frames, are laced to the wires stretched round the outside of the frames for this purpose (*see* Plate 33). The bottom edges are laced to the edges of the bottom parts and also to G longitudinal. At the bottom corners of each side sheet are fitted very strong canvas lugs, sewn on to canvas doubling pieces and carrying two eyelets each. On the side edges of the sheets at each transverse frame-joint a piece of canvas doubling about 20 by 10 cms. (7·8 inches by 3·9 inches) is sewn on to the fabric and is fitted with five lacing eyelets closely spaced. It would appear that the method of fitting the side sheets in place is as follows:—

(1) Hook the top edges to the A longitudinal.
(2) Pull the bottom corners into position close up to the G girder joints by means of cords attached to the joints and passing through the eyelets in the lugs mentioned above.
(3) Pull the side edges into position at the transverse frame joints.
(4) Lace all edges.
(5) Secure the sheets to longitudinals, wires, &c. by the tapes sewn on to the sheets, as described below.

It is important to note the following means adopted to prevent the outer cover from bagging in the lower part:—

(a) The outer cover sheets are attached to the main longitudinals by tapes composed of double pieces of outer cover fabric which are sewn on to the inside of the cover at intervals of about 80 cms. (31·2 inches) and tied to the inner channels of the longitudinals.
(b) The outer cover sheets are tied to the diagonal bracing wires by means of similar, but shorter, tapes at intervals of about 33 cms. (12·8 inches) along the wires. The remains of these tape attachments were found only at a few parts which had partly escaped the fire, and it is not known how far up the side they were carried, but it may be supposed that they extended up to F longitudinal at least.

Covering strips of outer-cover fabric are stuck over all the edge lacings of the sheets, and it may be presumed that a wide strip was fitted over the A girder and stuck down to the sheets at each side, so as to maintain weathertightness.

In the outer-cover sheets the warp of the fabric is generally fore and aft.

Forward of frame 1, where the surface to be covered is nearly hemispherical, the sheets are made up of gores, with the seams following great circles. The outer edges of these sheets are laced to No. 1 frame, but there is no information as to the number of sheets on the bow or of their other attachments. As stated in Chapter III., a network of wires is stretched between the longitudinals forward of frame 2 to support the outer cover against wind pressure.

The fin covers are of ordinary outer-cover fabric and the sheets covering the opposite sides of each fin are laced together at the outer edges. The manner in which the fin covers are attached to the outer-cover sheets is not known, but there is some evidence that the outer-cover sheets are interrupted at each fin and that their edges, and also the inner edges of the fin covers, are laced to the longitudinal at the root of the fin.

The rudder and elevator covers are also of ordinary outer-cover fabric, and are attached in the manner shown in Plate 75, Figs. 192, 194, and 195.

*Fabric.*—The outer-cover fabric consists of a cotton fabric of similar weight and strength over the whole ship. All the specimens examined except that from SL 11, which had a lighter cloth, have been very similar to the Admiralty A quality. In the case of fabric from LZ 77, which was examined at the Municipal School of Technology, Manchester, the resemblance is stated to be too striking to be accidental. The opinion was expressed that the cloth was woven of Sea Island cotton yarns, and was manufactured in Lancashire.

The weights and strengths found are:—

|  | Weight, Gms/m². | | Strength.—Kilogrammes per Metre. | |
| --- | --- | --- | --- | --- |
|  | Fabric. | Dope. | Warp. | Weft. |
| LZ 77 | 130 | 42 | — | — |
| L 15 | 123 | — | 1,400 | 1,320. |
| L 33 | 135 | 40 | 1,140 | — |
| SL 11 | 95 | 20 | 550 | Somewhat damaged. |

All the specimens, except those from L 15, are printed with a characteristic pattern of small dots, which gives the cloth a uniform grey appearance at a slight distance. The pigment used is ultramarine. This pattern is stated by the prisoners to be for protective purposes, but it has been suggested that the pattern is printed on to disguise the nature of the cloth, in order to facilitate exportation from this country.

The specimens of outer cover obtained from L 15 were left the natural colour of the cotton, or a sort of cream.

*Proofing and Dope.*—The outer covers are all doped. The aim in doping is stated by prisoners to be to render the upper half of the cover proof against the weather, while leaving it as permeable to gas as possible. The lower half, on the other hand, requires to be fairly impermeable and fireproof.

The dope used is a cellulose acetate dope, and is referred to by the prisoners as "Cellon."

With reference to the weight of dope shown in the above table, it is to be noted that all the airships were burnt except L 15, and, in consequence, the portions of outer cover obtained from the wrecks probably all came from the bottom parts of the outer cover, where, if the prisoners' statements were correct, the dope would be of the fireproof type, and heavier than on the upper parts.

It is, however, to be remarked that no evidence has been obtained of the presence of any special fireproofing ingredients, and all specimens examined are moderately inflammable. It may perhaps be inferred that the Germans have not yet been able to produce an efficient fireproofing dope. No provision is made for protecting the outer cover from the flash of machine guns beyond the plating of the gun platform forward.

*Lacing.*—The arrangement of canvas, and lacing eyes at the edges and corners of the outer cover sheets in L 33 is illustrated in Plates 76 and 79. The arrangement of lacing at the outer edge of a fin is shown in Plate 76, No. 3.

The canvas used is generally similar to that used for the water-ballast bags, and is always very thoroughly sewn on to the cover.

The lacing cord is of 3-ply cotton, weighing about 4·3 grammes per metre, and having a breaking strength of 75 kilogrammes.

**Water-ballast Bags in "L 33."**

The general arrangement of the water-ballast bags is shown in Plates D to G (in packet). Details of construction and dimensions of the larger type are shown in Plate H, which also shows the method of attachment to the box girders above.

Each water-ballast bag consists of an inner and outer bag. The inner bag is of parallel 2-ply cotton fabric weighing 430 gms. per square metre, made up of two plies of fabric at 130 gms. per square metre, and one layer of rubber weighing 170 gms. per square metre. The strength is 2,560 kilogrammes per metre. The joints are stuck, with an overlap of 25 cms., and covered on the inside with strips, 25 cms. wide, of rubbered fabric, stuck on.

The outer bag is made of canvas, weighing 700 gms. per square metre. Strength, 3,120 kilogrammes per metre.

The seams in the sides are vertical, the overlap being about 20 cms., with three rows of stitches. A red thread is woven into the canvas 20 cms. from the selvedge, and forms a useful guide in sewing.

The canvas of the outer bag and the fabric of the inner bag are doubled at the lower part near the valve.

The connection of the valve to the bags is made by clamping the fabric of both inner and outer bags down on to a flange surrounding the valve body. Suitable holes in the bags are made to enable this to be done, and clamping is effected by means of a flat ring with numerous bolts passing through the flange (*see* Plate 21).

The discharge pipes are made of thin aluminium, solid drawn, internal diameter about 8 cms., and discharge outside the G girder, except in the case of those bags abaft the after car, which are fitted with short vertical pipes discharging below the bags. Ends of the pipes projecting through the outer cover are of streamline form.

The valve and body is aluminium or aluminium alloy, with a rubber seating. The valve is operated by a wire passing up through the top of the bag to guide sheaves on the supporting girders, and thence to the control cars.

The outer bag is provided at the top edges with stiffeners carrying lacing eyelets corresponding to those in the supporting box girders. The manner in which lacing is effected is shown in Plate 21, Fig. 84.

There is an opening in the top of the bag for filling purposes and two small openings with short lengths of piping, providing alternative passages for the valve operating wire.

As shown in Plate H, the outside of the bag is clearly marked with level lines and the corresponding weight of water, in kilogrammes, the capacity of the larger bag being 1,000 kilogrammes.

The capacity of the smaller water-ballast bags fitted under the 100-kg. bomb positions is not known.

The emergency water-ballast bags are of the form shown in Plate 75. The materials used are similar to those described above, and the fabric of the inner bag extends down to the mouth of the pipe. Each of these bags is laced to a tube above it. The capacity of each bag is 80 kilogrammes of water, and the internal diameter of the discharge pipe is $3\frac{1}{4}$ inches.

### Sundry Fabrics.

In L 33 the covers for the kingposts and trusses are cotton fabric of 130 gms. per square metre.

Canvas of various thickness is used for the fairing of cars. It is painted with "cellon" containing a grey pigment. No attempt appears to be made to fireproof the material.

A small piece of fabric was obtained from the airtight bumping bags in the landing buffers in L 33. This was single-ply cotton weighing 130 gms. per square metre, rubbered 174 gms. per square metre, giving a total of 304 gms. per square metre. A portion of 3-ply rubbered fabric, one ply being canvas, from LZ 77 was said to be from the bumping bag.

A cotton fabric weighing 123 $Gm/m^2$, containing in addition about 30 $Gm/m^2$ of wax, was obtained from L 15. Its purpose is not definitely known. It is pretty clear, however, that it was not generally used to form bulkheads at the transverse frames. The Zeppelins do not appear to have any fabric bulkheads except that abaft the extreme after gasbag from which this particular sample appears to have been taken.

In the L 33 class the outer cover fabric is doubled in way of the propellers placed under the middle line of the hull, and it is supported by closely spaced girders. For this purpose this would appear to take the place of the 3-ply wood ice-shield opposite the wing propellers to which reference is made in Chapter III.

### Gasbag Valves.

In all recent German rigid airships two types of gas valves are fitted. The first type is entirely automatic, and one of them is fitted to every gasbag. The second type is hand-operated, and is fitted only to certain gasbags, *e.g.*, in L 33 there were 19 gasbags, and only nine of these had valves of this type: these are called "Manœuvring valves."

*Type I.*—The design of these appears to vary very little in different ships. A typical example, as found in L 33, is shown in Plate 80. The valve disc consists of an aluminium spinning, forming a rim, with light duralumin channel stiffeners fitted at right angles across it. Gas-tight fabric fills in the centre part of the rim, and is stuck to the latter, as shown in the figure. The outer ring, which constitutes the valve body, is formed of two aluminium spinnings with their edges rolled together in such a way as to form a very rigid hollow ring. On each side of this ring, a light framework of duralumin is fitted; that on the side which is within the gasbag being designed to carry the central spring, which closes the valve, and also to support a ring of aluminium tube which forms one boundary of the valve seating; that on the side which is without the gasbag, being fitted to ensure a free space for the lift of the valve disc, and to enable the gas to escape.

The valve seating is of slightly conical form, and consists of a membrane of doped two-ply fabric, stretched between the outer ring, or valve body, and the tubular aluminium ring mentioned above. The face of this fabric, on which the valve disc bears, is heavily doped so as to present a glazed surface. The valve disc is kept in a central position relative to the outer ring by means of three small stranded steel wires attached to its centre on the outside, and to the outer rim, as shown in the drawing.

The joint of the outer ring with the neck of the gasbag is formed as follows :—

A strip of soft rubber about $\frac{1}{16}$-inch thick is first stuck round the outside of the ring. The gasbag neck is then secured tightly to the rubber by two duralumin securing rings fitted with tightening screws. These two securing rings are in electrical connection with the outer ring by means of small steel wires specially fitted for this purpose, and the outer ring itself is connected electrically to the framework of the ship by means of a small stranded wire. In L 33 these wires were all attached to the "E" longitudinals.

The positions of these valves in this ship appears to be between "F" and "F'" longitudinals, and usually from 3 to 5 feet from one end of the gasbags in which they are fitted. The valves are placed some on one side of the ship and some on the other, without any apparent order. In L 33, the automatic valves amidships are set to blow at 22 mm. pressure, and those towards the end of the ship at 14 mm. pressure.

*Type 2.*—Plates 81 and 77 show one of these valves found in L 33, which is generally similar to those found in previous Zeppelins.

In the valve shown on Plate 81, the valve body consists of an aluminium alloy casting carrying the india-rubber seating-ring as shown, and having three radial arms carrying a guide sleeve for the valve stem. The valve disc is composed of two aluminium sheets dished for stiffness and carrying the valve stem in the middle. The notable feature of this valve is the arrangement of springs and levers for closing it. The springs are four in number. One is carried on the valve stem and exerts a force which will increase in proportion to the lift of the valve. The other three springs act upon levers arranged in the form of toggles, the arrangement being such that the effort they exert to close the valve falls off rapidly as the lift increases and at the full lift their effort in this direction is exceedingly small.

The net effect of the whole combination of springs and levers is that the force required to open the valve is fairly constant throughout its lift. The valve is operated through a simple lever as shown.

Manœuvring valves in L 33 are fitted to gasbags Nos. 2, 5, 6, 7, 8, 11, 12, 14, 15. The valve bodies are fixed to the structure by means of duralumin lugs. The positions in the ship are shown in Plate A, and a typical arrangement of the girders to which the valves were attached as shown in Plate 77. It will be seen that a short girder is introduced across the corner between "A" longitudinal and the main transverse frame in order fully to support the valve, and to carry the sheaves in connection with the operating gear. The arrangement of control wires is described in Chapter III.

## CHAPTER VI.—BOMB-DROPPING GEAR AND ELECTRICAL INSTALLATION.

L 33 is fitted to drop 40 explosive bombs weighing 100 kilogrammes (220 lbs.) and eight weighing 300 kilogrammes (661 lbs.). The releases to which the 106-kilogramme bombs are attached are fitted approximately $16\frac{1}{2}$ inches apart, while those for the 300-kilogramme bombs are placed about $17\frac{1}{2}$ inches apart.

*Bomb-Dropping Positions.*—There are two bomb-dropping positions, one forward of the midship cars between frames Nos. 15 and 17 and one aft between frames Nos. 21 and 23. The forward position is arranged with two sets of apparatus, numbered 11 to 20, for 100-kilogramme (220-lb.) bombs, and 21 and 22 for 300-kilogramme (661-lb.) bombs—one set fixed on the port side of the keel gangway and one set on the starboard side. The after position is exactly similar, but the sets of apparatus are numbered 1 to 10 of 100-kilogramme and 23 and 24 of 300-kilogramme bombs. There are also two spaces for incendiary bomb frames, but only one space contains a frame. This frame is fitted to carry 30 incendiary bombs weighing 10 kilogrammes each.

*Bomb Releases.*—The releases each consist of two duralumin side plates of the shape shown in the illustration, Plate 82, of a total length of $6\frac{1}{2}$ inches, $2\frac{7}{8}$ inches deep and $\frac{9}{16}$ inch apart. They are flanged outwards on top and bolted to a dished steel plate $5\frac{1}{2}$ inches long by $3\frac{5}{16}$ inches wide. The forward edges of the side plates are cut away to accommodate the eye on the stranded wire rope, by which the bomb is suspended (*see* A, Fig. 9).

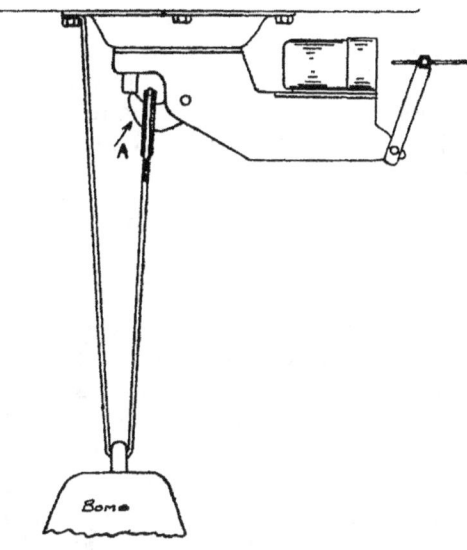

Fig. 9.

The tumbler (*see* A, Plate 82), pivoted at the centre (*see* B, Plate 82), is in the form of a hook at the lower end—the other end being shaped to form a transverse knife edge. Behind this are pivoted two arms forming a sear (*see* C, Plate 82), the upper ends of which are riveted by a cross-piece formed to engage with the upper end of the tumbler when in the "cocked" position (*see* Fig. 205), the hooked end being riveted by a cross-piece (*see* D, Fig. 204) engaging with a bent in the horizontal trigger (*see* E, Fig. 205).

On the right side of the sear is an arm (*see* F, Figs. 205-6), which bears against a cylinder of insulating material containing mercury (*see* G, Figs. 204-6). The trigger is pivoted through the head in which is the bent. The top of the side plates are flanged outwards above the trigger, and on to the flanges is screwed the solenoid. The armature is situated vertically behind the solenoid, pivoted about its axis, the lower end being morticed to receive the tail of the trigger (*see* K, Fig. 205). To the forward end of the solenoid is screwed a bracket (*see* M, Fig. 204), to which is pivoted and sprung the cylinder containing mercury. To cock the gear the sear is rotated until the tail cross-piece engages with the knife edge on the tail of the tumbler. The bent on the trigger engages with the cross-piece in the hook end of the sear; the tail end of the trigger slips into the mortice in the lower end of the armature. As soon as the current passes, by attracting the head of the armature the tail of the trigger is released from the mortice, the sear swings and clears the tumbler, which is allowed to rotate, and the bomb drops. Fig. 204 shows the gear in this position; the dotted arrows indicate the necessary movements to cock the gear. Fig. 206 shows the gear "cocked" and the safety lever in the "safe" position. The dotted lines in Fig. 205 show the position of the pieces—the safety lever at "fire."

So long as the gear is cocked, the cylinder containing mercury rests horizontally on the arm of the sear, as shown in Figs. 204-5 at G. The mercury completes the lamp circuit, indicating the bomb is on the rack. As soon as the bomb is released, the arm on the sear dropping, allows the spring on the cylinder to depress it, thus breaking the circuit and extinguishing the indicating lamp on the switchboard, to show that the bomb has gone. This is seen in Fig. 204. To prevent the sear swinging and the light, in consequence, flickering, a catch on the left side plate is arranged to hold the sear forward (*see* P in Fig. 204). As the cocking of the tumbler requires a certain amount of manipulation, two horns are situated in front of the hook completing the recess for the eye (on the stranded wire rope used for suspending the bomb). (*See* S in Figs. 204-6.) These are sprung so that the gear may be cocked and the bomb subsequently placed on the rack without interfering with the tumbler, sear, or trigger. The safety device consists of a lever (*see* V in Figs. 204-6), with a projection on it situated at the rear. When this lever is vertically upright, as in Fig. 206, the projection on it bears against the lower end of the armature, preventing it from freeing the tail of the trigger. A quarter turn either way of the lever leaves the armature free. These levers are each connected to a flexible steel wire which runs to the forward car. Each set of 12 releasing apparatus port and starboard, forward and aft, are connected to a separate wire, making in all four control wires. It was not possible to trace the arrangement of the control of these wires in the forward gondola. Two of the 6-core cables are split up into their separate cores at the No. 20 apparatus, and six similar wires are jointed to the single-core cable. One core of each of these cables is then taken to each apparatus, Nos. 17, 18, 19, 20, 21, and 22, as shown on Plate J (in packet). The common wire is connected to the centre terminal of the apparatus.

*Electrical Installation for Bomb Release.*—The electrical installation for operating the bomb releases in L 33 consists of two switchboards (Plates 84–7, 89 and K), fitted in the forward control car—one port and one starboard—from each of which eight six-core and one single-core cables are led through two flexible aluminium tubes to the " G " longitudinal girders and are then taken along the port and starboard sides of the keel walking way to the bomb chambers. The cables are attached to the framework by means of wood clips which are bolted to the girders. The bolt is inserted in one of the holes usually used for the rivet holding the parts of the framework together. Some of the clips are grooved to take the cables, in which case an aluminium saddle lined with fibre is used for fixing, but in the majority of cases the cables are rove through a hole in the clip. All particulars of the wiring that could be obtained are shown in Plate J. The single core cables port and starboard are spliced at frame No. 16.

The switchboards are each fitted with six main switches, five push and one lever type.

Two of the push switches are distinguished by the inscription " Abwurf t " and " Abwurf T," a third by " Licht," the last two by " S " and " L "; and the lever switch by " Aus " and " Ein." For reference purposes these are labelled I., II., III., IV., V., VI. respectively, in Plate K (in packet).

In addition there are 25 single switches arranged in two rows at the top of each board, 24 of which are connected with a fuse in the circuit of the solenoid which operates the release while the centre switch controls the lighting of a lamp for illuminating the board.

The latter is distinguished from the others of its kind by a handle which is made of a different material.

The 24 switches are identified by numbers engraved above the first row.

Switch I. is the final control for releasing the bombs either singly or in batteries, and is used in conjunction with the 24 lever switches.

Switch II. is used for the very rapid release of bombs singly, in rotation. When depressed it closes the solenoid circuit and just afterwards rotates a drum, upon which is engraved a series of numbers from 1 to 24, and also a series of 24 contact fingers are mounted, which in turn come into contact with contact pieces—these contacts being connected to one side of the 24 lever switches. A serial number automatically appears as each bomb is dropped.

Switch III. closes the indicating lamp circuit and when it is depressed all lamps which are connected to an undischarged bomb light up.

Switches IV. and V., if depressed simultaneously with switch VI. in the off position, operate a voltmeter fitted on the switchboard.

Switch V., when switch VI. is in the off position, closes the circuits connecting the resistances (shown in Plate J) to the voltmeter, and the resistance of the solenoid circuit can then be measured in ohms by direct readings on the voltmeter scale. To measure the resistance of solenoid circuit No. 12, open switch VI., depress switch II. until the arm actuated by the ratchet is in contact with No. 12 release gear, and then depress switch V., when the voltmeter will give a direct indication of the resistance in ohms of No. 12 circuit. A sliding index on the instrument found in L 33 is set at 12 volts.

*To Release Bombs Rapidly.*—Push main switch marked VI. over to " ein," then alternately depress and release switch II. at the same speed as it is desired to drop the bombs. The number appearing just above the switch indicates the number dropped.

*To Release Bombs Singly.*—With main switch over to " ein," push down any of the lever switches connected to a loaded bomb release (*i.e.*, one against which a lamp lights when switch III. is depressed) ; then depress switch I. and the bomb falls.

*To Release Bombs in Batteries.*—Up to the number carried on board the ship. Push down as many of the levers connected to loaded bomb releases as it is desired to drop bombs (say, ten), then depress switch I., and the ten bombs are dropped.

*Indicator Lamp.*—To find the number of releases which are loaded, it is necessary to push main switch marked VI. over to "ein," then depress switch III., and the lamps in the circuits which have loaded releases will be lit. Lamps which remain unlit indicate dropped bombs.

By keeping the switch depressed while operating the bomb release switches, the lamps can be observed to go out as the bombs are dropped.

*Bomb Release Gear for Incendiary Bombs.*—Mechanical bomb release gear, which is apparently for incendiary bombs, is slung on three parallel tubes in L 33. These consist of a sheet of duralumin pressed out to form a circular collar (*see* A, Fig. 10) to fit round the tubing from which they depend, and kept in place by a pin, and into two vertical side plates arched at the bottom to take the eyebolt of the bomb. They are $3\frac{9}{16}$ inches over-all height, $2\frac{5}{16}$ inches long, and the side plates are $\frac{7}{16}$ inch apart. To the bottom of the side plates, either end of the recess, are pivoted two arms (*see* S and S', Fig. 10), one $1\frac{5}{16}$ inch long and the other $1\frac{7}{8}$ inches long, scarfed to meet each other—the other end of the longer (*see* S', Fig. 10) being cut to form a bent (*see* B, Fig. 10). The release lever (*see* L, Fig. 10) is pivoted

Fig. 10.

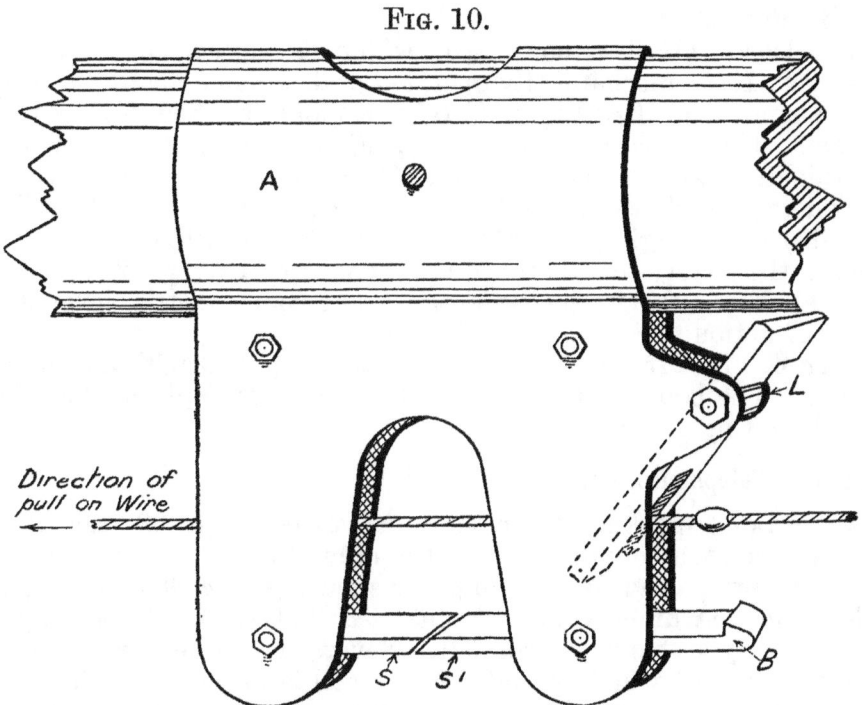

INCENDIARY BOMB RELEASE GEAR L 33 (FULL SIZE).

above the centre of the long arm. When the bottom end of the lever is engaged with the bent, the arms are prevented from falling; as soon as it is pulled away the weight of the bomb pulls down on the arms, which allow it to drop. The levers are all longitudinally cut away in the centre below their axes and one wire passes through all these slots. On to the wire at different distances are soldered small steel balls, which, when the wire is pulled, bear against the levers, and so release the bombs in turn (*see* Plate 40).

*Comparison with Earlier Dropping Gear.*—The electrically-operated bomb-dropping gear found in L 33 is evidently of the latest pattern, though similar to that in LZ 77, which came down at Revigny, in France. The Bomb Release is an improved design compared with those found in previous wrecks (shown in Fig. 207). Not only is the installation mechanically superior but the method of indicating the release of a bomb is absolutely certain, whereas in previous types, the masking of the light (as shown in Fig. 208) merely showed that the plunger had been depressed. In L 33 the light is extinguished, doing away with the unsatisfactory shutter arrangement which was previously used.

To render clear a comparison of plates (Plate 82, L 33, Plate 90, LZ 77), it may be explained that the same parts are indicated by different reference letters as follows:—

In Plate 82, L 33, Part A = Part c in Plate 90, LZ 77.
,, ,, B = ,, c′ ,, ,,
,, ,, C = ,, d ,, ,,
,, ,, D = ,, g ,, ,,
,, ,, E = ,, f ,, ,,
,, ,, F = ,, z ,, ,,
,, ,, G = ,, n ,, ,,
,, ,, K = ,, h ,, ,,
,, ,, P = ,, s ,, ,,
,, ,, S = ,, r ,, ,,
,, ,, V = ,, t ,, ,,

There are indications that the previous switchboards arranged for the release of bombs in batteries, but this is open to doubt.

The present type of bomb-release switch is controlled by a main plunger switch which renders unnecessary the magnetic control switch found in LZ 85 (shown in Figs. 210–2), and which was used to obviate the possibility of accidentally depressing the switch plungers (shown in Fig. 209). Fig. 213 shows a back view of the switchboard shown in Fig. 209.

A bomb release salved from L 15, which is shown in Plate 88, represents an earlier type. It is mechanically operated and weighs about $3\frac{1}{2}$ lbs. Its main features are a zinc body (see Fig. 216) 6 inches long at the top and 4 inches long at the bottom. The release lever (see L, Figs. 215–6), made of duralumin, is $\frac{3}{8}$ inch in thickness and 9 inches long. The "tumbler" (see T, Figs. 215–6) is $\frac{11}{16}$ inch wide, $\frac{7}{16}$ inch thick, and 3 inches long, and has a "bent" at the back (see N, Fig. 215) for engaging the safety sear (see S, Fig. 215).

As soon as the safety sear is released a hook (see H, Figs. 215–6) engages with a block (see K, Figs. 215–6) screwed on to one side of the lever to ensure its remaining out of action.

A link (see Z, Fig. 215) allows the lever to be pulled without interfering with the safety sear. The bomb is then released by a direct pull on the wires attached to the top of the lever.

**Electric Lighting, Telephones, &c.**

*Telephones.*—Loud-speaking telephones are fitted in L 33 (see Plate 91), but according to information obtained from the crew, these were not satisfactory, and nearly all orders were passed by messenger or shouting (see Plate L, in packet).

Two telephone instruments were found (see Plate 91). These appear to be of the loud-speaking type hand instrument. Several novel features are embodied in these instruments, one of which is that the line coil is fitted inside the hand grip. A ringing push button, coloured black, is fitted on one side of the grip and at the opposite side is a push button, coloured white, for operation by the thumb, which it is considered can be used for "morsing" in case speaking is inaudible. A felt pad is fitted for the ear, and a folding mouthpiece.

A loud-calling telephone buzzer was found, of the usual diaphragm type fitted with a trumpet. A small condenser is fitted in the base of this, and connected across the break.

A telephone head-gear with leather strap sliding adjustment and one telephone receiver was found—it is considered that this may have been for the telephone or wireless operator.

The receiver (only one being used) is of the watch type, with a 1,000-ohm winding. The method of clamping the diaphragm is entirely new, and the necessary adjustment is made by rotating the earpiece, which can be locked after adjustment. From a visual examination the receiver appears highly efficient.

The head-piece is in the form of a strap made of flexible leather, and is more comfortable than the aluminium band type.

To prevent interference a pad is attached by an adjustable clamp to cover the operator's other ear.

The two-way plug has a special pin of efficient construction.

*Electric Lighting.*—The electric lighting installation of the forward and after cars of L 33 was so burnt away that it is not possible to give details of the whole of the equipment. The lighting of the port and starboard cars was practically undamaged (*see* Plate M, in packet).

The supply for the lighting cannot be definitely traced, but it is considered to be derived from the generator on the same bed as the Alternator described in Chapter VII. in connection with the wireless equipment. This D.C. generator gives 24 volts when running at approximately 4,000 revolutions per minute. When this is not running the supply can be obtained from the accumulators also referred to in connection with the wireless equipment. No switchboard has been found.

From the forward car three cables are led away along the framework. The cables are encased throughout in braided tubing, which is fixed to the framework by means of tape similar to Blackley tape. The cable is further protected with pieces of large braided tubing where bent round sharp corners. One cable is led along the keel walking way, one cable along the starboard and the other along the port framework. Tappings are taken off for the midship cars, spliced joints being made—the joints being covered with tape similar to Blackley tape—and enclosed in aluminium boxes. The supply cables are taken along on opposite sides of the girder framework through tubes to the forward cars. Here they pass through a fuse box, fitted with fuses rated at two ampères, to a junction box.

The supply for the port car is tapped off the port and middle line cables and that for the starboard car off the starboard and middle line cables. The three cables pass along to the after car but could not be traced to any connections inside this car.

The cables did not continue beyond the after car.

All the incandescent lamps were broken and the sockets found were marked 4 K. (C.P.) 14 V, except the lamps for illuminating the steering wheel dials, which were marked 1 K. (C.P.) 14 V. The fittings used are of the simplest type, and consist chiefly of miniature B.J. lampholders fastened to the framework.

In the forward car a bow light is fitted, of which the candle-power is not known, as the bulb is missing. The port and starboard navigation lights consist of a miniature B.J. lampholder secured to the framework of the port and starboard midship cars, which are fitted with pockets for this purpose, the inside being painted red or green. A switch is fitted to control each of these lights with an aluminium clip sprung over the switch-knob, which has to be lifted before the switch can be operated. Under the switch is fitted a tally "Bordelichte" (Navigation Lights). The lighting of the port and starboard midship cars consists of four lights: one navigation light, one light for illuminating the engine telegraph (the telegraph is mechanically operated), and two portable leads of approximately 7 feet in length fitted with portable electric hand lamps. These hand lamps are of brass of simple construction. The overall length of the fitting is approximately 6 inches; it consists of a brass tube, forming the handle, fitted at one end with a hemispherical reflector arranged with plain glass front and wire guard secured by a screwed ring. A miniature screw bulb is fitted in the centre of the reflector. At the other end of the handle is a milled knob fitted with a hook and gland to receive the flexible cable. Each light is controlled by a switch—the switch for the telegraph light being marked "Telegraf." The light illuminates the telegraph dial externally, and is hung from the centre framework. All the fittings were missing. Self-contained hand torches of the usual commercial variety, fitted with dry cells, were found. It is understood from the crew that these electric torches are used generally for moving about the ship.

Another dry battery light fitting was found, consisting of a bull's-eye type front, a small dry battery behind, and a small switch, all mounted on a board. It appears that this is an auxiliary lighting fitting, and is probably used to illuminate a gauge, &c. on the engines or control gear.

The crew space, so far as can be seen, is not electrically illuminated, although a twin flexible cable was found on the outside of the framework between the forward car and the crew space. This was found lying on the framework and not attached in any way, and its purpose could not be ascertained. So far as could be seen a stern light is not fitted.

Two steering wheels are fitted in the forward car, and the dials are electrically illuminated. This is a very simple arrangement, consisting of an ordinary miniature B.J. lampholder fixed inside the dial by means of a clip. The lamp sockets are engraved 1 K. (C.P.) 14 V.

A set of five switches are arranged with a brass cover engraved to indicate the lights they control. The particulars engraved are as follows: telegraph, reserve, chart table, altitude steering wheel, side steering wheel.

Two 4-pin plugs are also fitted in this car, one port and one starboard, from which four stranded cables of about $\frac{5}{32}$ inch diameter are led to a junction box from which four similar cables are led forward to the position of the wireless gear. These plugs are used for supplying current to a signalling light. This signalling light consists of an aluminium alloy lamp about 13 inches diameter, fitted with a flashing shutter operated by means of a Bowden wire from a hand plunger. The parabolic reflector is made of polished aluminium. The bulb, which is missing, is fed from a 4-core plug and flexible leather-covered cable. A sighting tube with rifle sights is fitted on top of the lamp, and the whole fitting is mounted on a universal joint pivot which fitted into a socket on the car.

It is understood that the electric lighting is turned off when reaching the enemy's shore and that each of the crew is supplied with an electric torch for use as necessary on moving about.

## CHAPTER VII.—WIRELESS TELEGRAPHY APPARATUS.

From examination of the Wireless Telegraphy apparatus and parts found in the destroyed German Rigid Airships, it has been possible to obtain the following information and drawings. On referring to the drawings, it should be observed that the lines shown in full are definite, whilst the dotted lines are only assumed, but in each case there is good reason for assuming the dotted lines to be fairly correct (*see* Plates 95–5 and L, in packet).

Each ship appears to have been fitted with two transmitting sets in the forward car, one being a standard 1 kilowatt Telefunken set and the other a 40 to 50 watt set. Both sets work with a quenched spark gap, and the plates used in the gap are copper, with silver centres. No cooling device appears to have been in use. The condenser used in the transmitter is made up of copper foil with mica dielectric, and has a capacity of 17 jars.

Three alternators were used in L 33, of which one was fitted in the forward car, one in the starboard midship car, and one in the after car.

*Forward Car Alternator.*—The alternator in the forward car is fitted on a common bedplate with a generator, which can also be used as a motor. On top of the generator-casting is fitted a switch, marked "Motor" and "Lichtmaschine." The generator and alternator may be run as separate machines, or they can be connected by means of a stop pin, and the generator run as a motor driving the alternator.

Three pulleys are fitted, one for the generator, one free, and one for the alternator. These are driven by means of a belt from a pulley on the main engine shaft. The control gear for these machines was not seen.

The name plate on the generator gave the following particulars:—

    Type M.G., No. 2,511.
    Volts, 24.
    R.P.M. { Revol / Touren } 3800/4500.
    Eisemann.

The name plate on the alternator gave the following particulars (*see* Plate 92):—

"Telefunken," Type T.F., 15/16, A. 3, W. 270, C. Revs. 4,000, Freq. 534.

The commutator of the generator has 14 segments, and fixed carbon brushes are used—$\frac{9}{16}$ inch by $\frac{1}{32}$ inch by 1 inch.

A twin steel-braided cable is led away from the alternator, and was traced to a position near the wireless gear.

There are six cables leading from the generator, and the terminals are marked D +, M +, R 2, R 1, or R.A., or R 4, D −, and M −.

The cables from the terminals M + and M − are flexible, and of about $\frac{5}{32}$ inch diameter core, and branch off into two cables by means of a spliced joint about 9 inches from the machine. These are again spliced together at a point near the wireless gear.

The bedplate of the generator and alternator appear to be aluminium. The bedplate is made with a sliding base similar to the slide rest of a lathe. This is apparently for adjusting the belt.

A set of 12 accumulators of about 30 Ampère-hour capacity were found near the wireless gear, and it is considered were used for driving the motor alternator when necessary, and probably also used for the electric lighting.

There are 12 cells, and each cell has five plates, $5\frac{1}{2}$ inches by $3\frac{7}{8}$ inches.

*Results of Tests.*—A test was carried out on one of the small machines, and although it was difficult to have a prolonged test owing to the insulation having been damaged, the following information was obtained.

The alternator was driven at 3,900 r.p.m., at which speed the D.C. excitation was ·4 ampères, at 16·5 volts. The alternator output was 16·5 volts on open circuit. This voltage fell quickly as the load was applied, the most satisfactory position being 2 ampères at 20 volts. When a greater current than 2 ampères was taken, the voltage dropped very rapidly. The machine was running slightly below the speed marked on the label (a copy of which is shown in the drawing on Plate 92), and consequently the A.C. output was only 40 watts. The frequency of the two A.C. machines appears to have been almost the same, and was very high for a machine of the size of the smaller set.

A further test was carried out on the small combined set, running it as a motor generator from a battery of accumulators. The speed maintained was 3,900 r.p.m. The motor at first took 7 ampères at 23·5 volts without any load on the alternator. As the load was put on, the motor current rose to 9 ampères. Owing to the damaged condition of the windings it was difficult to run at full test, but it was obvious from the figures obtained that only brief messages could be sent by this set if the ship's engines were out of action, and she had to depend on the accumulator battery for communication. This agrees with the statement made by the prisoners of L 33. An automatic no-volt release was used between the battery and the dynamo.

In the charging circuit this was of an ordinary electro-magnetic design as used in motor car lighting circuits, but of rather heavier design. The weight of the motor generator and alternator is 72 lbs.

*Starboard Midship and After Car Alternators.* — The alternators in the starboard midship and after cars are similar to each other but of a larger size than the one in the forward car. They are pipe ventilated, the inlet being taken to a position outside the car and flanged to catch the air. Gauze partitions were fitted in the ventilating pipes as a fire precaution. These alternators are arranged for driving by belt only, and the belt tightening gear consists of a jockey pulley held in position by means of a spring. The drive is obtained by means of a belt run from a pulley fixed on the main shaft.

These alternators are fitted with change of speed gear, which appears to be two-speed, thus allowing the alternator to be run at full speed with the engine running at half speed. The nameplate on one of the alternators gave the following particulars (*see* Plate 92):—

**A. 22 Revs. 3,000. Frequency 506, W. 1,500. Type T.L. 35/26 Telefunken.**

In the after car there are three propelling engines, two abreast forward and one on the middle line aft, and each is fitted with a pulley for driving an alternator. Only one alternator was found in this car. The weight of the large alternator is approximately $52\frac{1}{2}$ lbs. and that of the two-speed gear box 12 lbs.

All the alternators are arranged to fit into four brackets and are held in position by a pin through each bracket and tightened up by adjusting screws. Above the after engine four brackets are fixed to which an alternator can be readily attached by means of pins. This appears to be an alternative position for an alternator to be fitted in case of breakdown of the other engines. A pulley for driving is fitted on the engine shaft. Connections are provided at this position and two steel braided cables run to a switch box in the forward part of this car.

Two steel braided flexible cables also run to this switch from the alternator fixed between the forward engines in this car. The switch is then connected to a fuse box similar to that fitted in the starboard midship car, and from the fuse box two steel braided cables run along the framework on the port side of the ship to the forward car. The switch is burnt out.

The above cables consisted of a three-core cable of 1/20 wires and a twin cable of stranded wires of approximately $\frac{5}{32}$-inch diameter. These cables run along inside the framework and are secured by means of aluminium clips about every 4 feet.

Two steel braided cables also run from the fuse box for the alternator fitted in the starboard midship car along the framework on the starboard side of the ship to the forward car.

*W/T Receiver.*—The various drawings on Plate 93 show the receiver from L. 32 so far as its construction and connections could be ascertained. The dotted lines in the diagrams represent what are believed to be connections which are missing in the actual sets.

The circuit was so arranged that two alternatives existed, and reception was possible either by (1) crystal only, or (2) crystal and valve amplifiers. The crystal circuit consisted of a loose coupled circuit, a secondary variable condenser, a silicon crystal, and low-resistance telephones in conjunction with a telephone transformer.

The drawing on Plate 93 also shows a change-over switch, used apparently to change from crystal to valve and crystal. The valve circuit consisted of two Lieben valves acting as amplifiers, as shown in the diagrams on the right-hand side of Plate 93.

The separate inductance shown in the top centre of the drawing would appear to be for use in tuning up the transmitter and for listening to its note in the receiver. This would obviate the use of a separate wavemeter for rough measurement.

The receiver weighed approximately 23 lbs.

*Aerial.*—Owing to the very damaged condition of the aerial gear difficulty has been experienced in arriving at a definite conclusion. The arrangements appear to be as follows :—

One wire of stranded silicon bronze or phosphor bronze about 117 feet in length extended from the deck insulator. At the end of this two other wires of the same material are attached to form a three-wire tail 278 feet long. Lead weights weighing respectively $6\frac{1}{2}$ lbs., $4\frac{1}{2}$ lbs., and 2 lbs. are attached to the ends of these three wires.

The ship's motion through the air would tend to spread these wires like a fan with the lightest weight above and the heaviest below.

Small chains of porcelain insulators were found which might possibly have linked these weights to prevent them twisting up, and also to maintain a constant spread, but no definite information has been obtained on this point.

An alternative arrangement has been suggested for the aerial on these ships, and that is as shown in the following sketch :—

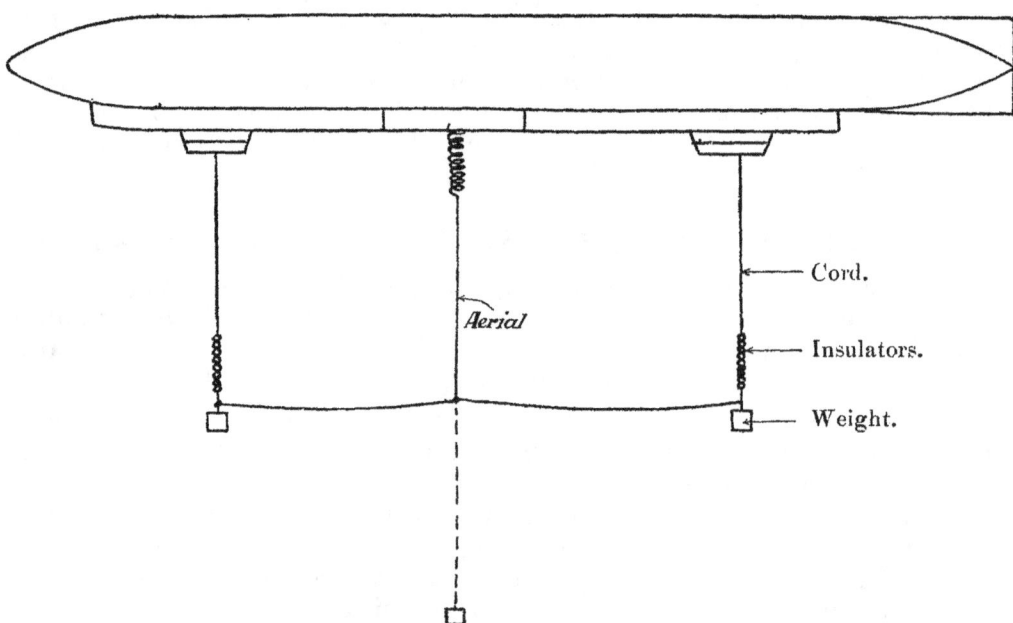

A single wire extends from the middle of the keel to a wire stretched between the forward and after car in the form of an inverted "T." The wire stretched between the cars being suspended by lengths of rope and chairs of porcelain insulators. It is also possible that the centre wire hung down below the horizontal wire, as shown dotted.

Several difficulties are, however, seen to this arrangement:—

(1) That all W/T has been found in the forward car, whereas this Aerial would point to it being amidships.
(2) Difficulty in reeling in, and the fact that no reels were found for the cord in the forward and after cars.

It is, therefore, considered that the first arrangement is the most likely one of the two.

The best pieces of the Zeppelin's aerials wire recovered gave a maximum breaking strain of 320 lbs., and consists of a seven-strand wire, each strand of 13 wires No. 38 S.W.G. hard-drawn, phosphor bronze. The lay is—

Wires, 24 turns per foot, right-handed.
Strand, 18 turns per foot, left-handed.

The metal is alloyed as follows:—

| | |
|---|---|
| Copper | 93·2 per cent. |
| Tin | 6·0 ,, |
| Iron | 0·4 ,, |
| Phosphorus | Small quantity. |

The average breaking strain of samples tested was 130 lbs.

*Deck Insulator.*—So far as can be gathered from the fragments found, the deck insulator appears as shown in the drawings on Plates 94 and 95, and was made of high-grade porcelain and fitted into the ship's deck with wooden clamps. A heavy metal tube is fitted into the centre to form a guide for the aerial wire to pass through. At the top of the insulator there is a cup in the brass tube, and at the top end of the aerial a brass stop is fitted which rested in the cup and thereby fixed the length of aerial used.

The main length of aerial was connected by a heart-shaped thimble to the bottom of this stop, and a short length of aerial wire connected to the top of this stop also by a heart-shaped thimble was connected to the drum on which the aerial wire was wound. To ensure good contact, a short wire was soldered to the stop and to the aerial wire above the heart-shaped thimble in each case, as shown in the drawing on Plate 94.

*Aerial Winding Gear.*—This gear was of a very heavy type, consisting of a thin metal drum mounted on a shaft supported on four porcelain insulators. The shaft was in ball bearings and was geared down and hand-operated by a chain driven on sprockets. A foot-operated brake was fitted on a drum mounted beside the reel, and a toothed wheel and pawl controlled its movement. Contact was made by brushes on the shaft.

The general arrangement was as shown in the drawing on Plate 94. One spare length of aerial wire was carried. The process of winding-in would appear to be very slow.

*Wireless Cabin.*—All the above-mentioned gear except the motor generators was contained in a duralumin cabin which measured approximately 5 feet 6 inches in length, 2 feet 1 inch in height, and 1 feet 8 inches in breadth. A prisoner stated that this cabin was insulated from the ship, but the only insulation found was small composition feet which could not possibly insulate electrically but which might possibly have been fitted for the purpose of breaking sound conduction and thereby rendering the cabin less liable to external vibration and noise. The cabin was only large enough to hold the apparatus and could not have formed more than an open cupboard for it. There was no sign of a silent cabin or any enclosed space for the telegraphist, and no signs of this cabin being in any way enclosed. A steel shelf was fitted in this cabin on which the accumulators were kept. This cabin is shown in the drawing on Plate 94.

*Earth.*—The earth connection was made with four strands of aerial wire to the deck and side of the car. A prisoner stated that the earth connection was made by an insulated cable run to the engines and bolted to them, but no trace of this was found in any ship.

*General Arrangement.*—The wireless apparatus and cabin was fitted in the forward car, which also appeared to be the navigator's position and chart room.

# CHAPTER VIII.—BOMBS, FLARES, MACHINE GUNS AND AMMUNITION.

## Bombs Carried.

German naval airships of the new series, commencing with the L 30, have carried larger types of bombs than those used by the older ships. There had, it is true, been several instances in which bombs of 220 lbs. weight had been employed previously, and in one case a bomb weighing 650 lbs. was used, but they do not seem to have formed part of the standard armament of an airship until the raid of the 24th–25th August 1916. Both these types of bombs are similar as regards shape and construction to that shown in Fig. 11,* with the exception that they are considerably larger and of heavier construction throughout. For example, the walls of the 650-lb. bomb are $\frac{1}{2}$ inch thick, instead of $\frac{3}{16}$ inch; the base plug is $8\frac{1}{2}$ inches in diameter and weighs 27 lbs. In the 220-lb. bomb the base plug is 7 inches in diameter and weighs 17 lbs., as against a base plug of 4 inches in diameter and $6\frac{1}{2}$ lbs. weight in the smaller bomb.

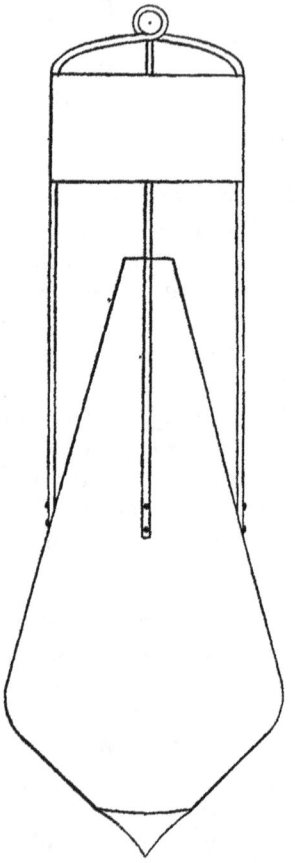

Fig. 11.
Scale, approximately $\frac{1}{8}$.

Pear-shaped Bomb.
50 and 100 kg.

A low standard of armament of the new series of German naval airships may be said to consist of the following:—

2—300-kilogram (660-lb.) bombs.
10—100-kilogram (220-lb.) bombs.
15 or 16 — 50-kilogram (110-lb.) bombs.
20 incendiary bombs.
Flares.

This would bring the weight of the total armament up to 5,780 lbs., or about $2\frac{1}{2}$ tons; but, as will be seen in Chapter III., on Construction, it is possible to carry a much greater load.

## High-Explosive Bombs.

The bombs which were dropped in the earlier raids on this country, for example by the L 3 and L 4 on 19th January 1915 (at Yarmouth), were pear-shaped, about 36 inches long and 12 inches in diameter, with a spherical bottom, but this pattern has now been considerably improved upon.

*Pear-shaped Bomb.*—See illustration, Fig. 11.

Height.—Varies between 24 inches and 36 inches.

Diameter.—Varies from 12 inches to 18 inches.

Weight.—Usually 50 kg. (110 lbs.), sometimes 100 kg. (220 lbs.). A smaller pattern of this bomb, weighing from 50 to 60 lbs., has been employed by airships cruising over the sea.

---

* The information and illustrations of bombs and flares are reproduced from Intelligence Circular No. 8 of G.H.Q., Home Forces.

*General Appearance.*—Pear-shaped, made of iron, painted greenish grey, with four iron handles joined to a loop at the top, and a cylindrical sheet-steel guard at the top. Usually the bomb is found with the propeller missing.

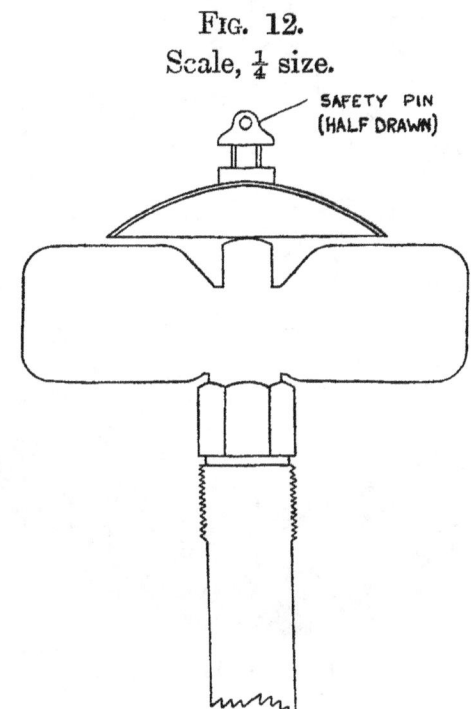

Fig. 12.
Scale, ¼ size.

PEAR-SHAPED HIGH-EXPLOSIVE BOMB FUZE.

*Fuze.*— See illustration, Fig. 12. Percussion, screwed into tail of bomb; the spinning of the propeller disengages a needle weight, which is then held off the percussion cap by a light spring. On impact the inertia of the weight overcomes the spring and the needle fires the detonator. In addition to the usual safety pin, a device for preventing the propeller revolving until the bomb has fallen some distance is provided by an inverted brass saucer; air pressure lifts this about ½ inch, and two pins which, when the saucer is down, prevent the propeller from revolving, disengage and permit it to rotate. This saucer must considerably increase the air resistance to the bomb.

*Pieces.*—These vary in size and shape from 6 inches square downwards; they are $\frac{3}{16}$ inch thick and possess a characteristic light brassy bronze colour. The edges are very jagged, and have particularly good cutting properties. The line of fracture is almost always at an angle of 45° to the surface of the piece.

Iron bars ½ inch in diameter, varying in length, but usually about 18 inches long and generally flattened at one or both ends, are often found near the seat of the explosion. They are portions of the handle of this bomb.

*New, Extra Large, Pear-shaped H.E. Bomb.*—A full and accurate description of this bomb cannot be attempted until an unexploded specimen is found.

Weight.—About 660 lbs.

Pieces.—The pieces resemble, in general outline and cutting properties, those of the pear-shaped large high-explosive bomb described above, with the exception that they may be larger and that they are ½ inch thick. The base-plug is of similar construction to that of the bomb above described. It is, however, much larger, its weight being about 27 lbs., and its diameter 8¼ inches.

*H.E. Spherical Bombs.*—The bombs carried by the German military airships are spherical in shape; they vary in weight and size, ranging from 50 lbs. to 230 lbs. in weight and from 8½ to 14 inches in diameter. The bombs are painted black, and they resemble large cannon balls (*see* Fig. 13), with a fuze projecting at one place. The fuzes are clockwork and have a delay action of 0·5 second (*see* Fig. 14). The casing is of oil-hardened steel, which an explosion breaks up into knife-edged pieces with the line of fracture at about 45° to the surface. This type of bomb is intended solely for the destruction of property and life.

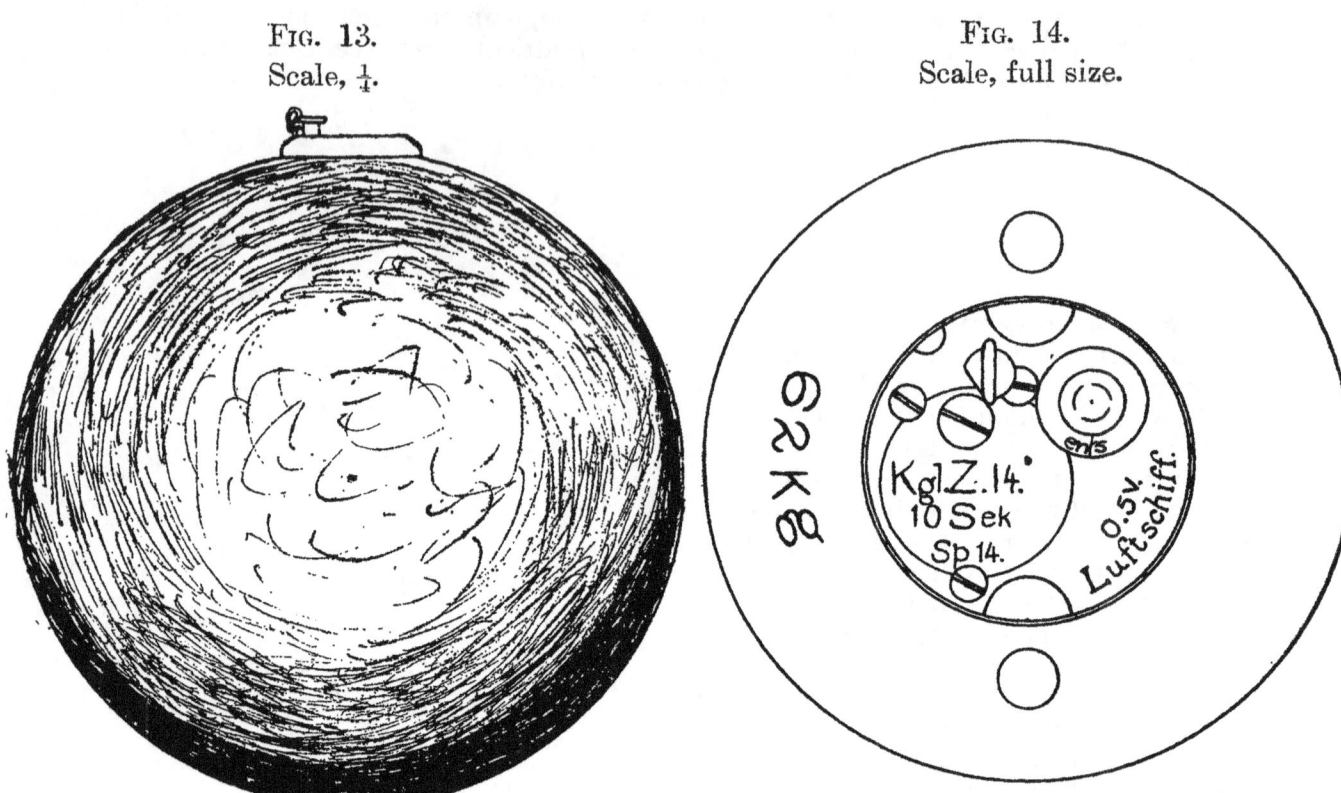

Fig. 13. Scale, ¼.

Fig. 14. Scale, full size.

MILITARY SPHERICAL H.E. BOMB.   DELAY ACTION FUZE FOR SPHERICAL H.E. BOMB.

**Incendiary Bombs.**

Height.—Excluding looped iron wire handle, 17½ inches.
Diameter.—Overall, 7 inches.
Weight.—20–25 lbs.

*General Appearance.*—Resembles a drum of tarred rope with a handle at one end (*see* Fig. 15). There is usually attached to the handle a calico streamer about 4 inches wide and 36 inches in length, to ensure the bomb remaining vertical during its fall

Fig. 15. Scale, ⅛.

Fig. 16. Scale, ½ size.

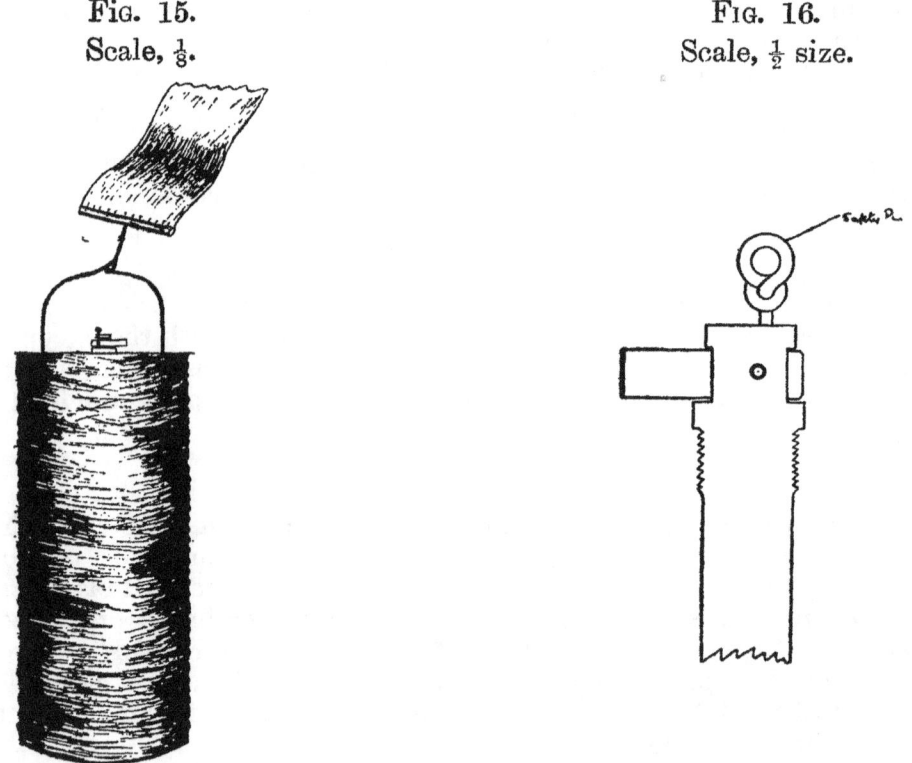

INCENDIARY BOMB.                INCENDIARY BOMB FUZE PERCUSSION.

*Fuze.*—See illustrations, Figs. 16 and 17. Percussion, kept out of action by a vertical brass pin, which prevents the flat spring at the top of the bomb from expanding horizontally.

Fig. 17.

Scale, full size.

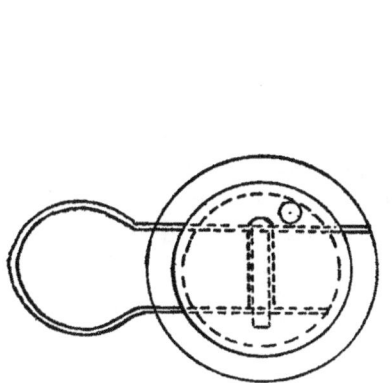 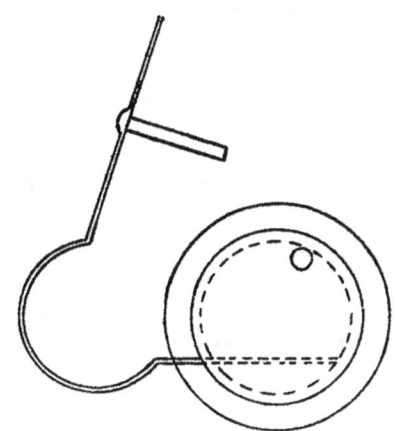

PLAN SHOWING SAFETY PIN IN PLACE.    PLAN SHOWING SAFETY PIN REMOVED.

*Effect.*—On firing, this bomb emits an intensely hot flame for 20 to 30 seconds; then blazes well for 10 minutes. The burnt-out framework consists of an iron centre tube, sometimes enclosed by a partially-burnt or melted cylinder of thin sheet iron. The centre tube has a flat perforated iron disc at the top and one at the lower end about 7 inches in diameter, resembling a saucer.

**Flares.**

*The Illuminating Flare.*—This illuminating flare is used by naval and military airships in order to illuminate the ground beneath them, doubtless for the purpose of ascertaining the exact position of bomb-targets.

Height - - - - - - - 25 inches over all.
Diameter - - - - - - - 4 inches.

*Appearance.*—Made of thin sheet iron, painted grey, with red paper band 2 inches wide near the base (Plate 96, Fig. 226). If exploded, nothing remains but an empty case; if partially exploded, it may have the cap at the tail end missing and a parachute of black material (gingham), 8 feet 6 inches in diameter, still attached. Sometimes has a label marked either "Leuchtbombe" (trans.: "Illuminating bomb") or "Blendbombe" (trans.: "Blinding or dazzling bomb") pasted on.

*Fuze.*—Fitted at the nose with an aluminium time fuze, of shell pattern, marked "Dopp Z. 08," and bearing a paper scale graduated to 3,000 metres.

*Action.*—The flare is thrown overboard timed to burst at the requisite height above the ground in order to provide the best illumination. When the fuze has burnt for the required time, it ignites a bursting charge in the nose of the bomb, which at the same time ignites the star candles, and blows them together with the parachute out of the casing. The candles then remain floating in the air, suspended by the parachute, at approximately the spot at which the bomb burst.

The light given out is very bright and intense, effectively illuminating—for the purpose of observation from the ship—the ground below for a radius of about half a mile.

A slightly modified type of this flare has since been discovered. In this modification, although the external appearance remained exactly the same, the flare did not contain a parachute; it contained, instead of five long candles, six sets of five shorter ones, about $3\frac{1}{4}$ inches long, each set of five candles being superimposed on another. By the action of the fuze these are blown out of the bomb and fall to, and burn on, the ground.

*Water Flares.*—Water flares are either lowered from the ship by a stranded wire or are thrown overboard. As the flare touches the water, the calcium phosphide in the container at the bottom of the drum gives off phosphoretted hydrogen, which is spontaneously inflammable in contact with air, and thus gives sufficient flame to act as an indicator. It is only used at sea and by night.

*Water Flare, Patt. No.* 1. *See* Plate 96, Fig. 227.

Height - - - - - - 28 inches.
Diameter - - - - - - 14 inches.

*General Appearance.*—Resembles a galvanised iron drum with a conical top, which is 10 inches high. It may have a thin two-stranded wire coiled round it. The flare ignites on touching water, thus indicating whether the ship is over land or not.

*Water Flare, Patt. No.* 2.—*See* Plate 96, Figs. 228–9 :—

Height - - - - - - 34 inches.
Diameter - - - - - - $3\frac{1}{2}$ inches.

*General Appearance.*—This has a long fish-shaped wooden body with a galvanised iron nose-cap nailed to the wood. The nose-cap is hollow and contains about $\frac{1}{4}$ lb. calcium phosphide. At the upper end of the nose-cap is hinged a brass plate at an angle of approximately 45°; this plate is about 2 inches square, and has fastened to the under side a strip of thin brass foil, the other end of which is lightly soldered over a $\frac{1}{4}$-inch hole in the nose-cap.

Fig. 18.
Scale, $\frac{1}{2}$.

WATER FLARE—ALTERNATIVE FORM.

*Action.*—When thrown from the ship, if the flare falls into water, the force of impact tears the strip of light metal away from the hole communicating with the interior of the nose-cap; water flows in and generates phosphoretted hydrogen, which ignites spontaneously in contact with air.

This type of flare has recently been found with a new form of wooden float. Instead of the long fish-shaped wooden body, the new type is furnished with a piece of wood about 6 inches square and $\frac{7}{8}$ inch thick. To the under side of this is nailed a disc of wood $3\frac{1}{4}$ inches in diameter and $\frac{7}{8}$ inch thick. Over this block is fitted the nose-cap secured by four nails. The nose-cap is exactly similar to that shown in Fig. 229, with the exception that it is made of sheet zinc instead of galvanised iron. Attached to the square wooden block is a cotton streamer $3\frac{3}{4}$ inches wide and about 6 feet long. The purpose of this is to ensure a steady fall of the flare nose first. For an illustration of this flare *see* Fig. 18.

*Water Flare, Patt. No 3*:—

| | |
|---|---|
| Height | 8 inches. |
| Diameter | $2\frac{7}{8}$ inches. |
| Weight | $1\frac{1}{2}$ to 2 lbs. |

*General Appearance.*—This flare is a small cylinder made of tinned sheet iron, painted post-office red, with a small triangular drop-handle of looped wire at one end. (*See* Fig. 19.)

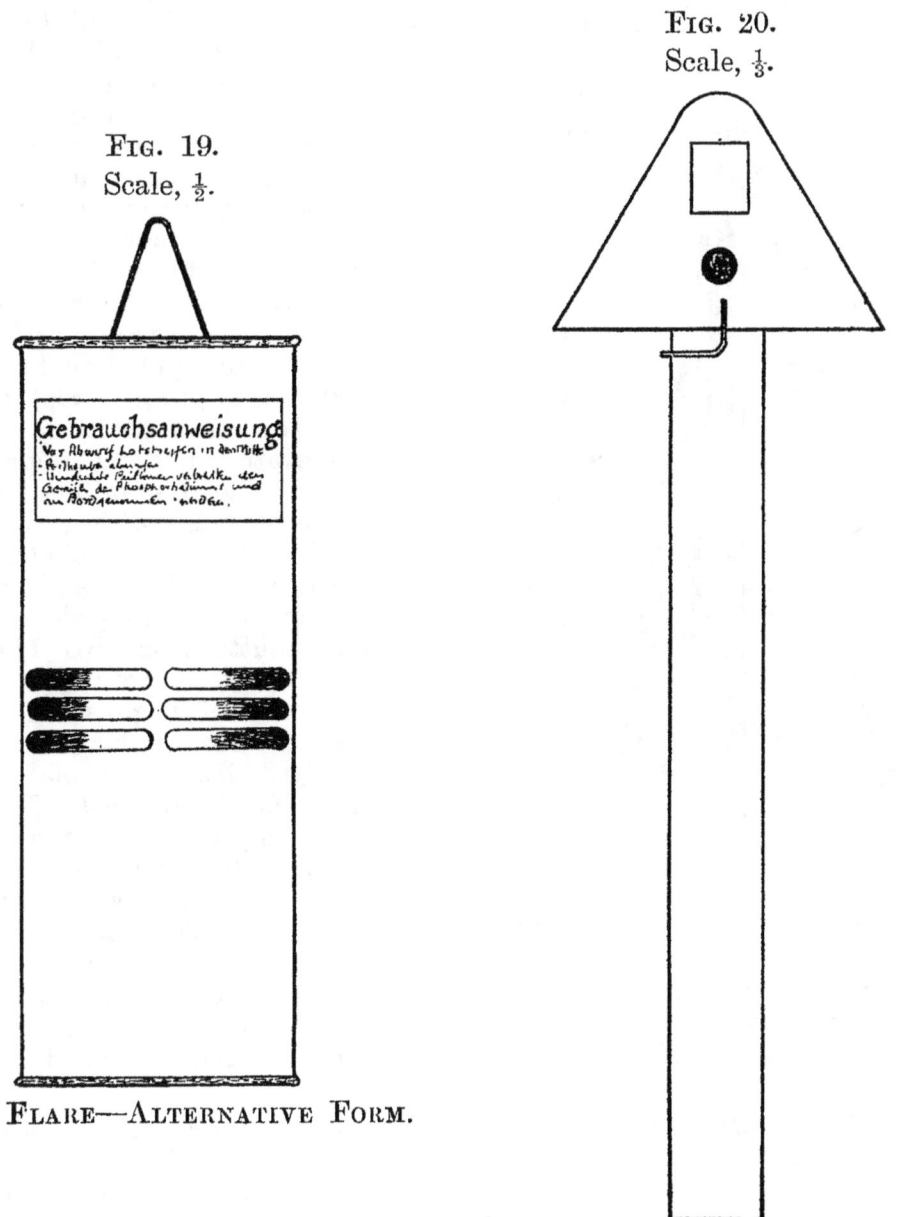

Fig. 19. Scale, $\frac{1}{2}$.

WATER FLARE—ALTERNATIVE FORM.

Fig. 20. Scale, $\frac{1}{3}$.

WATER FLARE—ALTERNATIVE FORM.

At one side, at the middle point of the surface length of the cylinder, is a small lead patch about 1½ inches square and fixed in position by solder. At the opposite side of the cylinder there will be found a tin patch, which may be torn off in a manner similar to that adopted for opening tins of canned food.

The purpose of this flare is exactly similar to that of the water flares described above. It is also used when travelling over the sea to obtain "fixes" for working out the dead reckoning.

*Water Flare, Patt. No. 4* :—

| | |
|---|---|
| Height | 18 inches. |
| Diameter | 1½ inches. |
| Weight | 1½ to 2 lbs. |

*General Appearance.*—This flare is a long cylindrical body with a conical head, roughly resembling a rocket (*see* Fig. 20). The head is about 3 inches in diameter at the base and tapers to a blunt point; its length is 3¾ inches. The whole flare is painted a light greyish-brown. Its purpose is exactly similar to that of the water flares described above.

### Machine Guns.

At least six machine guns, and probably seven or eight according to the work required, are carried in the L 33 class of Zeppelins. Two are mounted on tripod mountings (*see* Fig. 21) on the forward gun platform on top of the hull. This platform is connected with the keel-walk by a ladder shaft up which the guns and ammunition are hoisted. Two gun positions are fitted in the forward and, possibly, in the after cars, one on either beam. The two side cars have fitted outboard a pillar mounting for one gun in each car. Another gun is mounted on the top platform aft (*see* Plate 97).

In the L 33 class the guns are Maxim-Nordenfeldt, German army pattern, 8-mm. bore. They are sighted on tangent sight to 2,500 metres (2,734 yards), fitted with all-round traverse and with an arc marked to 80° elevation and 29° depression. A tray to hold the ammunition box is fitted to the gun to train with it and has a lever spring clip at the side to hold the ammunition box in place. The weight of the gun and tray, excluding the pedestal, is 61 lbs.

The Schütte-Lanz (SL 11), brought down at Cuffley, had both Parabellum and Maxim-Nordenfeldt guns. The Parabellums are, in form, similar to a Maxim, but weigh only 22 lbs., the whole of the construction being lightened. They are 48 inches in length, firing at the rate of 400 per minute—100 cartridges to a belt—and are mounted on a pivot similar to our standard Lewis. There is no sign, however, of these guns being carried by Zeppelins of the L 33 class. It is also noteworthy that whereas rifles were found in SL 11, none were found in the L 33 type of ships.

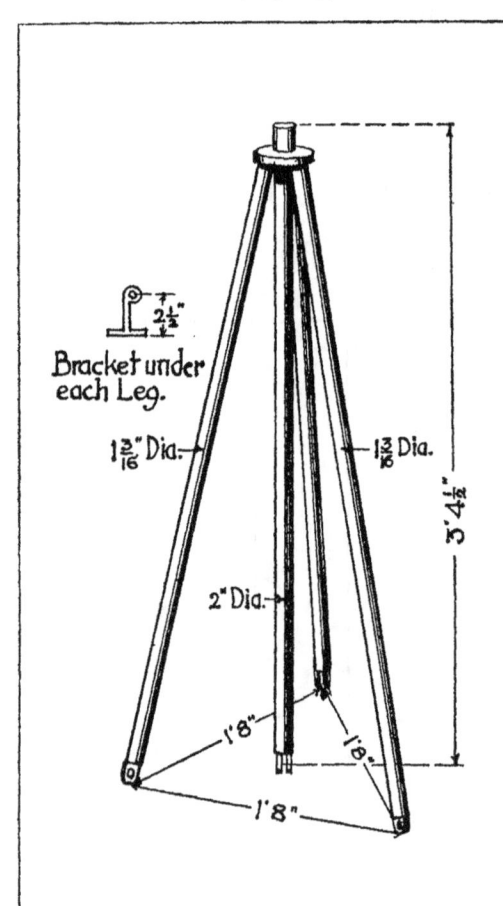

Fig. 21.

DIAGRAMMATIC SKETCH OF TRIPOD MACHINE GUN MOUNTING ON TOP FORWARD PLATFORM OF L 33 CLASS. *See* ILLUSTRATION, PLATE 28.

The Maxim-Nordenfeldts mounted on the platforms on top of the hull have fitted to them, on a steel bracket, a wooden shoulder piece. This shoulder piece is pivoted about its centre to enable it to be folded parallel to the axis of the bore to facilitate its passage up and down the shaft. When in position to fire, a clip holds it open.

The ammunition belts, 22 feet long, holding 250 cartridges and weighing 40 lbs., are stowed in boxes, two of which are issued to each gun. The box is of duralumin with the exception of three brass hinges and two spring bolts, and measures 14½ inches by 3½ inches by 6¾ inches, and is provided with the ordinary leather handle.

**Ammunition.**

The ammunition carried by the L 33 Zeppelins consists of both the armour-piercing and explosive types. Every tenth bullet in a belt is usually explosive. The following details of each type are extracted from reports furnished by the Research Department, Woolwich, through the Ministry of Munitions :—

*Explosive Type.*—The general construction of the bullet is shown in the sketch (Fig. 22). It is essentially an explosive system consisting of an ignition device

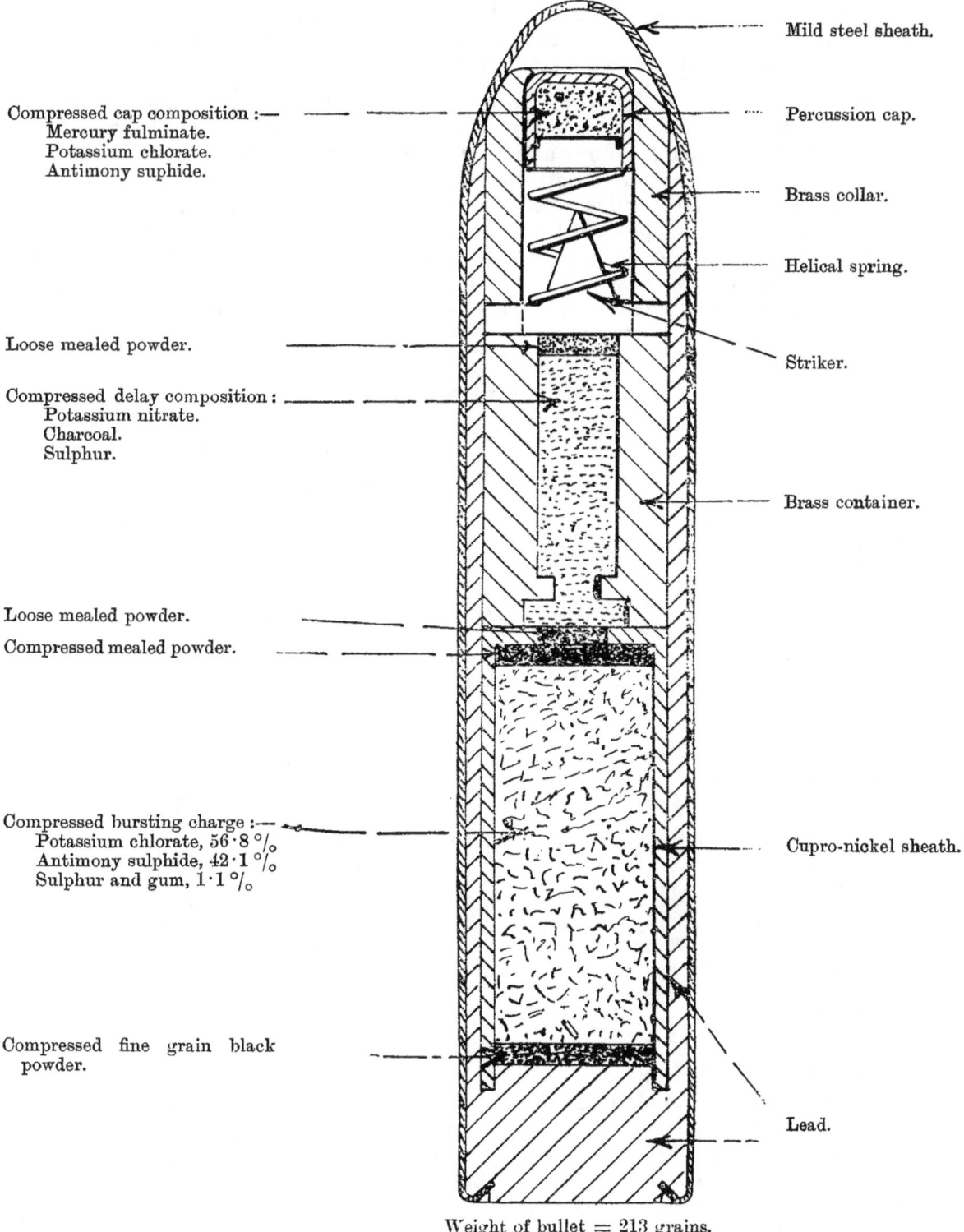

Weight of bullet = 213 grains.

FIG. 22.—DIAGRAMMATIC SKETCH OF EXPLOSIVE BULLETS FROM L 33.

which functions on shock of discharge, a delay pellet, and the bursting charge. The outer sheath of the bullet is steel. It is fitted with a lead lining and is perforated at the nose with a small hole about 0·05 inch in diameter. The igniting device is housed in a stout brass collar situated in the nose of the bullet, and consists of a small percussion cap supported over a bar striker by a spiral spring. Immediately below is a delay pellet which is made up of a stout brass collar filled with compressed gunpowder and primed with mealed powder. Under the delay pellet is a cupro-nickel sheath filled with a compressed mixture of potassium chlorate, antimony sulphide and sulphur, and primed at each end with mealed powder. This composition, which constitutes the main bursting charge of the bullet, is somewhat sensitive to shock and explodes violently when ignited. Apparently the shock of discharge causes the percussion cap to set back on to the striker and ignites the delay pellet, which in turn causes explosion of the sensitive composition in the base portion of the bullet. One bullet was fired at rest by driving the percussion cap on to the striker by a small weight and drift. After a slight delay the bullet exploded violently, shattering the sheath into many fragments.

*Armour-Piercing.*—Both types of bullets examined are pointed and forward of the cannelure resemble the ordinary German bullets in shape; while one of the bullets is of the same colour as the ordinary cupro-nickel coated bullet, the other is of a copper colour. Both bullets (*see* Fig. 23) consist of (*a*) an envelope of steel coated on both sides with a copper alloy, (*b*) a lead lining, and (*c*) a pointed steel core. Rearward of the cannelure is an approximately parallel portion about 0·3 inch long, followed by a portion about 0·2 inch long, which tapers slightly to the base. The core of each bullet is cylindrical, with an ogival point and a tapered base. The weights and compositions of the bullets and of their different parts are set out below:—

*Weights (in Grains).*

|  | Cupro-Nickel Coloured Bullet. | Copper Coloured Bullet. |
|---|---|---|
| Complete bullet | 179·5 | 178·4 |
| Envelope | 43·7 | 43·2 |
| Lead lining | 47·7 | 47·9 |
| Steel core | 88·1 | 87·3 |

*Chemical Analysis.*

|  | Cupro-Nickel Coloured Bullet. | Copper Coloured Bullet. |
|---|---|---|
| Envelope:— | | |
| Steel | 89·8 | 90 per cent. |
| Coating material | 10·2 | 10 ,, |
| Coating of envelope:— | | |
| Copper | 88 copper | 92 ,, |
| Nickel | 11 zinc. | 8 ,, |
| Lining. Antimonial lead containing nearly 4 per cent. antimony. | — | — |
| Steel core:— | | |
| Carbon | 1·1 | 1·3 per cent. |
| Silicon | 0·1 | 0·2 ,, |
| Manganese | 0·2 | 0·2 ,, |
| Tungsten | 2·9 | 3·4 ,, |
| Nickel | 0·4 | 0·2 ,, |
| Chromium | 0·3 | 0·2 ,, |

The cupro-nickel coated bullet was practically identical in all its characteristics with two German armour-piercing bullets previously examined in this Department, and the copper-coloured bullet resembled them in all essentials except (*a*) the coating of the steel envelope, and (*b*) the hardness of the steel core. The coating of the copper-coloured bullet was of brass containing about 92 per cent. copper; this alloy closely resembles in composition that of which the solid French bullet is made. No reason is seen for the change from a cupro-nickel to a brass coating.

The hard steel cores of the brass-coated and cupro-nickel coated bullets are of essentially the same composition, but it is clear from the hardness and from the microstructure of the two cores that the treatments have been different. The core of the brass-coated bullet appears to have been hardened by rapid cooling in air, while that of the other bullet has been fully hardened, probably by quenching in water. Both cores are sufficiently hard to have great penetrative power against metal plates.

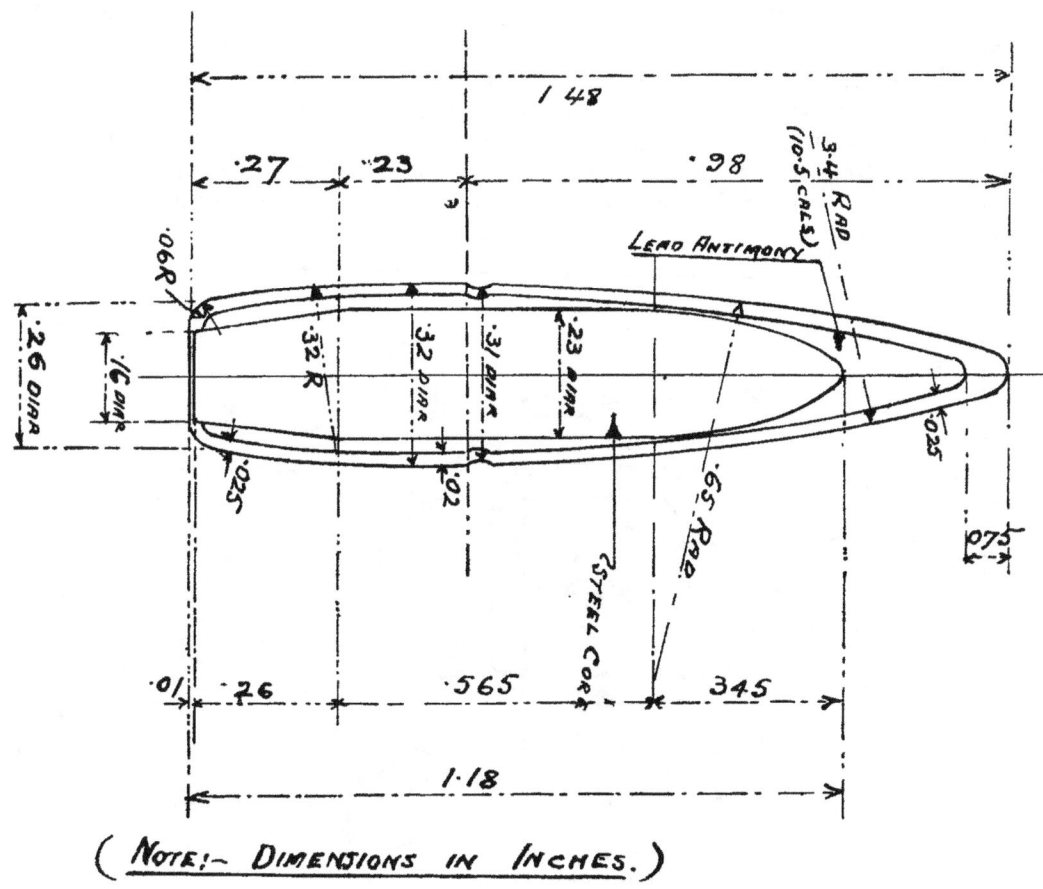

Fig. 23.—Details of Armour-piercing Bullets from L 33.
(Three times proper size.)

### General.

The prisoners stated that the guns on top of the ship are not brought into action when the ship is rising from fear of an accident due to possible films of gas. They further said that for the same reason they were much averse to firing the gun on the after platform.

The ship apparently has the following "blind" spots for machine gun fire: (i) Immediately underneath the ship over the whole of her length; (ii) Some distance aft on a line more than 20° below the longitudinal centre line of the ship; (iii) When the ship is rising rapidly in any line immediately above her longitudinal centre and not forward of the after car.

The 8-mm. German army pattern rifles which were carried in the SL 11 were presumably intended for use against attacking aeroplanes when the latter were in a "blind" position so far as its machine guns were concerned.

It is further stated that when crossing the North Sea, the ship flies at 200 to 300 metres high, and that the machine guns are then used for the purpose of destroying mines.

# CHAPTER IX.—BOMB AND GUN SIGHTS.

**Bomb Sight and Ground Speed Indicator.**

*Description.*—The Zeppelin bomb sight consists of a prismatic telescope directed downwards to the earth below. It is hung pendulum-wise on ball bearings, and the line of vision is deflected 15° forward of the vertical by means of a double prism arrangement. The support is made in the form of a turntable, also on ball bearings, to enable the instrument to be turned in azimuth to allow for the drift of the airship. The optical system consists of two prisms, an objective and an eyepiece; the two latter form an astronomical telescope, whilst the two prisms, in addition to deviating the light through the angle of 15°, also form an inverting system, thus rendering the provision of inverting lenses unnecessary. The general arrangement is shown in Fig. 24.

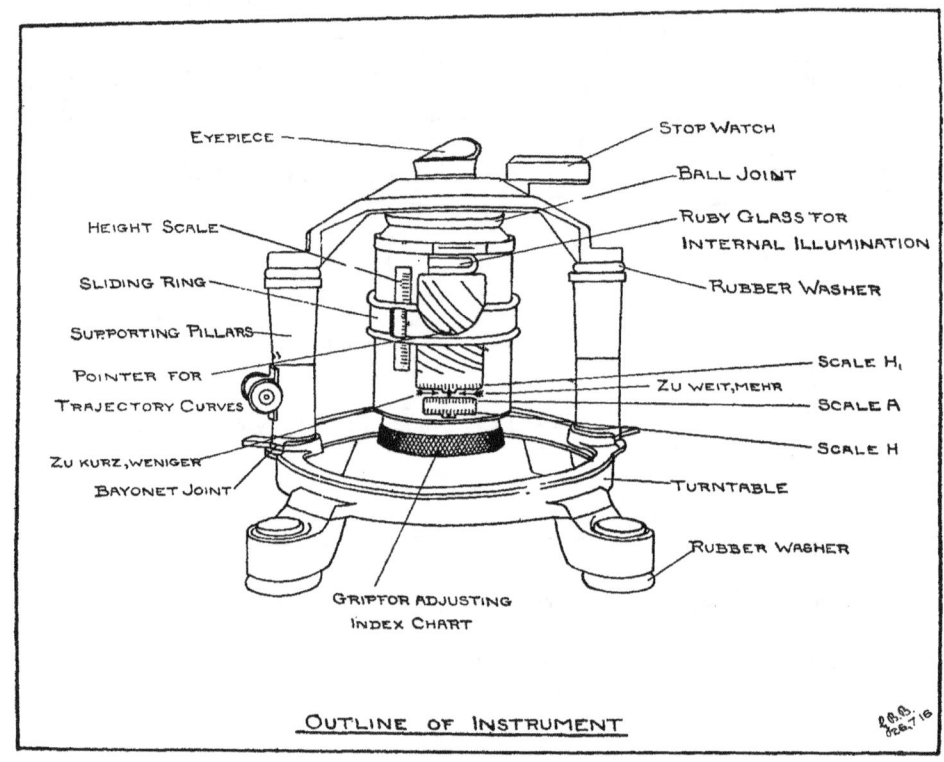

Fig. 24.
General view of Zeiss Prismatic Bomb Sight and Ground Speed Indicator.
(Its size is about 10″ × 10″ × 10″.)

In Figs. 233-4 (*see* Plate 98) are reproduced photographs of the main part of the instrument used in L. 31; this is identical with that used in L. 33, but the latter was too burnt to photograph well.

In the field of view is a graduated scale (*see* Fig. 25) carrying a cross line which indicates the point vertically below the instrument, and having a cross line on a movable graticule which can be set by the rotation of the milled drum seen at the foot of Fig. 234 (*see* Plate 98).

The milled drum is turned by hand in accordance with the scale of heights (*Messung*) seen in the lowest window, and the arrangements are such that for all heights the distance on the ground between these two cross wires is 300 metres (984·2 ft.). In the older type of instrument, used in L.Z. 85 (Salonica), the distance was only 140 metres (459·3 ft.). This constant distance is used for timing for ground-speed, and the watch (*see* Fig. 235, Plate 98—which shows the L.Z. 85 watch; others too damaged to photograph) employed for the purpose is graduated directly in metres per second to save arithmetic. This bomb sight, it will be seen, is an effective ground-speed measurer in addition to being a bomb sight (*see* Chapter X. on Navigation, page 78).

When the ground-speed has been obtained by this means, the movable cross wire is set to a new position by means of a sliding ring, clearly seen in Figs. 233-4, and the cylinder of curves, seen in Fig. 234. The drum on which these curves appear is easily removable, and it seems that three are usually carried with each sight, one for the 50-kg. (110-lb.) bomb, one for a vacuum trajectory, and the other probably for the 100-kg. (220-lb.) bomb. The curves for the 50-kg. bomb have been analysed and appear to be based on a terminal velocity of about 600 ft. per sec., but the error is made of correcting for the lag of the bomb by an amount proportional to the ground-speed instead of to the air-speed of the airship; for a height of 12,000 feet and a bomb of this size and shape the range error so produced would be about equal to the distance travelled by the wind in eight seconds. A further omission in this bomb sight is that the calibration does not take into account the effect on the trajectory of the tenuity of the atmosphere at the heights at which the sight is designed for use.

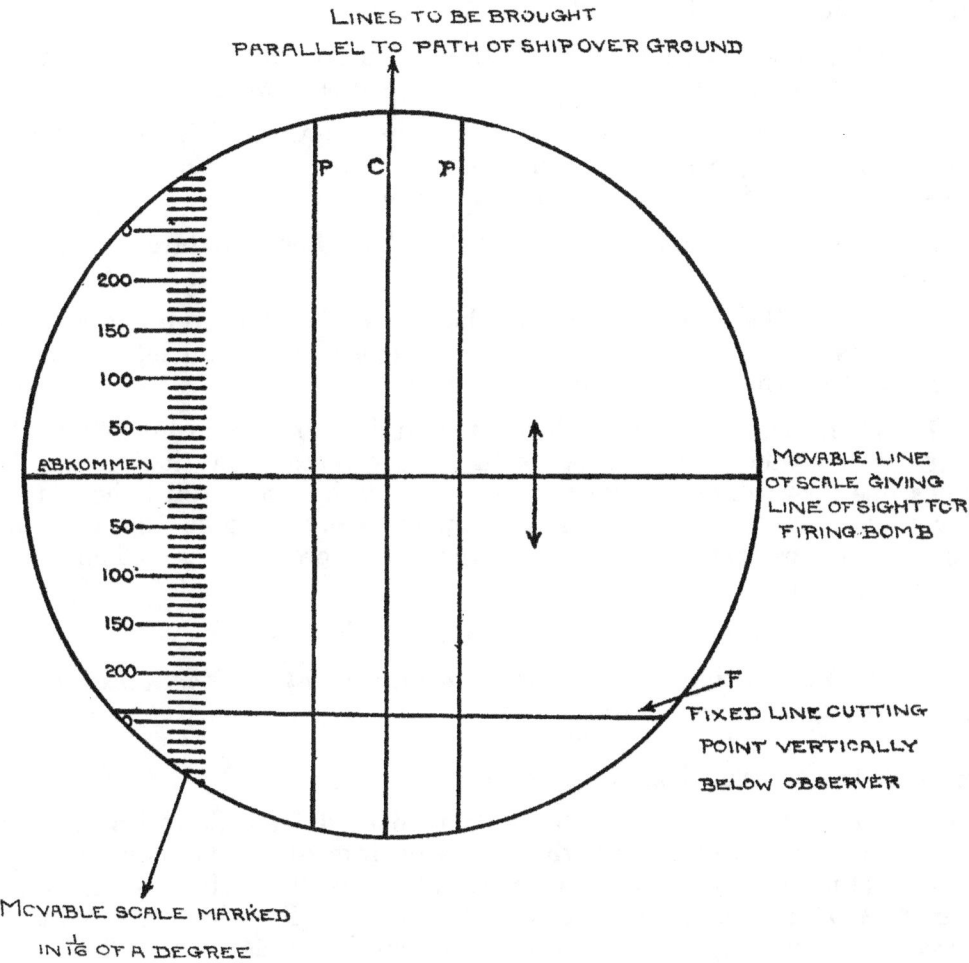

Fig. 25.
Lines in field of view.

The sight has a spotting correction. The operator watches for the point of impact of the bomb, notes how many divisions in the line of flight it falls away from the target, and turns the milled drum through a corresponding number of divisions, so making a "spotting correction," which adjusts the sight correctly for any further bombs which may be dropped at the same ground-speed and the same height. The wording "zu kurz" (short) and "zu weit" (over) is clearly seen in Fig. 233. The instrument is also provided with a separate scale (in the middle window) to enable the timing for the ground-speed to be done on the actual target instead of on an auxiliary target; this latter procedure is necessary when bombing at sea. To facilitate this a small bubble is placed in the field of view; this becomes central when the axis of the telescope is tilted through 35°. The distance timed over remains 300 metres (984 feet).

In order to enable the instrument to be used at night means are provided for illuminating the lines on the glass scale or graticule; a small electric lamp shines through a ruby glass and the cross wires stand out boldly against the dark background.*

---

* The glass graticule receives the light through a vertical edge; and the edge at the far end of the graticule is polished, silvered, and bevelled so as to retain the rays of light by total reflection from the surface of the glass.

The following instructions for use have been freely translated from a German leaflet which was found pasted in the lid of the box which had evidently contained the bomb sight in L. 15, which came down off the Kentish Knock on March 31st, 1916. The same sight is evidently referred to:—

### Instructions for Use.

1. Keep the telescope on its revolving stand continuously in the direction in which you are heading, so that the objects in the field of view move parallel to the longitudinal lines.

2. Turn the drum until the figure (red scale) denoting the instantaneous height coincides with the index.

3. Allow the telescope to hang freely. Take the time until the auxiliary target selected has crossed the space between the two cross wires.

4. Set the index of the ring to the instantaneous height. Read off the speed in metres per second from the stop watch and set the speed curve in question on the index by turning the drum.

5. Allow the telescope to hang freely; fire when the target appears on the aiming wire.

6. For further shots measure the correction on the previous shot in sixteenths of degrees on the scale in the field of view and set a corresponding correction on the spotting correction index.

7. When using the spirit level (for timing on the target) turn the drum until your present height appears on the green scale on the index. Press the telescope forward, or lift the telescope and its bearings, until the bubble of the level placed in the field of sight appears central; take the time until the *target* has crossed the space between two cross wires. Then follow the instructions 4 to 6.

8. For night use switch on the lamp.

9. *The lamp casing and lamp must always remain on the telescope even when the instrument is used by day.*\*

## Variation in design of Bomb Sights.

Examination of the various bomb sights recovered from Zeppelin airships brings to light certain minor differences which are of interest. The bomb sight, No. 54, from the Zeppelin destroyed at Salonica, L.Z. 85, has the height scale engraved up to 4,000 metres (13,123 feet), and the trajectory drum with which it was fitted makes no allowance for the lag of the bomb; whereas in those (Nos. 76, 77, and 80) recovered from the Zeppelins brought down at Potters Bar (L. 31), at Billericay (L. 32), and at Little Wigborough (L. 33) the height scale is increased to 4,500 metres (14,764 feet), and on the drum in use allowance is made for the lag of the 50-kg. pear-shaped bomb.

This indicates that the German pilots found by experience that a height scale up to 13,100 feet did not meet their needs, and that an extension of 1,600 feet was necessary. Moreover, the Little Wigborough (L. 33) sight was found set at 3,600 metres (11,800 feet) and 27 metres per second (60 miles an hour) ground-speed. This sight was badly burnt, and the scale shows no motion of the moving parts of the sight since the paper was charred; there is good reason, therefore, for thinking that the setting at which the scales were found was that of the last use of the sight. As regards the Billericay (L. 32) sight much the same remarks apply except that the sight was less burnt and may perhaps have been disturbed by the shock of striking the earth. The setting was about 4,000 metres for height (13,100 feet) and 31 metres per second (69 miles an hour) for speed relative to the ground.

Examination of the bomb sight taken from the Zeppelin (L. 31) which fell at Potters Bar indicates that it was probably last used at a height of 3,500 metres (11,500 feet) and a ground-speed of 40 metres per second (89 miles an hour). This ground-speed is much higher than those (60 m.p.h. and 69 m.p.h.) mentioned above;

---

\* Evidently with a view to the centre of gravity of the telescope being preserved in position.

inquiry was therefore made of the Meteorological Office as to the wind-speed and direction on September 23-24 (when L. 32 and L. 33 were destroyed) and on October 1-2 (when L. 31 was destroyed). It appears that on the former night there was at the height of attack "scarcely any wind at all"; whereas on the latter night there was a "probable wind at 10,000 feet in the neighbourhood of London of from 10 to 30 m.p.h. coming from the north-west." Since L. 31 was dropping her bombs near London while approaching from a generally northerly direction, it would come with a brisk following wind, and might therefore be expected to travel at a very high ground-speed.

A further point which is of interest is the fact that whereas the two height scales which each instrument possesses are graduated from 1,000 metres upwards in each of the three new Zeppelins (L. 31, 32, and 33); in the Salonica one (L.Z. 85), although the height scale for bombing begins at 1,000 metres, that for measuring ground-speed for navigation, &c. starts at 250 metres. Navigation at such low heights as this seems to have been abandoned; the lowest region is now 1,000 metres.

### Summary.

| Zeppelin. | Destroyed | Where. | Last Bomb Sight Setting. | | Probable Wind. m.p.h. |
|---|---|---|---|---|---|
| | | | Height. Feet. | Speed. m.p.h. | |
| L.Z. 85 | 1916. May 5 | Salonica | Not known | Not known | Not known. |
| L. 31 | Oct. 1-2 | Potters Bar | 11,500 | 89 | 10-30 following. |
| L. 32 | Sept. 23-24 | Billericay | 13,100 | 69 | Almost none. |
| L. 33 | ,, | Little Wigborough | 11,800 | 60 | ,, |

### Method of Attack.

It appears from the examination of the crew (carried out by G.H.Q., Home Forces) captured from L. 33 that the two bomb sights on the airship were operated by the Commander and the Second Officer, one working on the starboard and the other on the port side of the ship. Officers are given previous practice in bomb-dropping over a prepared course near Dresden.

On the night of 24th September 1916 an attack was made on London, and the positions of the fallen bombs have been carefully mapped. From an examination of these records certain conclusions can be drawn as to the accuracy which can be achieved by the use of this bomb sight. Most of the 50-kg. (110-lb.) bombs have evidently been sprinkled at random, much in the same way apparently as the incendiary bombs were thrown out. The two 300-kg. (660-lb.) bombs which have been traced were, however, clearly aimed at a definite target; the first one missed a suburban railway station (which the airship lit up by a flare) by only about 150 feet, and the second (which appears from the record to have been corrected by two spotting shots) missed another surburban station by about 100 feet—in each case the shot was "over." Direct hits on railway tracks were obtained with 50-kg. bombs, and it seems that the bomb sight was correctly set for this bomb—possibly the "50-kg." curve drum was in use, as in the sights recovered from L. 31, L. 32, and L. 33; if this were so, it would be natural for the much heavier bomb to carry further and fall beyond the target by one or two hundred feet. It appears, therefore, that considerable accuracy can be obtained by the use of this sight when desired, although general destructiveness is the more usual aim.

### Sights for the Machine Guns.

In addition to the ordinary tangent sight a telescopic prismatic sight is used with the machine guns. Photographs of the latter are seen in Figs. 236-7. (*See* Plate 99.) The magnification is 2·5 and the field of view 20°. Various coloured screens can be inserted in the eyepiece at will. The circular scale seen on the right of the sight allows for deflection. The divisions correspond very closely with metres right or left at a distance of 1,000 metres (3,280 feet) and are probably intended to do so exactly. This scale can be illuminated by a small electric light when desired.

Elevation is given by turning the horizontal milled ring immediately under the prism at the head of the instrument. The divisions evidently represent the range in hundreds of metres. The corresponding angles of elevation have been found by measurement with this instrument to be:—

| Range. | Angle of Elevation. |
|---|---|
| | ° |
| 500 metres (528 yards) | 0·37 |
| 1,000 ,, (1,093 yards) | 0·98 |
| 1,500 ,, (1,640 ,, ) | 2·35 |
| 2,000 ,, (2,187 ,, ) | 4·7 |
| 2,500 ,, (2,734 ,, ) | 8·5 |

These compare with our own figures for Mark VII. ammunition, which are as follows:—

| Range. | Angle of Elevation. |
|---|---|
| | ° |
| 500 yards | 0·45 |
| 1,000 ,, | 1·05 |
| 1,500 ,, | 2·25 |
| 2,000 ,, | 4·6 |

It is evident from the range and deflection scales that this sight was not constructed for use in the air; the ranges are too great and the deflection scale has no immediate application to air fighting. This sight was indeed stated by members of the crew of L. 33 to be too complicated for aircraft work, where both target and gun platform are moving, although excellent in trench warfare when firing at a moving target from a fixed position. Moreover the sight is heavy and clumsy for use on a machine gun which has to be operated so rapidly as air fighting often requires.

These objections appear to have been recognised in Germany, since an effort has been made to modify it to a form suitable for use in the air. A graticule has been added in the field of view, which consists of three circles with arrows pointing inwards around the circumference of the outermost circle; the diameters of these circles have been measured and found to correspond (assuming the mean velocity of the bullet to be 700 metres (2,296 feet) per second) with the correct deflection for a hostile aeroplane passing, at right angles, at the relative speeds of 20, 40, and 60 metres per second (44, 90, and 136 m.p.h.). Instead of using the elaborate deflection and elevation scales on the instrument, the machine gunner would merely sight so that the hostile machine was flying in the direction of the arrows, towards the centre of the circles, and would hold it on the circle which most nearly agreed with the estimated relative speed of the two machines perpendicular to the line of sight.

## CHAPTER X.—NAVIGATION.

**General.**

The information in this chapter was obtained by personal examination of four members of the crew of L. 33 by a naval officer in conjunction with officers from G.H.Q., Home Forces. Though their testimony cannot be implicitly relied on, it has proved correct in certain details which have been checked independently, and as a whole may be considered as being not far from the truth.

*Control Position.*—The navigation of the ship is normally conducted from the control car forward, but an auxiliary position for rudder and elevator control is arranged in the forward end of the after car.

*Officers and Crew.*—The commander assisted by the navigator, a warrant officer, determines and corrects the course. The directional steering is done by the quartermaster and a relief, while the steering for altitude devolves upon the chief boatswain with an assistant as relief.

Every member of the crew has two different sets of duties—one on which he is regularly employed, and another which he takes up in the time that he is off his raid duty. Second numbers usually are available for any odd jobs during their time off duty.

A watch is generally two hours in duration, although in certain cases it may be four. At the end of a watch the men relieved have half an hour's rest, after which they take up their secondary duties.

*Controls.*—The ship steers fairly well except in bad weather. The newest type (L. 30 onwards), however, does not seem to be so handy as its predecessors. The ship can be kept to her course within half a point under very favourable circumstances. Lateral as well as vertical steering is by hand, no power being used. The wheels controlling both are set at right angles to each other in the control car (*see* Plate 48). A reserve steering set is in the after car. The lateral steering requires little physical effort, but on the other hand the wheel controlling the elevating planes frequently demands the exercise of all the strength that a single helmsman can muster.

### Navigating Instruments.

*Instruments, Generally.*—The ships are equipped with the usual navigating instruments, *i.e.*, compass, pelorus or bearing plate, station pointer, drift indicator, &c. (*see* Chapter XI. on Instruments).

*Compass.*—The compass is magnetic, of the liquid type. No provision is made for adjusting it for deviation. It is checked on the trial flight of the ship by observations of terrestrial objects. The ship is not swung systematically for compass adjustment, though it is stated that a prescribed item in the ship's trials was a flight where the ship turned through a complete circle in order to test the free and accurate working of the compass. In any case the amount of the deviation is inconsiderable.

The coxswain (quartermaster), who was an intelligent man, stated that there was no need to adjust the compass, as it was placed nearly 30 feet from any disturbing magnetic influences and had practically no error.

An examination of the compass position in L. 33 by the compass department of the Admiralty showed that this statement is probably quite correct.

*Pelorus.*—The bearing plate or pelorus is used in the ordinary way for taking bearings in lieu of the standard compass from which an all-round view is unobtainable. It is made of aluminium, consequently has little weight, and is somewhat larger than usual, with an open centre and a skeleton radius bar fitted with longitudinal threads, thus allowing objects on the ground to be viewed through the middle of the instrument. It has also a reflector to the sight vane for observing celestial objects, and it is probably the case that when the ship is turned in azimuth, as described in the preceding paragraph, solar bearings are taken with it to check the compass. Two bearing plates are provided, resting on brackets which fit into sockets on the outside and on either beam of the control car. Not only are they used for taking bearings, but they also serve the purpose of a drift indicator and corrector. By observing objects on the ground as they pass along the longitudinal threads on the radius bar the amount of the drift can be calculated and then corrected.

*Station Pointer.*—The station pointer requires no particular notice. It is used in the ordinary way.

*Observing Sextant.*—The observing sextant is one with its own artificial horizon, and known in Great Britain as the bubble clinometer sextant. (One prisoner stated that it was called in Germany "Liebern's Quadrant.") With this instrument observations of the sun can only be taken at low altitudes, it being obviously impossible to take readings when the sun is high in the heavens, the outlook from the car being limited by the form of the hull directly overhead. Observations of celestial objects have only yielded fair results and the fixes so obtained have been far from satisfactory; the vibrations of the forward car preventing any real accuracy of observation. The upper machine gun platform above the hull, which should afford a steadier base, does not appear to have been used for these observations.

It is of course impossible to say how far the foregoing statements are true, but they may probably be taken as correct.

Of the four men examined, the navigator was the only one who said that the instrument had been used. He had personally given it a trial on a former voyage and had abandoned it as of no value on account of the poor results obtained. The others stated that they had never seen it used; they understood it to be an instrument of little value for the purpose intended, and in their own words described it as an "article of luxury." Experience in this country has shown that in aircraft bubble instruments of this description are far from accurate except under the most favourable conditions.

The ordinary sextant for taking horizontal angles between terrestrial objects does not appear to have been given a trial, though its value for position finding, in combination with the station pointer, is precisely the same in an airship as in a steamship.

*Drift Indicator.*—A drift indicator or mechanical plotter is used for calculating drift, wind, &c. The instrument recovered from L. 33 was in rather a battered condition, but its uses could be easily determined, and it is more fully described on page 83. This is only for calculating the allowance to be made for the drift observed or to work out the velocity of the wind, the other usual elements being known.

Drift is estimated by day, not only by the bearing plate, but, over sea, by wooden crosses dropped at short intervals—a rough idea thus being obtained of the leeway made, as in the case of a sailing ship and her wake. At night, tins of calcium phosphide, which burn for thirty minutes on reaching the water, are used instead of the wooden crosses. These are also utilised at night-time to indicate whether the ship is over the sea or land, no action taking place when they drop on dry land.

**Calculating Ground Speed and Fixing Positions:**

The ground speed is calculated :—

  (a) By observing the ship's shadow over the land; or
  (b) By rough calculations based on the engine speed and on the estimated velocity of the wind.

A much more accurate method than these is the use of the bomb sight, a most scientific and admirably conceived optical instrument, which, in addition to its primary function, serves a navigating purpose also. Two observations of a ground object are taken, a slight adjustment made by the help of a scale on the side compensates for the altitude of the airship, and by means of two cross wires, and a special stop watch on the instrument, the ground speed can be read off. (*See* Chapter IX.)

The prisoners, particularly the navigator, insisted that the graticulæ in the bomb sight were not used for this purpose. The latter stated that Lieut.-Comdr. Boecker quite recently experimented for some time by using the instrument in this way by observations of floating objects on a course laid near the German North Sea Islands, but abandoned the idea owing to the inaccurate results obtained. In this the man was probably departing from the truth, and it may be taken for granted that the bomb sight is the instrument chiefly relied on for accurately determining the ground speed.

It is certainly curious that the Germans have not further utilised the instrument as a drift indicator and have depended on two separate instruments, *i.e.*, bomb sight and bearing plate, for obtaining the ground speed and the drift, with a third instrument for working out the results. With a small addition the bomb sight can be made available for use as a drift indicator and corrector.

*Difficulty in Fixing Position.*—Of the foregoing navigating instruments chief reliance appears to be placed on the compass. The dead reckoning calculations are generally very unsatisfactory, in great measure owing to the varying amount of drift at different altitudes, which cannot be allowed for. In fact the usual nautical instruments and methods of navigation fail to fix the ship's position with any real accuracy when out of sight of land.

*Directional Wireless.*—Directional wireless is largely relied upon to correct the ship's track and determine her position. It is of varying accuracy, largely depending upon the relative positions of the ship and the directional stations. So long as the cut obtained from the directional stations is a good one, an accuracy as near as a five-mile square on a squared map can be expected; but as soon as the angle of cut is small the resulting position may be considerably in error.

Generally a position is indicated within 10 or 20 miles. When travelling at a great height or near the English coast an equally favourable result is not to be expected.

The experience of the navigator was that the error of the directional wireless is so variable as to preclude a mathematical rule being formulated for its correction. For instance, on the occasion of the Edinburgh raid, when off Dunbar, bearings were received which gave the ship's position as alternatively either 60 miles to the north or 60 miles to the south of where she actually was at the time.

### General Data.

*Weather Reports.*—One weather report at least is regularly received on the journey. The Hamburg Meteorological Observatory and the Ostend weather station furnish these, which are based on information received from Sweden. The isobars are only completed on the southern part of the weather chart. Information is received regarding these from the directional stations at Bruges, Borkum, Sylt, and Nordholz. All wireless bearings and reports are handed direct to the commander, who jointly with the quartermaster plots the ship's position and the isobars (*see* Chapter XII. on Weather).

*Courses of Rigid Airships.*—In the case of L. 14 and L. 33, a due westerly course was usually taken on leaving the shed and continued for a couple of hours. The ship was then turned south-west and skirted the German coast until west of Borkum when the westerly course was started, which was adjusted on the voyage according to the landfall desired.

Other commanders, however, skirt the coast until west of Ameland, then proceeding as set above.

The commander of L. 14 and L. 33 particularly avoided the chance of being observed from neutral territory. He did not wish to run the risk of his approach being advised by English agents or hostile observers.

Commanders of airships are expected to avoid crossing neutral territory if only for the reason explained above. On the return journey, however, the same importance is not attached to being observed by neutrals; and, indeed, the circumstances are sometimes such that the ship is not in a position to be able to select her course.

It was stated that the Dutch authorities had recently started opening fire on airships passing over their territory with artillery as well as with machine guns, causing thereby considerable surprise and annoyance to the German authorities, who had been accustomed to see the duty of punishing an infringement of neutrality performed in a perfunctory manner.

It was stated that the crew are not always warned in advance when a trip to England is to be undertaken, precise instructions being received by the commander in mid-air by wireless.

*Making a Landfall.*—The degree of accuracy with which a landfall is made varies considerably. At a height of 800 metres (2,625 feet) in clear weather the radius of observation is approximately 60 sea miles; at night, under the most favourable circumstances, this is reduced to at most 20 miles. On arriving within sight of the coast the course is adjusted to make the desired landfall, though frequently by reason of inaccurate navigation it is not possible to make that desired.

On the occasion of the Hull raid on the night of 5th–6th March 1916 the L. 14 came on to Hull by a fluke. The objective was Middlesbrough and Skinningrove, and the landfall was intended in the north of Yorkshire. Recognition of the coast on making a landfall is by no means easy. It is necessary to pick up a clearly distinguishable feature, such as an estuary, a promontory, or a distinctively marked bay. The attacks on the Midlands are usually made *viâ* The Wash, which, however, it would appear is by no means easy to recognise.

*Cruising and Full Speed.*—The North Sea is passed at a cruising speed which in the latest type of ships is about 45 miles per hour. This reduced speed is obtained by diminishing the number of r.p.m. to about 1,200, all motors as a rule being used. The full speed of these ships under normal circumstances is about 56–60 miles per hour. It is, of course, materially affected by head or following wind.

*Heights attained.*—The height at which the passage across the North Sea is made varies considerably according to the weather conditions. Previously this was between 300 to 800 metres (984 to 2,625 feet). L. 33, however, in her last trip made the passage at nearly 1,500 metres (4,920 feet) high. The greatest height to which this ship has ascended is about 4,000 metres (13,120 feet). On crossing the English coast the vessel is elevated to a minimum height of about 2,000 metres (6,560 feet) in order to avoid the anti-aircraft defences on the coast. The ship can rise with the assistance of her planes to a height of about 1,200 metres (3,937 feet) in between three and four minutes. The average height of flight when over England was stated to be about 3,000 metres (8,747 feet), provided no gunfire is expected. In entering over Folkestone on August 24th, 1916, however, L 32 flew at 12,000 feet.

A view of the earth's surface can be obtained from the control car through horizontal windows which resemble an ordinary window ledge, which is transparent by reason of cellon panes being fitted.

*Airship Lights.*—On arriving within sight of the English coast the greatest care is taken that no lights are visible. All unnecessary lights are extinguished, and those which are essential are carefully shaded.

*Lightships.*—The following lightships would appear to have been often picked up in making the passage:—Terschelling, Haas, Texel, and Galloper. It was stated that the position of the English lightships is mistrusted, as it is thought the British authorities have shifted them to frustrate their possible use by the enemy. This statement is very doubtful, and it is probable that full use is made of the lighthouses and lightships for purposes of night landfalls, and, moreover, with favourable results, as the airships' logs indicate.

*Position at Night.*—The position at night when over land cannot always be accurately determined. Outstanding features, such as the shape of the coast, can in good weather be recognised. Rivers, lakes, &c., are quickly picked up, but at the height at which ships now have to travel the lights used on land are seldom recognised. This was not the case before the present reduction of lighting came into force.

Railways, it would appear, are not easy to recognise. The Thames at London is always visible. The assistant coxswain stated that when over London on the last trip he was able to pick up the sweep of the Thames round the Isle of Dogs, but could not distinguish the docks, and did not recognise the gnomon formed by the South West India Docks and the Millwall Docks.

*Course during Bomb Dropping.*—When bombs are being dropped, the course of the ship must not be altered if accuracy of aim is required.

*Course for Attack on the Midlands.*—From an independent but reliable source the following information was obtained, which is more or less confirmed by the foregoing statements made by the prisoners :—

The German airships when making attacks on the Midlands almost invariably make a landfall near Cromer. To enable them to do this, the Frisian islands are so disposed geographically as to form an admirable leading line, some 50 miles in length, which runs almost due west (magnetic), 273° to be exact, across the North Sea to one of the lightships off the Norfolk coast.

Flying over these islands the track can be corrected for drift before losing sight of land, and any small error in direction can be checked, after leaving Borkum, by the Dutch coast lights and the Terschelling light-vessel; these final corrections making a total length of leading line of about 120 miles. The remaining oversea passage to the Norfolk coast is about the same length, and with the course so accurately set, and the direction and force of the wind so clearly determined, it should not be difficult to make a good landfall.

The accompanying chart (Plate 100) will explain the above.

## How German Rigid Airships land.

According to the examination of prisoners of L. 33, made by G.H.Q., Home Forces, the landing of German rigid airships is effected in the following manner :—

*Landing by Day.*—In calm weather the ship, when within a kilometre (1,094 yards) of the sheds, drops to a height of about 100 feet. She is then brought nose down gradually until within a distance of about 500 metres (547 yards) of the shed when she is 50 or 60 feet above the ground. At this time all engines, except those working the two wing propellers, are shut off, although this procedure varies according to the individual commander. In shutting off the engines, great care must be taken that the propellers are braked and locked in the horizontal position, so that when the ship touches the ground they may not be broken.

The ship is ballasted so as to remain steady and horizontal. The two smaller landing ropes are thrown overboard, and the landing party haul down the bow of the ship, while certain of the crew stand by the water ballast bags at her bows ready to release water ballast at a moment's notice in case of emergency.

The bow of the ship is now hauled to ground level, and a section of the landing party stand underneath the forward car with upturned hands ready to take her weight as it comes within reach.

Men of the landing party then swarm up into the forward car and traverse the keelway to the stern of the ship, thus bringing it down gradually. She is then man-handled into the shed by the landing party.

In breezy weather the airship appears to be brought to ground in much the same manner, with the addition that the main landing ropes are thrown out at a much earlier period of the ship's descent. To each of these is attached a special many-tailed rope kept on the landing ground. These enable a large number of men to control the ship and move her as desired. The number of men required to handle the ship on these occasions is approximately 700 as against 100 or even fewer in still weather.

In a wind the ship is brought down at an angle to the shed, at a considerable distance from it. Alertness of the crew and landing party must be considerably increased, and the ship must be brought into the shed upon trollies running on rails.

The principal handling ropes (for further details, *see* page 24) are as follows :—

One rope forward at the anchor point 125 metres long (410 feet).
Two near the same position 70 metres (249 feet) long, and 1 inch in diameter.
Three behind the rear car attached to the keelway.

It was stated that none of the airship sheds were provided with wind screens, and that it was risky to try to get a ship into its shed with a cross wind of more than 12 m.p.h.

In rough weather it is customary, when there is too much side wind and the shed is not a revolving one, to inquire of other sheds by wireless the weather conditions, and if they are such as to permit of an easier landing, the ship berths where this is available.

If the conditions are such that the ship cannot continue her journey, she is moored in the open. In this case all the landing ropes are used, and the men hold the ship down by sheer weight. The number of men employed for this purpose may be as many as 1,000.

*Landing by Night.*—Landing at night is carried out in precisely the same manner as by day, with the addition that the ground is illuminated by searchlights cast along the ground, while guiding lights are employed on the shed. The ship's searchlight is also used.

*Mooring on Water.*—Experiments appear to have been made at Friedrichshafen and on the Alster Lake, Hamburg, with mooring ships on the water. This is possible in the case of calm weather on still water, such as the surface of a lake, but is not practicable in a seaway.

*General.*—Many double sheds capable of berthing two ships are in existence, but when possible they are only used as single sheds.

A ship under normal conditions is ready for service almost immediately after a raid, 8 to 10 hours being required to top-up the ship. Should she immediately proceed on another journey the strain on the crew is considerable.

In case of need, ballast is stated to be jettisoned according to a prearranged scheme—water ballast and machine guns first, then surplus petrol and finally everything available, even to the extra clothing of the crew.

The sailmaker of L. 33 affirmed that the general understanding was that it was most inadvisable to force the ship up dynamically at a greater angle than about 20° when the gasbags were full, or at more than 15° when they were slack.

## CHAPTER XI.—INSTRUMENTS AND FITTINGS.

**Zeppelin Compasses.**

One compass only was recovered from L. 33; it was manufactured by "Carl Bamberg," Berlin, and is his 1908 model of ship's compass. It would appear that no effort has been made to save weight in the construction of the type of compasses used in these airships, nor is there any anti-vibrational arrangement. The compass is also gymballed each way. The peculiarity, and indeed the only point of especial interest in this compass, is the arrangement by which the expansion of the liquid, on a change of temperature, is absorbed (*see* Fig. 26).

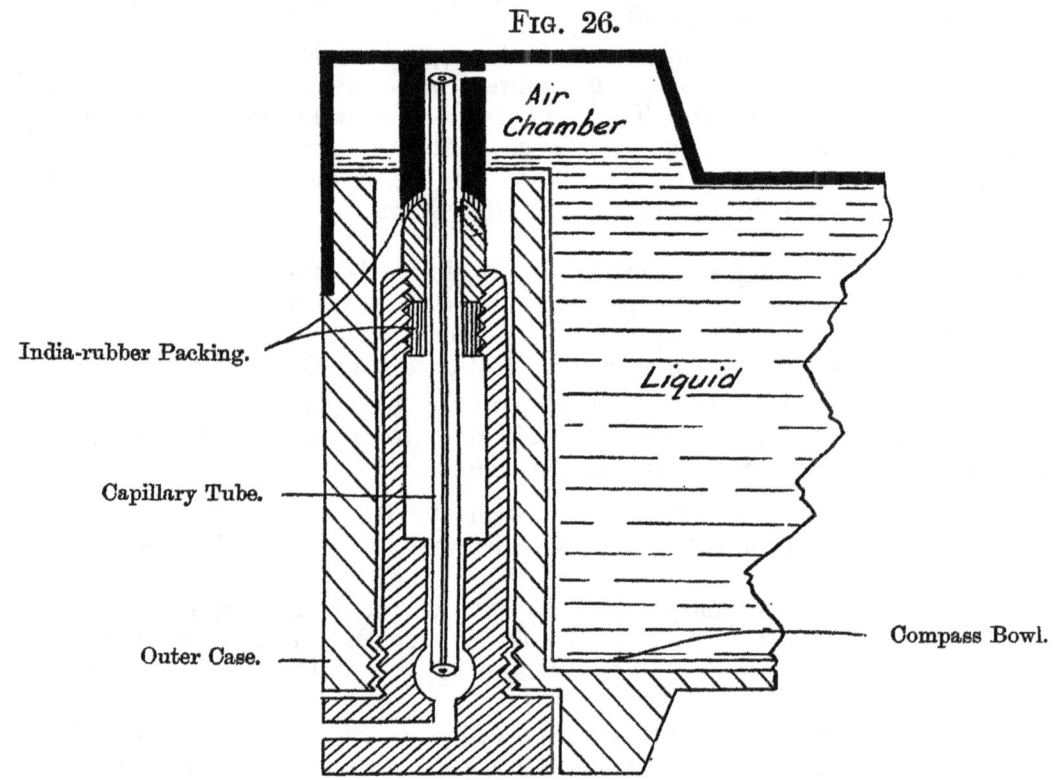

Fig. 26.

DIAGRAMMATIC SKETCH.—CAPILLARY TUBE EXPANSION ARRANGEMENT.

The feature of this arrangement is the air chamber round the top of the compass above the level of the glass, so that when a bubble forms from any cause it immediately passes into the air chamber and is not apparent in the compass. On expansion of the liquid the air in the air chamber, instead of being compressed, is forced through the capillary tube, and similarly, on contraction of the liquid, air is drawn through the capillary tube. The air chamber is sufficiently large to allow for the probable maximum expansion of the liquid so that no liquid will be forced into the capillary tube. In this particular compass a piece of ordinary clinical thermometer tube is used. This arrangement is not new, valves of various sorts having been employed in compasses by this and other makers as far back as 1903. The system of needles is crude, but follows the type employed by this maker for the past 26 years.

Inside the float are four magnets, each consisting of three laminations, a further six magnets being attached to the bottom of the float outside. The result is that a very strong directive force is obtained, but at the expense of lightness, since in order to float this great weight of magnets sufficiently to reduce the weight suspended by the pivot, the float has to be very large. Its weight complete is 290 grammes (10·2 ozs.). This large float means that a proportionately large amount of liquid must be employed. The objection to the large directive force is that no form of soft iron correctors can be employed for adjusting purposes, which, however, are not necessary, as the magnetic conditions prevailing at the compass position are ideal, no magnetic material of any size being placed within 24 feet of the compass.

It is not apparent what the necessity is for this very heavy type of compass, since the resulting period, 25 seconds (the ordinary marine liquid compass period), can be equally well obtained in a compass of one-half the total weight and with the same size of card.

The card used is a metal one graduated in degrees from 0° to 360°.

The lighting arrangements are an electric light placed under the compass bowl, the bottom of which is made of glass, and since the card is of metal no light is seen through it, but light is reflected from the sides of the bowl on to the card. The primary lighting is run off the main dynamo, while an alternative dry battery light is arranged for use when the main dynamo is stopped.

The construction of the bowl leaves much to be desired. It consists of two parts:—

(1) The bowl proper.
(2) An outer strengthening case.

It would appear that the latter is an unnecessary complication, due perhaps to a shortage of brass, as the bowl itself is very thin and not sufficiently strong to support the gymbal roller bearing; the outer strengthening case is a skeleton frame made of aluminium casting, the two being joined at the bottom by a watertight screwed joint.

There is no bearing attachment, the only facilities for obtaining bearings from the compass direct being a shadow-pin fitting.

The cap and pivot arrangement is the upright pivot and inverted cap; the cap is of sapphire and the pivot of manganese bronze.

No effort appears to have been made in the following directions:—

(a) Reduction of weight;
(b) Magnification of reading;
(c) Alternative lighting in case of electrical failure;
(d) Anti-vibrational;

and there is no progress as regards the special requirements desirable for airship conditions.

The compass recovered from the Observation car which was dropped in one of the Home Counties, and to which reference is made on page 31, Chapter III., was of the same size and type as that above described.

The other compasses recovered from Zeppelins L. 31, L. 32, and S.L. 11 were all of another pattern manufactured by Ludolph of Bremerhaven. They are very third-rate instruments, and have no special feature except extreme weight.

*Compass Position.*—The main navigating position in the L. 30 class is situated in the bow of the foremost car, and the greatest care has been taken to arrange that no magnetic material of any size is used in the vicinity of the compass. The nearest mass of magnetic material, *i.e.*, the foremost engine, is some 30 feet away. The W.T. instruments were also some distance off. The reason for this care is obvious, as the accurate adjustment of the compass of one of these large vessels would be extremely difficult, if not impossible.

No arrangements for adjusting were provided, but as it is very doubtful if the errors in this carefully arranged compass position were as much as 1° to 2°, there would be no necessity for adjusting arrangements.

In the secondary or alternative navigating position no compass was observed in the case of L. 33, but L. 31 and L. 32 each had two compasses, as also did S.L. 11.

*Bearing Plate.*—The bearing plate is of the portable type, having three legs which slip into shoes. It is not known how many alternative positions were arranged—but probably three. This is a combined pelorus and drift indicator, having three cross wires which, when set parallel to the drift lines, indicate the direction of drift on the graduated arc. It is fitted with a bearing sight for obtaining bearings, and also with a reflector for use when observing bearings of the sun.

### Drift Indicator.

The instrument consists of three graduated arms and two graduated discs (*see* Plate 103).

*Assembly:*—

Arm A graduated in three scales, and used for course steered and air speed.

*Scale* 1.—Metres per second from 0 to 40, every ·5 of a metre being marked. This is the standard scale.

*Scale 2.*—Kilometres per hour, marked from 0 to 150, every one kilometre marked. For use when flying over land.

*Scale 3.*—Sea miles per hour, marked from 0 to 80, every one mile marked. For use when flying over sea.

*Arm B* graduated in metres per second, from 0 to 20, every ·5 of a metre marked. This arm is used for wind.

*Arm C* graduated in metres per second, from 0 to 35, every ·5 of a metre marked. This arm is used for ground speed and course.

The three scales on arm A are proportionate, *i.e.*, opposite 17·5 metres per second read 64 kilometres per hour or 34 knots, by means of which the readings on arms B and C can be converted into kilometres per hour or knots as desired.

Two arms A and B and disc X are pivoted at a point; travelling along arm B is a pointer carrying arm C, travelling along arm A is pointer carrying disc Y, and through which arm C may slide.

*Method of Using :—*

(a) Disc X is rotated until the course steered appears in the slot cut in arm A.
(b) Arm B is set to direction from which wind is blowing on disc X.
(c) Travelling pointer on arm B is set to speed of wind in metres per second and clamped there by means of a screw underneath.
(d) Travelling pointer on arm A is set to air speed of aircraft in any one of the three scales.
(e) Disc Y is set and clamped by means of an eccentric clamp to course steered by aircraft.
(f) Arm C cuts disc Y at course made good.
(g) There is no means of reading on arm C the speed made good.

It appears that the instrument is used mainly for obtaining the course made good, no means being arranged for reading the speed made good on arm C, which is graduated in metres per second. This is probably explained by the fact that the speed made good can be obtained from the bomb-dropping sight (pages 72 and 78). The instrument is almost an exact reproduction of the "Drift Calculator" designed by Captain F. Creagh-Osborne, R.N., in 1913, two of which were issued for trial in the R.N.A.S. in that year; the German instrument is rather more clumsy and complicated to use than the earlier type.

The only problems which can be conveniently solved by this instrument are (1) To obtain course made good, having observed the speed made good, and knowing the speed and direction of the wind; (2) To obtain the allowance to make for wind.

The instrument is not adaptable to manœuvring problems. It is constructed by Harteman and Braun, of Frankfurt (well-known instrument makers), and is numbered "4," which seems to indicate that the production of this instrument has been recent.

There would appear to be no good reason for graduating arm C, since it is not possible to read it when set. Arm B, on the other hand, is graduated on both sides, and can consequently be set and read when the pointer is under the disc X. It seems probable that this instrument was used more as an adjunct to the bomb-dropping sight than as a navigational instrument, being inconveniently arranged for navigational purposes.

**Aneroid.**

The photograph reproduced on Plate 101 (Fig. 239) shows the mechanism of the instrument recovered from L. 33. The mechanism is of unusual design. In most instruments the spring connected to the aneroid box which supports the pressure of the atmosphere is mounted externally. In the Zeppelin instrument the spring is placed inside the aneroid box, and is therefore protected from rusting. In this respect this instrument conforms closely to our practice as supplied to recording aneroids.

The adjustment for setting the index to zero is usually carried out by altering the height of the bridge to which the fixed end of the spring is attached. In this instrument the adjustment is carried out by shifting the plate to which the lower side of the aneroid box is fixed. The arrangement of the levers resembles closely our practice, but a rather more perfect form of adjusting the gear for setting the amplitude of the movement of the pointer to the scale, is fitted in the Zeppelin instrument.

The main lever, which is connected direct to the top of the aneroid box, has its motion conveyed to the rotating balanced lever by means of a small connecting link. In our instrument a series of holes are drilled in this lever, to one of which the link is attached, the lower end of the link being firmly attached to the rotating lever. In the Zeppelin instrument the holes are drilled in the rotating lever and the link attached thereto, but in addition to this the upright end of the link can be slid along the straight lever so as to obtain a fine adjustment. The differences of the two instruments are shown in the following diagrams :—

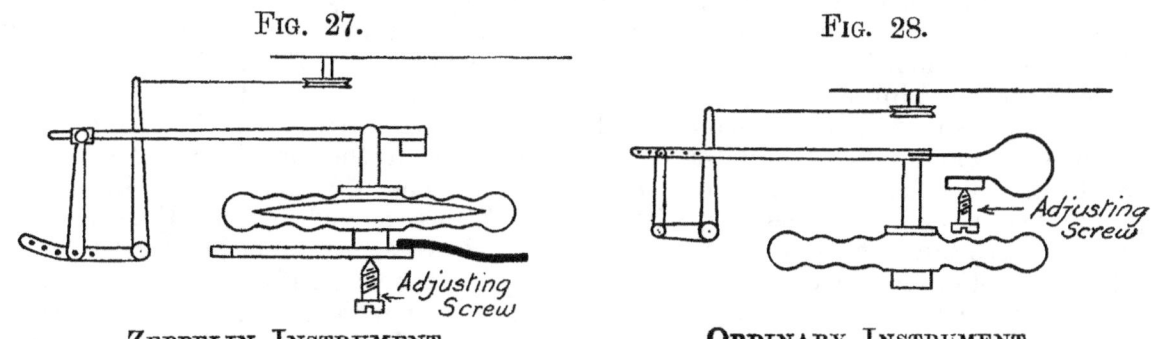

FIG. 27.   ZEPPELIN INSTRUMENT.

FIG. 28.   ORDINARY INSTRUMENT.

### Statascope.

The following is a description of the statascope, to indicate the rate of rise or fall, recovered from L. 15. (*See* Plate 101, Figs. 242 to 245.)

It consists of a thermos flask, into the neck of which is fitted an india-rubber cork. Through two holes in this cork are placed two glass tubes A1 and B1. Tube A1 is carried through the cork only. Tube B1 is connected by a rubber tube inside the flask to a fine capillary thermometer tube extending nearly to the bottom of the flask. The upper ends of tubes A1 and B1 are connected to the two ends A and B of a U tube bent as shown.

Both the upper ends of this U tube are fitted with a non-spill chamber. Into one one of these chambers is fitted a stopper for filling the U tube with liquid. Leading into the tube B1 just above the cork of thermos flask is another tube C, the lower end of which is open to the atmosphere. The entrance of this tube is designed with a view to making it dust-proof. As a further precaution against dust a glass-wool plug is fitted inside the tube.

The portion of the U tube to which the scale is fitted is fixed at 15° to the horizontal and has the internal diameter decreased considerably. This increases the movement of the liquid along this section. Any change of pressure of the atmosphere is communicated along tube "C" to tube "B" of the U tube, the liquid in the narrow portion of the tube immediately registering the rise or fall of pressure as the case may be. In the meantime the pressure in "B" is gradually communicated through the fine capillary tube to the thermos flask "A." The amount by which the pressure in the thermos flask lags behind that in the atmosphere is dependent on the rate of rise or fall which is thereby indicated by the reading of the U tube.

The whole of the apparatus is enclosed in a light metal casing, an opening in the cover exposing to view only the narrowed portion of the U tube with the rise and fall scale attached. An electric light is fitted inside the casing. The chief difference between this instrument and our statascope is that whereas on the British instrument each unit of the scale represents a given number of feet rise, on the German instrument each unit of the scale represents a given rate of rise. A similar instrument was found in LZ. 85 at Salonika.

### Zeppelin Rate of Rise Indicator.

The remains of the statascope found in L. 33 were almost completely destroyed.

The instrument shown on Plate 101, Fig. 241, was got from L. 32. The principle on which it depends for action is exactly the same as that of the L.15 instrument described above. The principal difference is that a mechanical pressure gauge takes the place of the liquid one in the previous instrument. Referring to Fig. 29 of the cross section, the instrument will be seen to consist of an outside case, containing an inside metal container, between the two being placed a thick layer of felt. An aperture, fitted with a fine gauze diaphragm, allows the inside container to communicate with the outside air. Immediately above this aperture, and attached to the bottom of the inside container, is mounted a corrugated diaphragm which would appear to be of German silver. It is very thin, being approximately ·06 mm. in thickness. On its centre is mounted a light brass plate, to which is attached a short link connecting with the indicator gear. This consists of two levers, A and B. Motion is communicated to the short end of lever A, and is communicated by the long end to lever B through fine silk cord, the motion being magnified in the ratio of 1 to 5. The fine silk cord is attached to lever B at a distance of approximately 1·5 mm. from its axis. 1 mm. movement of the diaphragm would therefore cause a movement of index of approximately 220 mm. The cord is kept taut by means of a hair spring mounted on the shaft of lever B. On the top of this same shaft is mounted a light steel pointer which indicates on to a cardboard scale backed with looking glass. The scale is mounted immediately over an aperture cut in the top of the outer container, and is rendered airtight by means of a flange holding a mica plate. The bearings for lever B are mounted on a suitable frame fixed rigidly to the container. The bearings for lever A are, however, mounted on a frame, one end of which is attached to the casing, the other end connecting with a nut which is moved by an adjusting screw which passes through the casing from outside. By rotating the adjusting screw the relative positions of lever A and the diaphragm can be altered and the index set at zero on the scale. The part of the inside container above the diaphragm is therefore airtight, with the exception of two small holes which are drilled through the side. These connect with two tubes—one passing round the circumference of the inside container, and being about 10 cm. long; the other being approximately 2 cm. long, passing towards the front of the inside.

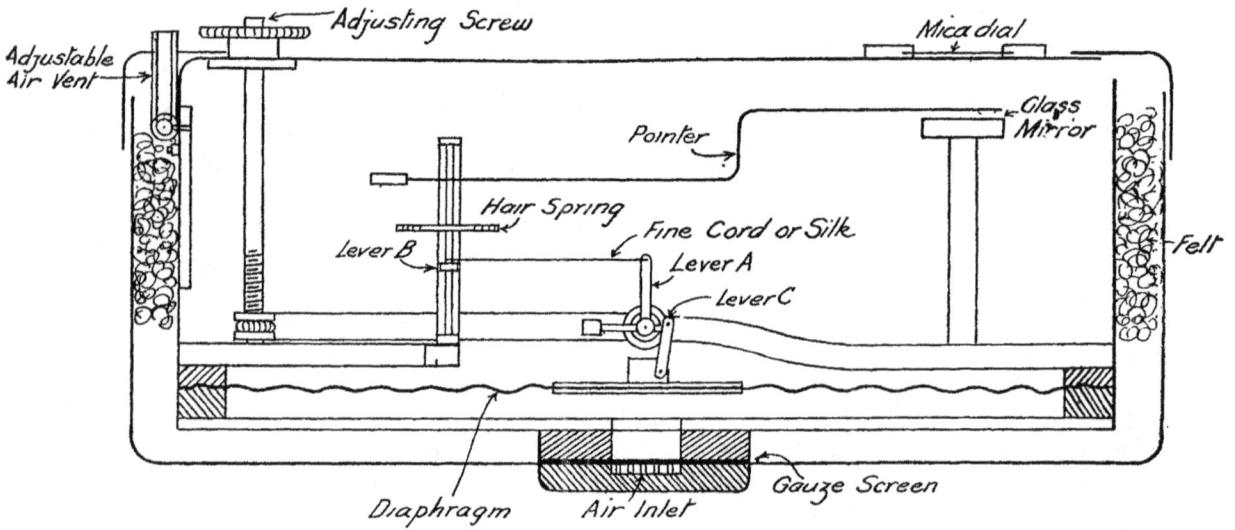

Fig. 29.—Rate of Rise Indicator from L 32.

This latter tube has been plugged. It is used probably for testing and calibrating instruments. The longer tube has a loose-fitting wire partially obstructing its bore for a certain length, the flange being secured in place by means of a small set screw. This tube takes the place of a small glass capillary tube which was used in the earlier instrument. The action of the apparatus is as follows:—

The change of barometric pressure due to height above sea level of the apparatus is immediately communicated to one side of the diaphragm by the recording gear, causing a movement of the index. It also communicates, but more slowly, through a partially plugged pipe with the other side of the diaphragm. Since the rate of leakage through the tube is proportional to the difference in the pressure of the air inside and outside the central container, the indicator shows the rate at which change of external pressure is occurring, and thus the rate of change of level of the airship.

**Electric Thermometers.**

An electric thermometer was fitted to indicate the temperatures at two points, probably one in the gas and one in the outer air. The instrument in L. 33 was badly burnt. The instrument shown in Plate 102, Fig. 251, is similar and was recovered from L. 15, having been immersed in seawater for two months. Many of the steel parts of the instrument had entirely disappeared. The aluminium was in process of corrosion, and fine copper wires had corroded through. The scale was in fairly good condition, except in the centre, where the pointer had probably laid. Here there was evidence of considerable electrolytic action, the aluminium backing to the scale being almost eaten through. The woodwork was in fairly good condition.

The instrument, when opened, was found to consist of the following parts:—

- A 4-volt battery of "Ever Ready" type, the size used in small pocket flash lamps.
- A moving iron voltmeter that apparently acted as an automatic cut-out, as there was no space provided in the outside case through which a pointer could be visible.
- A large, carefully-made, moving-coil galvanometer, provided with two finely-wound coils which passed round the central iron core placed between the poles of two compound horseshoe permanent magnets. The coils crossed over one another, making an angle of 60°, their ends being connected to separate terminals, by, presumably, fine spirals of copper wire.

It has been a matter of difficulty to follow the exact electrical connections, as all the finer wires had disappeared, but the arrangement of the circuit is almost certainly as shown on the diagram (*see* Fig. 30).

Fig. 30.

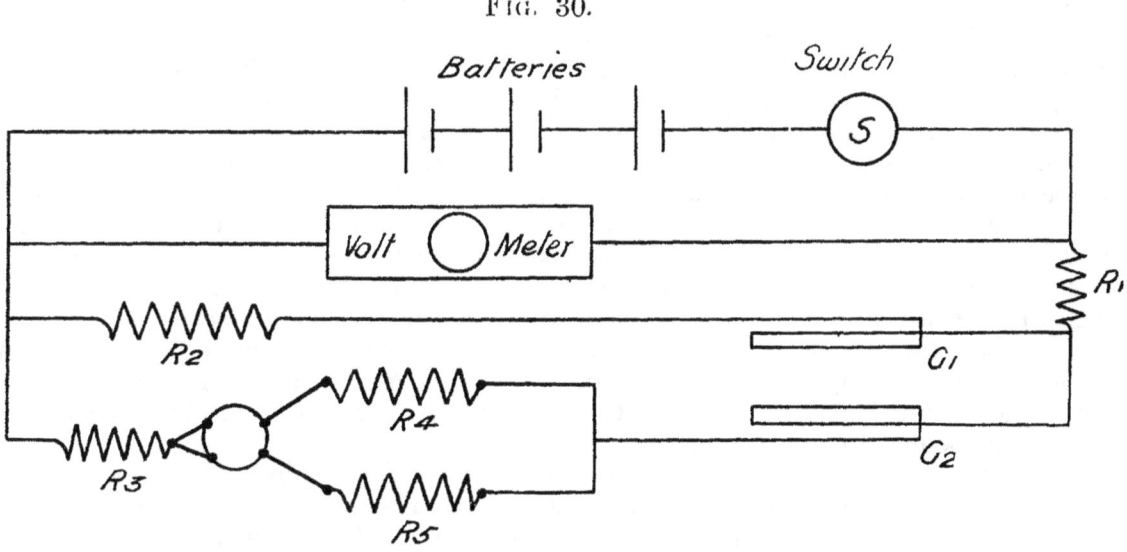

CONNECTIONS OF GALVANOMETER.

The resistances were wound as open coils, there being no bobbins, and they were non-inductively wound.

The action of the apparatus appears to have been as follows:—

Two currents flow in opposite directions around the iron core. When these two currents are equal, the pointer of the instrument remains at zero; when they are unequal, the pointer will move by an amount which will depend upon the ratios of their strengths. The current flowing round one coil depends solely on the initial voltage of the battery and the resistance of the two coils which are marked R. 1 and R. 2 in the diagram. The current through the other coil, however, depends on the E.M.F. of the battery, as before, but also on the magnitudes of the resistances R. 1 and R. 3, and the resistance of the other two external leads connected to the 4-way plug. If the resistances of the external circuits connected to the terminals of the 4-way plug vary, the throw of the galvanometer will vary, and owing to the nature of the arrangement the throw will vary almost exactly with the value of the resistance and will be almost independent of the electromotive force of the battery. This is the watt-meter principle, and as a method of measuring resistances has the advantage over the Wheatstone bridge that it is not affected by the initial electromotive force to the same extent.

Past experience of small hand-lamp batteries of this type has shown that they undergo considerable voltage drop towards the end of their life.

The action of the voltmeter referred to above was probably automatically to break the circuit when the E.M.F. dropped below a certain pre-arranged value, this value being that at which the pointer of the galvanometer ceased to give readings proportional to the resistance in the external circuit. As further evidence that this was the case may be noted the fact that the battery was coupled direct to the resistances and galvanometer, there being no potentiometer or variable series resistance that could be operated by hand.

The scale of the instrument is graduated from minus 25° to plus 50° C.

### Engine Revolution Indicator.

The instrument (*see* Plate 102, Figs. 246 to 248) is found on examination to be of the governor type. Motion is transmitted to the instrument through a flexible cable, there being a small gear box interposed to increase the speed of rotation of the governor to four times that of the shaft. The governor, which is of the rotating-ring type, can be seen both in the photograph of the assembled instrument and also in the photograph of the parts of the instrument. As the rate of rotation increases, centrifugal force causes the ring, which is initially tilted, to become more and more horizontal. In so doing, the circular spring attached to the shaft is set in torsion. This motion is conveyed to the mechanism by means of a small fixed shaft which passes down the hollow governor spindle. This shaft is attached at its upper end to a small lever mounted on the index mechanism. The lower end of this small shaft has an annular groove which engages with a small sliding piece, which is in turn attached to a lever connected to the rim of the governor. When the governor tilts, therefore, the small sliding piece is drawn downwards, which causes the shaft attached to it to be drawn downwards, therefore causing the index mechanism to be moved in proportion to the tilt.

The index mechanism consists of three rotating shafts, attached to which are four gear wheels, the motion being conveyed to the first shaft by means of the small lever referred to above, and is conveyed to the second shaft through a pair of gear wheels which increase the motion in a ratio of about 1 to 9. To this second shaft the hand is attached. Motion is conveyed from this second shaft to a third shaft by means of a pair of gear wheels which increase the rate of rotation in a ratio of 1 to 13. To this third shaft are attached four small air vanes which act as dampers.

It will be seen that this mechanism differs materially from the ordinary type of speedometer in the way in which the motion is conveyed from the governor to the index. The whole apparatus appears to be extremely sensitive, and the method of air damping seems to be particularly good.

The flexible shaft used to convey the motion from the engine to the instrument is constructed as follows (*see* Plate 102, Fig. 248):—The outside casing is ordinary flexible metallic tube wound on a right-hand thread (B). Packed between the spirals is ordinary asbestos string (D). Inside this is wound a ribbon of mild steel $\frac{3}{8}$ inch wide, there being no overlap between successive spirals (C). The flexible shaft which rotates inside this casing is built up of a succession of flat and cylindrical links, which are riveted together in such a way that they form a series of universal joints (A). Details of the mild steel strip and the central shaft are shown in Fig. 248.

### Petrol Gauge of Storage Petrol Tanks.

A gauge is fitted in German rigid airships to indicate the level of petrol in the tanks in the keel. The gauge illustrated on Plate 102, Fig. 252, is constructed almost entirely of aluminium; it is $4\frac{1}{4}$ inches in diameter and $1\frac{5}{8}$ inch depth. It consists of a four-spoked aluminium pulley, with grooved flange, and a fine silk cord is fastened in this groove, wrapped several times round the pulley and led through the outlet pipe to the petrol tank. To the other end of this cord is fastened a weighted float. The string is held taut on the pulley by means of a watch spring, fastened to the rear of the pulley. The motion of the pulley is transmitted to a registering hand by brass reduction gears, the gearing being 3 to 1. The gauge is made petrol-tight, so that no gland is required on the cord. The illustration on Plate 102, Fig. 252, shows the instrument without the dial, which was graduated up to 220 kilos (485 lbs.) of petrol.

### Fire Extinguisher.

The general arrangements of the fire extinguisher from L. 15 is seen in the illustration, Plate 101, Fig. 240. The diameter of the container is 5 inches, and the depth to the point at which it tapers towards the handle is about 6 inches. The lever lid top has to be pulled off before the contents can be shaken through the sprinkler. This appears to be a distinct disadvantage to the design. The contents consist mainly of sodium bicarbonate with a little sodium carbonate. Ferric oxide has been added, and gives the contents a brownish-red colour. The whole weighs about 4 lbs. Prisoners from L. 33 state that fire extinguishers have not been found of much use, and are usually the first thing to be jettisoned.

**Life Belt.**

A life belt recovered from one of the Zeppelins consists primarily of a set of braces made of canvas tape, which strap tightly round the chest and back of the wearer (*see* Plate 102, Figs. 249 and 250). The main framework of these braces is rectangular in shape both front and rear, the two portions being connected together by a pair of shoulder straps, and also a strap at the top of the rectangular frame which passes under the left armpit of the wearer. At the bottom of the rectangular frame on the left side, and at the top and bottom on the right side, are attached buckles and straps—three in all—which allow the braces to be strapped firmly to the body and permit of adjustment to various chest dimensions.

Two rubbered canvas bags, rectangular in shape, are connected loosely by means of thin tape to the braces, one bag being in front and one at the back. These air bags are connected together by a narrow neck of the same material, which passes under the left armpit and allows air to pass from one bag to the other, thereby ensuring an even pressure of air in both bags. The front air bag is fitted with a valve, placed at the top of the bag, convenient to the mouth of the wearer, as shown in Fig. 249, allowing the bag to be inflated at will.

The chief feature of this life belt is the method of fastening the air bags loosely to the body braces, so that although the braces may be strapped tightly around the body and remain in position, the air bags, when inflated, will not exert a pressure on the chest and back of the wearer.

The valve used for inflating by means of lung power is of the ordinary type, as used in air cushions, &c. The life belt is stamped: " Size 1. For chest measurements from 80–100 cms."

## CHAPTER XII.— METEOROLOGICAL CONDITIONS OF RAIDS.

An investigation into the meteorological conditions in connection with Zeppelin raids up to the end of September 1915 was made in the Air Department, Admiralty, in October 1915. Nineteen raids were dealt with, and the conclusions arrived at were as follows :—

*Wind.*—In 16 out of the 19 raids the wind was less than 20 miles per hour on the Continent at the time of starting, and less than 25 miles per hour over here during the raids. The direction does not seem to have mattered, as all parts of the compass were represented.

On the other three days, May 10th and 26th, and June 15th, 1915, the wind varied from 25 to 40 miles per hour, and the direction in all three cases was somewhere between north and east.

From pilot balloon ascents we know that, on May 10th and 26th, 1915, the high N.E. wind did not extend to any great altitude; in fact, at 10,000 feet it was either calm or light westerly, but on June 15th the strong easterly breeze extended as high as 10,000 feet, where it was still blowing at 25 miles per hour.

*Cloud.*—In most cases the state of the sky on the Continent before starting was fairly clear—a little cloud about. In four cases it was quite clear; in three others rather cloudy. In one case only was it overcast. Over here, during the raid, on five occasions the sky was clear; on seven there was a little cloud; on seven it was rather cloudy, but on no occasion was it overcast during the whole raid.

*Mist or Fog.*—On six occasions mist or fog was reported at places on the Continent—North Holland—at times that would correspond to the start of Zeppelins on night raids. On 10 occasions during raids, mist or fog was reported over here. In no case, neither on the Continent nor over here, was the mist or fog general, but in patches, and mostly on the coast.

*Rain.*—Light rain occurred twice on the Continent just previous to raids, and twice over here during raids. In all four cases it was light drizzling or misty rain.

*Thunder.*—August 12th, 1915, was the only raid day on which thunder was reported. Thunder was reported at Spurn, and the raid was over Suffolk.

*Moon.*—On 16 raids there was no, or practically no moon, but on the other three—April 30th, May 21st and 31st, 1915, it was practically full moon. (Nor was there any cloud worth considering on these three occasions.)

*Barometer.*—In 13 of the 19 raids it was steady; in three it was rising; in two it was falling slightly; in only one was it rather unsteady, but in every raid it was either normal or high.

(In referring to the barometer, high means above 1,020 millibars; normal means between 1,010 and 1,020 millibars; low means below 1,010 millibars. Corrected to mean sea level.)

*Time of Year.*—January, one raid. February, none. March, none. April, three. May, four. June, three. July, none. August, three. September, five.

### Comments on the Foregoing.

*Wind.*—There is usually a westerly wind above 10,000 feet. Generally on the Continent if wind is favourable at low altitudes for starting, it will be more favourable at high altitudes for returning.

*Pilot Balloons.*—The importance of these in connection with aircraft is fully appreciated by the Germans. Information at hand states that they make very extensive use of them.

*Cloud.*—The enemy appear to think an overcast night unfavourable for the purpose.

*Fog.*—Unless it be general it should not, and does not, hamper the enemy. In fact if there are sufficient clear spaces to pick up coastlines and rivers it should help.

*Thunder.*—It seems fairly certain that prevalence of thunder prevented raids during July 1915. Thunder was reported in England on 15 days in July.

*Rain.*—Evidently avoided if possible, as on no occasion was there heavy rain over here during a raid.

*Moon.*—The fact that on three raids it was practically full moon with a clear sky is rather surprising and seems to have no meteorological explanation. It may however be interesting to note that during the last 10 raids in 1915 there was practically no moon.

*Barometer.*—The height and steadiness of the barometer seem to have been the most important factors in deciding on suitable raid days. In fact the enemy appears to have waited for a high and steady period.

*Summary and Conclusion.*—In general the conditions were:—Barometer high and steady; no depressions approaching; wind less than 25 miles per hour; no rain about; little or no cloud; no moon; no thunder; there was often some coast fog.

### Zeppelin Weather During 1915–16.

The tables set out across the two pages 92 and 93 show a similar investigation into the weather during raids between the end of September 1915 and the end of November 1916. The barometer readings given are those for Blaavands Huk, which place is sufficiently near Cuxhaven to make very little difference.

### Raids, Barometer, and available Darkness Graphs.

The graphs, shown in Plates 104 to 106, are practically self-explanatory. The actual barometer readings in Germany not being available, those for Blaavands Huk, in Denmark, have been used. A good general idea of the main conditions selected for Zeppelin raids is given, but for a more complete idea of the actual conditions the weather charts should be consulted.

Considering meteorological conditions alone from the graphs, the following dates stand out as probable chances missed :—

*October 5th*, 1916.—Apparently good, but there was a shallow depression over Germany, which caused rain in the North Sea.

*October 6th.*—Really favourable. Probably the same shallow depression prevented a raid.

*October 7th.*—The conditions in Germany would be favourable. A depression west of Ireland towards evening caused rain from North Spain to N.W. Scotland, including most of the Channel but missing the North Sea.

*October 10th.*—Apparently good in Germany, but there was a depression causing rain and high wind over North Spain.

*October 11th.*—Again, apart from a little rain in the North Sea, it would appear favourable, but there were rather high winds in South Norway; also rain in North Spain.

*May 29th–June 2nd*, 1916.—During this period the barometer was fairly steady, but there was a lot of rain about. Thunderstorms also were probable and were reported in Southern Europe.

Other chances missed may suggest themselves, but the foregoing are the most apparent.

It must not be forgotten that surface weather is often very different from that above 2,000 feet. Although the weather at the surface here may be dangerous, yet for ships flying above 3,000 feet it may be quite safe. The enemy is mainly concerned with his own local or surface weather for purposes of starting and landing, and with the general weather above 2,000 feet. The former he knows, and the latter he can generally calculate from the data obtained from his numerous observation stations. Local bad weather in this country is therefore no guarantee that a raid will not be attempted.

**Results of the Investigation.**

*Wind.*—There is no evidence that in any of the raids the surface wind, at starting and returning, exceeded 20 miles per hour. It seems most likely that, at starting, it was considerably less than 15 miles per hour. There were areas of fairly high winds, but the enemy's information was generally sufficient to enable him to avoid these areas.

Doubtless the enemy has a well-organised system of stations for wind observation by means of surface anemometers and pilot balloons, and is therefore able to inform the returning airship of the best place to land. (*See* Chapter X. on Navigation.)

*Cloud.*—In almost every raid there was absence of heavy cloud at low altitudes. In several it was cloudy, but in the majority the sky was perfectly clear.

*Fog.*—Fog or mist was reported on more than half of the raids. This must be expected as times of high barometric pressure are chosen for raids.

*Rain.*—Very seldom was rain reported during a raid. The presence of rain even in Spain seems to have prevented a raid.

*Moon.*—On every raid night during this second period of investigation there was either no moon at all, or practically none.

*Barometer.*—On two occasions only was the barometric pressure low, and on one of those it was rising. On the other 21 occasions it was either normal or high—mostly high.

*Summary and Conclusions.*—The conditions selected by the enemy for Zeppelin raids are very definite :—

    (1) Anti-cyclonic weather, giving calm or light breezes. Very little cloud. No rain or snow. (Fog, which is often present with these conditions, does not greatly hinder operations, and if patchy may help.)

    (2) No moonlight.

    (3) No thunder.

The enemy can obtain information from Scandinavia and Spain. This, together with his own meteorological system, would appear to be sufficient for his purpose. There is no indication that information is received from the Atlantic.

From this investigation there is no reason to believe that the enemy's airships are more "weather worthy" now than in 1915.

| Date. | Time of reaching English Coast. | Time of Leaving. | State of Barometer. | Time of Sunset. | State of Moon during Raid. |
|---|---|---|---|---|---|
| 1915. Oct. 13 to 14 | 6.20 p.m. | 1.50 a.m. | High, rising (1024 m.b.). | 5.12 | Set at 7.33 p.m., first quarter. |
| 1916. Jan. 31 to Feb. 1. | 4.50 p.m. | 5.25 a.m. | High, falling slightly (1039 m.b.). | 4.43 | None |
| March 5 to 6 | 9.45 p.m. | 2.25 a.m. | Low, rising slowly (1005 m.b.). | 5.43 | None |
| March 31 to April 1. | 7.45 p.m. | 2.50 a.m. | High, steady (1025 m.b.). | 6.29 | None |
| April 1 to 12 | 11.5 p.m. | 1.50 a.m. | High, rising slightly (1026 m.b.). | 6.31 | None |
| April 2 to 3 | 10.30 p.m. | 1.0 a.m. | High, falling slightly (1024 m.b.). | 6.32 | None |
| April 4 | 1.45 a.m. | 3.15 a.m. | Normal, falling (1015 m.b.). | 6.34 | None |
| April 5 to 6 | 9.10 p.m. | 2.50 a.m. | Normal, rising (1015 m.b.). | 6.37 | Set at 10.39 p.m., almost new. |
| April 24 to 25 | 10.15 p.m. | 2.5 a.m. | High, rising (1025 m.b.). | 7.8 | None |
| April 25 to 26 | 9.55 p.m. | 1.50 a.m. | High, rising (1028 m.b.). | 7.10 | None |
| April 26 | 10.25 p.m. | 10.50 p.m. | High, steady (1027 m.b.). | 7.12 | None |
| May 2 to 3 | 9.15 p.m. | 2.30 a.m. | Normal, falling (1014 m.b.). | 7.22 | None |
| July 29 | 12.10 a.m. | 3.25 a.m. | High, rising (1024 m.b.). | 7.53 | None |
| July 31 to August 1. | 10.30 p.m. | 3.0 a.m. | High, rising (1020 m.b.). | 7.49 | None |
| August 2 to 3 | 11.40 p.m. | 2.35 a.m. | High, falling (1021 m.b.). | 7.45 | None |
| August 9 | 12.10 a.m. | 2.35 a.m. | High, falling (1025 m.b.). | 7.35 | None |
| August 23 to 24 | 11.55 p.m. | 12.12 a.m. | Normal, falling (1012 m.b.). | 7.6 | Practically none |
| August 24 to 25 | 11.5 p.m. | 2.35 a.m. | Low, falling (1006 m.b.). | 7.3 | Practically none. Rises at 1.17 a.m., but past last quarter. |
| Sept. 2 to 3 | 9.50 p.m. | 4.20 a.m. | Normal, rising (1014 m.b.). | 6.44 | None |
| Sept. 23 to 24 | 8.55 p.m. | 2 a.m. | High falling (1020 m.b.). | 5.56 | None |
| Sept. 25 to 26 | 8.50 p.m. | 3.5 a.m. | High, rising (1021 m.b.). | 5.52 | None |
| Oct. 1 to 2 | 8.5 p.m. | 3.35 a.m. | High, falling (1020 m.b.). | 5.38 | None |
| Nov. 27 to 28 | 9.10 p.m. | 6.30 a.m. | Normal, rising rapidly (1015 m.b.). | 3.56 | None |

| Probable Weather at Cuxhaven at Time of Start. | Weather over North Sea and England. | Remarks. |
|---|---|---|
| Fine, calm - - - | Some fog. Calm to S.W., wind increasing during the night. | Ideal at start. |
| Some fog, S.E. wind about 10 m.p.h. | Some mist and fog. Very light indefinite wind. | Ideal. |
| Fair. Wind light, indefinite. | Fairly strong northerly wind. Some snow. | Rather a risky night for a raid. Perhaps there were special military reasons. About time of Irish rebellion. |
| Cloudy. Light W. wind. | Fine. Some mist. Very light westerly. | Ideal. |
| Cloudy. Light W.N.W. wind. | Fine. Some mist. Light indefinite wind. | Ideal. |
| Fine. Light indefinite wind. | Fine. Light indefinite wind. Some fog. | Ideal. |
| Rather misty. Light indefinite wind. | Rather misty. Light indefinite wind. | Evidently counting upon still anti-cyclonic. |
| Rather foggy. Light N.N.W. wind veering and decreasing. | Light indefinite wind - | Ideal. Scandinavian information. |
| Fine. Very light S. wind | Southerly wind, increasing on English side of North Sea. Cloudy, getting worse. | Depression apparently spreading in from Atlantic. |
| Fine. Very light indefinite wind. | Fine. Light S.E. wind - | Ideal. Depression did not spread and affect North Sea. |
| Fine. Light S.E. wind - | Fine. Light S.E. wind decreasing. | Ideal. No meteorological reason for the short raid. |
| Fine. Light indefinite wind. | Fine. Light indefinite wind. Foggy. | Ideal. |
| Rather dull, but very little wind. | Rather misty. Very light N.W. | Really quite good, but at Cuxhaven would appear rather risky. Reports from Belgium would give a good idea of the actual conditions. |
| Rather high N.W. wind, about 20 m.p.h. | Fine, hardly any wind, misty later. | Here again the conditions were actually good over the North Sea and south England, though at Cuxhaven they would not appear so. In fact, towards the end of the raid the wind over Norway and Scotland had in places reached 40 m.p.h. Belgium and the south of the North Sea were not affected. |
| Rather cloudy. Light, but increasing W.N.W. wind. | Fine, very light north-westerly wind. Not so fine in north of England. | Similar remarks to those on the raid of July 31st apply to this raid. |
| Fine. Light indefinite wind. | Fine. Light indefinite wind, increasing from E. | Ideal. During the night in the south of England the surface wind increased from the east to about 20 m.p.h. and decreased again towards morning. At high altitudes it was much less. Calm over the North Sea. |
| Rather dull. Light W.N.W. wind. | Fair to dull, some mist. Light W. wind. | A fairly calm region between two depressions. One off the west coast of Ireland and the other over Sweden. |
| Rather dull, some rain. Light W. wind. | Dull, becoming worse. Light southerly wind, increasing. | Still a region of practically calm weather. The enemy would know of the existence of the depression over Sweden. |
| Fine. Very light W.N.W. wind. | Dull and misty. Light indefinite wind. | Would appear ideal from enemy point of view. |
| Fine. Very light S.W. wind. | Fine. Light S.S.W. wind. | Ideal. |
| Fine. Light indefinite wind. | Fine. Light southerly wind. | Ideal. |
| Cloudy. Very light N.W. wind. | Rather dull, some mist. Light indefinite wind. | Almost ideal. Depression off S.W. of Ireland, but did not affect the North Sea. |
| Fine. Light N.N.W. wind. | Fine. Light indefinite wind south of Wash. Further north, increasing. Almost S.W. gale in Shetlands. | Strong westerly wind over Scandinavia, obviously moderating. From the information at disposal of the enemy the outlook would appear quite good, but was rather risky in the north. This they might know also. |

## APPENDIX A.—THE PRODUCTION OF HYDROGEN IN GERMANY.

Hydrogen is usually obtained in Germany by three main methods; two other methods are also used for active service:—

(1) By bottles.
(2) By plant at the airship station.
(3) By building an airship station near some factory where hydrogen is given off as a by-product.
(4) The silicon process is also used for war purposes and in the field.
(5) The oil cracking or Pictet process is also believed to be supplied on railway trucks as portable plants.

(1) *By Bottles.*—The bottles used for transport are short enough to be packed vertically in railway trucks with the valves upwards, with hydrogen compressed probably to 175 atmospheres, the bottles being charged and the hydrogen delivered direct from the trucks to a holder or an airship. The hydrogen is procured from the nearest large plant, or works where it is available as a by-product.

(2) *By Plant.*—The most popular type of plant now in use appears to be the Messerschmidt system, in which the hydrogen is produced by the alternate reduction and oxidation of spongy iron by water-gas and steam in a large retort with a central chamber and two annular chambers, the hydrogen being given off when the iron is oxidised by the steam. Plants of this kind are known to be installed at—

| | |
|---|---|
| Friedrichshafen | 7,500 cubic feet per hour, |
| Cologne | 5,500 ,, ,, |
| Königsberg | 9,000 ,, ,, |
| Mannheim | 15,000 ,, ,, |
| Schneidemuhl | 15,000 ,, ,, |

and probably some others. Hydrogen by this method is very cheap, as the only raw material used is coke. The Germans specialise in hydrogen manufacture and have taken out an enormous number of patents for various processes, all based on this type of chemical reaction, while changes and improvements are being made constantly. The Badische process is known to be largely used for the production of synthetic ammonia, though it is not thought to be ever used for airships on account of the high percentage of nitrogen contained in the gas.

The Linde Frank, Caro process is also used, as this process has been developed entirely in Germany and is very cheap. It consists of making ordinary blue-water gas and freezing out all the gases except the hydrogen, these other gases being used to run the machinery required to work the plant. There is an installation of this kind at Tegel.

(3) *By-product.*—Hydrogen is given off in large quantities as a by-product from the electrolytic production of caustic soda from brine at the works of the Griesheim Electric Company at Griesheim and Bitterfeld, in Germany. The Zeppelin garage at Frankfurt is supplied with hydrogen by means of a high pressure main from Griesheim, the distance being about 15 miles. The Griesheim Works produce daily about 700,000 cubic feet.

(4) *Silicon Process.*—So far as can be ascertained the silicon process does not differ materially from the same process in common use in France, Italy, and in this country.

(5) *Oil Cracking.*—This process consists in vaporising oil and heating the vapour to an enormously high temperature, say, 1,400° C., when it is "cracked" or split up into hydrogen and carbon, the carbon being a valuable by-product in the form of lampblack. It has not been adopted in this country as no tubes can be found which can stand the enormous temperature for any length of time, and it is not thought that its application in Germany is very large.

# APPENDIX B.—STATISTICAL DETAILS OF MACHINERY, &c.

## Weights of Machinery of L 33.

|  | Lbs. | Ozs. |
|---|---|---|
| Engine and flywheel | 970 | 0 |
| Clutch | 92 | 0 |
| *Engine Details:*— | | |
|     Top half of crank case | 70 | 6 |
|     Bottom half of crank case | 32 | 4 |
|     Cylinder complete | 37 | 0 |
|     Piston complete | 10 | 12 |
|     Connecting rod complete | 8 | 8 |
|     Crankshaft complete | 103 | 0 |
|     Carburetter | 11 | 8 |
|     Ex. manifold | 50 | 0 |
|     Camshaft | 12 | 0 |
|     Magneto (damaged by fire) | 9 | 4 |
|     Flywheel | 68 | 8 |
|     Main bearing cap and brass | 3 | 11 |
|     Front gear cover and pump | 8 | 4 |
|     Rear gear cover | 2 | 13 |
|     Main bearing bolt and clamp | 2 | 6 |
|     Tappet and guide | 1 | 0 |
|     Push rod | 0 | 6¼ |
| Petrol float chamber, strainer and pipes | 10 | 0 |
| Air pipe and manifold for crank case | 7 | 0 |
| Starting pump and piping | 14 | 0 |
| Stern drive transmission | 185 | 0 |
| Wing drive transmission | 782 | 0 |
| Radiator and pipes external to engine | 231 | 0 |
| Auxiliary water tank | 8 | 0 |
| Oil tank | 11 | 0 |
| Oil sump | 21 | 0 |
| Oil cooler, pipes, valves, fittings, and dashboard | 50 | 0 |
| Petrol tank (with slings) | 29 | 0 |
| Silencer, exhaust pipe, and casing from engine manifold | 67 | 0 |

NOTE.—All above weights are without petrol, oil, and water.

|  | Lbs. | Ozs. |
|---|---|---|
| *Oil:*— | | |
|     Tank | 265 | 0 |
|     Cooler and pipes | 30 | 0 |
|     Sump | 36 | 0 |
| *Water:*— | | |
|     Engine | 49 | 0 |
|     Radiator | 54 | 0 |
|     Tank | 155 | 0 |
| *Petrol:*— | | |
|     Tank (90 per cent. full) | 460 | 0 |

## Results of Analysis and Mechanical Tests of L 33 Machinery Parts.

### Engine.

*Main Bearing Bolt:*—

| | |
|---|---|
| Sectional area of specimen | ·248 square inch. |
| Elastic limit | 52·2 tons per square inch. |
| Maximum tensile strength | 56·9 ,, ,, |
| Elongation | 18·0 per cent. in 2 inches. |
| Contraction | 54·3 per cent. |
| Fracture | Fibrous. |
| Brinell | 255 |
| Sclero | 35 |

Analysis :—

| | |
|---|---|
| Carbon | ·31 |
| Silicon | ·26 |
| Sulphur | ·033 |
| Phosphorus | ·063 |
| Manganese | ·34 |
| Nickel | 4·06 |
| Chromium | ·50 |

*Main Bearing Cap :—*

| | |
|---|---|
| Copper | ·52 |
| Zinc | 12·95 |
| Iron | — |
| Manganese | ·02 |
| Silicon | — |
| Nickel | Nil. |
| Aluminium | 86·51 (by difference). |
| | 100·00 |

*Main Bearing Bush :—*

| | |
|---|---|
| Copper | 55·50 |
| Tin | Trace. |
| Zinc | 39·90 |
| Lead | 1·39 |
| Iron | ·98 |
| Manganese | 2·08 |
| Phosphorus | Trace. |
| | 99·85 |

*Camshaft Bearing :—*

| | |
|---|---|
| Copper | 86·50 |
| Tin | 9·37 |
| Zinc | 2·49 |
| Lead | ·90 |
| Iron | ·14 |
| Manganese | Nil. |
| Phosphorus | Trace. |
| | 99·40 |

*Valve Plunger Guide :—*

| | |
|---|---|
| Copper | 85·80 |
| Tin | 10·07 |
| Zinc | 2·88 |
| Lead | ·84 |
| Iron | ·14 |
| Manganese | Nil. |
| Phosphorus | Trace. |
| | 99·73 |

*Inlet Pipe :—*

| | |
|---|---|
| Copper | ·84 |
| Zinc | 11·70 |
| Iron | 1·10 |
| Silicon | ·45 |
| Manganese | ·02 |
| Nickel | Nil. |
| Alluminium | 85·89 (by difference). |
| | 100·00 |

*Cylinder Clamp :—*
- Carbon — ·44
- Silicon — ·08
- Sulphur — ·052
- Phosphorus — ·034
- Manganese — ·36
- Nickel — Nil.
- Chromium — Nil.

*Oil Base :—*
- Analysis :—
  - Copper — 1·1
  - Zinc — 12·1
  - Iron — —
  - Silicon — —
  - Manganese — ·02
  - Nickel — Nil.
  - Aluminium — 86·78 (by difference).
  - 100·00

*Water Jacket Casting :—*
- Silicon — 1·83
- Sulphur — ·131
- Phosphorus — ·42
- Manganese — ·64
- Graphite — 2·79
- Combined carbon — ·54
- Total carbon — 3·33

*Crank Case Bolt* (with right- and left-hand threads) :—
- Carbon — ·34
- Silicon — ·33
- Sulphur — —
- Phosphorus — —
- Manganese — ·46
- Nickel — 4·90
- Chromium — 1·04

**Clutch.**

*Clutch Centre Shaft :—*
- Hardness between clutch fins — Brinell, 363; Sclero, 57.
- Hardness on driving shaft — Brinell, 293; Sclero, 47.
- Analysis :—
  - Carbon — ·13
  - Silicon — ·36
  - Sulphur — ·044
  - Phosphorus — ·043
  - Manganese — ·40
  - Nickel — 4·76
  - Chromium — 1·10

*Clutch Shaft Key :—*
- Mark — 49
- Hardness — Brinell, 197; Sclero, 35.
- Analysis :—
  - Carbon — ·08
  - Silicon — ·02
  - Sulphur — —
  - Phosphorus — —
  - Manganese — ·50
  - Nickel — Nil.
  - Chromium — Nil.

*Nut of Clutch Shaft:—*

| | |
|---|---|
| Mark | N49/z.N.2120. |
| Hardness | Brinell, 190; Sclero, 33. |

Analysis:—

| | |
|---|---|
| Carbon | ·36 |
| Silicon | ·14 |
| Sulphur | — |
| Phosphorus | — |
| Manganese | ·73 |
| Nickel | Nil. |
| Chromium | Nil. |

*10-mm. Bolt ex Clutch Case:—*

| | |
|---|---|
| Hardness | Brinell, 286; Sclero, 38. |

Analysis:—

| | |
|---|---|
| Carbon | ·30 |
| Silicon | ·17 |
| Sulphur | — |
| Phosphorus | — |
| Manganese | ·54 |
| Nickel | 3·62 |
| Chromium | ·65 |

*Clutch operating Lever:—*

| | |
|---|---|
| Mark | N/49z.5666. |
| Hardness | Brinell, 197; Sclero, 35. |

Analysis:—

| | |
|---|---|
| Carbon | ·17 |
| Silicon | ·07 |
| Sulphur | ·019 |
| Phosphorus | ·032 |
| Manganese | ·40 |
| Nickel | 4·50 |
| Chromium | ·98 |

*Clutch Driving Plate:—*

| | |
|---|---|
| Mark | N49/z.5760. |
| Hardness | Brinell, 311–321; Sclero, 43. |

Analysis:—

| | |
|---|---|
| Carbon | ·44 |
| Silicon | ·32 |
| Sulphur | ·041 |
| Phosphorus | ·066 |
| Manganese | ·39 |
| Nickel | 3·20 |
| Chromium | ·62 |

*Thrust Collar:—*

| | |
|---|---|
| Mark | N49/z.698. |
| Hardness | Brinell, 296; Sclero, 49. |

Analysis:—

| | |
|---|---|
| Carbon | ·35 |
| Silicon | ·31 |
| Sulphur | ·027 |
| Phosphorus | ·060 |
| Manganese | ·31 |
| Nickel | 4·10 |
| Chromium | ·64 |

*Housing for Norma Ball-Bearing (on Clutch Shaft)* :—

    Mark - - - - - - - - N49/z.7180/2.
    Hardness - - - - - - - Brinell, 229; Sclero, 36.
      Analysis :—
        Carbon - - - - - - - ·27
        Silicon - - - - - - - ·10
        Sulphur - - - - - - - ·030
        Phosphorus - - - - - - ·049
        Manganese - - - - - - ·68
        Nickel - - - - - - - 4·68
        Chromium - - - - - - 1·18

*Universal Coupling (Male)* :—

    Mark - - - - - - - - zN.22915.
    Hardness - - - - - - - Brinell (German Impression), 401; Our Brinell, 388; Sclero, 55.
      Analysis :—
        *Carbon - - - - - - - ·21
        Silicon - - - - - - - ·30
        Sulphur - - - - - - - ·030
        Phosphorus - - - - - - ·041
        Manganese - - - - - - ·41
        Nickel - - - - - - - 4·40
        Chromium - - - - - - ·83

*Universal Coupling (Female)* :—

    Mark - - - - - - - - zN.26227.
    Hardness - - - - - - - Our Brinell, 352; Sclero, 58.
      Analysis :—
        *Carbon - - - - - - - ·23
        Silicon - - - - - - - ·11
        Sulphur - - - - - - - ·019
        Phosphorus - - - - - - ·032
        Manganese - - - - - - ·35
        Nickel - - - - - - - 4·68
        Chromium - - - - - - 1·19

*Clutch Plates* :—

### Inner Plate (Steel).

It was not possible to get a correct Sclerometer reading from these components owing to the shape of the teeth.

Two teeth were therefore cut from one of the plates, and the hardness figure for both pieces was found to be :—

    Brinell (500 kg.) - - - - - - - 100
    Sclero - - - - - - - - 25
      Analysis :—
        Carbon - - - - - - - ·08
        Silicon - - - - - - - ·02
        Sulphur - - - - - - - ·40
        Phosphorus - - - - - - ·076
        Manganese - - - - - - ·066
        Nickel - - - - - - - Nil.
        Chromium - - - - - - Nil.

### Outer Clutch Plate.

The same remarks apply in this case as in the steel plate. One of the teeth was cut off and gave the following hardness :—

    Brinell (500 kg.) - - - - - - - 74
    Sclero - - - - - - - - 21

---

\* NOTE.—These carbons are probably too high, as components appear to be case-hardened, and other steels used, which contain similar amounts of nickel and chromium, are about ·17 carbon.

Analysis:—

| | |
|---|---|
| Tin | 6·69 |
| Zinc | 1·29 |
| Lead | Nil. |
| Copper | 91·8 |
| Phosphorus | ·14 |
| | 99·92 |

*Outside Flange for Universal:*—
- Mark - No. 75z.26228.
- Hardness - Brinell, 229; Sclero, 37.

Analysis:—

| | |
|---|---|
| Carbon | ·15 |
| Silicon | ·24 |
| Sulphur | ·044 |
| Phosphorus | ·063 |
| Manganese | ·34 |
| Nickel | 4·40 |
| Chromium | 1·10 |

*Aluminium Driving Rim of Clutch-Case:*—

| | |
|---|---|
| Copper | 1·8 |
| Zinc | 10·7 |
| Iron | — |
| Silicon | — |
| Manganese | ·02 |
| Nickel | Nil. |
| Aluminium | 87·48 (by difference). |
| | 100·00 |

*Clutch Spring:*—

| | |
|---|---|
| Number of coils | 5 |
| Free length | $4\frac{1}{8}$ inches (approx.). |
| Outside diameter | $3\frac{1}{8}$ ,, |
| Inside diameter | $2\frac{1}{8}$ ,, |
| Section of coil | $\frac{1}{2}$ inch by $\frac{7}{16}$ inch. |

With load of 2,000 lbs. the spring compressed to $2\frac{3}{4}$ inches.

When this spring was stood on end on a flat surface it did not stand exactly perpendicular, owing to the ends not being true. This may have had some effect on the length when compressed.

Analysis:—

| | |
|---|---|
| Carbon | ·49 |
| Silicon | ·49 |
| Sulphur | ·027 |
| Phosphorus | ·063 |
| Manganese | 1·07 |
| Nickel | Nil. |
| Chromium | Nil. |

**Reversing Box.**

*Gear Box, Main Shaft* (with coupling and ball-bearing attached):—
- Bearing - D.W.F. type. No. 115E. 12.M.
- Shaft - No. 3z.26555.
  Two brass packing rings for ball bearing.

Brinell of shaft:—

Readings taken from the opposite end to coupling at 6-inch intervals down the shaft.

| | Brinell. | Sclero. |
|---|---|---|
| On spline | 415 | 60 |
| (a) | 415 | 60 |
| (b) | 415 | 60 |
| (c) | 415 | 60 |
| Spline | 401 | 60 |
| Flanged coupling | 201 | 35 |

*Lay Shaft* :—

    Shaft marked - - - - - - - No. z.21432. 64.

This shaft was fitted with two ball-bearings, one thrust bearing with spherical seating, and locknut on end :—

    Large ball-bearing numbered 311.E.
    Small    ,,    ,,    210.E.  10.N.
    Thrust bearing, z.2511, ball cage, No. 1011.
    Lock-nut, numbered z.6866.

Brinell of shaft :—

|  | Brinell. | Sclero. |
|---|---|---|
| Reading taken on the spline | 293 | 45 |
| ,,  ,,  ball race diameter | 302 | 47 |

*Small Bevel Wheel* :—

    Marks, &c. ☆C.29746. $\overline{\Lambda}7$. 267. NC.4. z.9368.
    Also, marked on back of teeth. S. 16 11 15.
    Fitted with one brass and one steel packing ring.

Brinell test :—

|  | Brinell. | Number. |
|---|---|---|
| Thick end of teeth | 415 | 415 German impression. |
| ,,  ,, | 415 | 415 Our impression. |
| Sclero, thick end of teeth | 75 | 80 |
| ,,  on line of teeth | 70 | 75 |

*Large Bevel Wheel* (with thrust and ball-bearings, dog, and ball-bearing inside the bevel).

    Mark on end of dog. z.6109.
    ,,  ,,  back of bevel wheel. V.C. 28589.  $\overline{\Lambda}7$  35z.  NC 4.
                                                      1  CN
    ,,  ,,  ,,  teeth. 4 10 15.
    Ball-bearing inside bevel marked. 214D. DWF. 90.

Brinell test :—

    Our Brinell on top of dog. 363  Sclero 55'.
    German  do.  back of teeth. 477, 514, 477.
    Our    do.    do.    477.  Sclero 80.
    Sclero on top of teeth. 70–75.

*Second Large Bevel Wheel* (with dog clutch, two ball-bearings, thrust bearing and securing ring).

    Mark on back of bevel. 1.DN.
    ,,  ,,  ,,  teeth. C.28589. S. 1 10 15.
    Ball-bearing inside bevel. Marked 214 D. 90. DWF.

Brinell test :—

    Our Brinell on top of dog. 352.
    German  do.  back of teeth.  461  429.
    Our    do.    do.          444  429.

Sclero test :—

    Top of dog. 55.
    Top of tooth. 70–75.  Back of tooth. 30.

*Internal Dog Clutches* ( (2) with ball-bearing to each, in aluminium case).

    Number on first dog. z.1056, and on opposite side, z.258.
      ,,     ,,  aluminium case. z.2390.

Brinell :—German impression. 331. Our impression. 321.
Sclero :—50.

    Number on second dog. z.9296.
      ,,     ,,  aluminium case. 2590. z.2589.
    Our Brinell (top of dog). 341. Sclero. 50.
    No German Brinell impression on second dog.

*Tensile Test of Aluminium in Case :—*

    Sectional area of specimen - - - - - ·0744 square inches.
    Elastic limit - - - - - - 9·47 tons per square inch.
    Maximum tensile strength - - - - - 10·46 ,, ,, ,,
    Elongation - - - - - - - 3·22 per cent.
    Fracture - - - - - - - Coarse.

**Wing Propeller Box.**

*Shaft :—*

    Brinell - - - - - - - - 277
    Sclero - - - - - - - - 43–45

*Bevel Pinion :—*

    Marks - - - - - VC.29019. $\overline{K}$7. 172.NC4. CO8 30 15
    Sclero on top of teeth - - - - - 75–80
      ,, ,, boss 60–75 - - - - - Brinell, 375–401
    German Brinell impression on end of tooth - 514, 601, 514
    Our impressions - - - - - - 514, 601
    Our Sclero reading - - - - - 80–85 and 85–90

*Wing Propeller Box Casting :—*

    Sectional area of specimen - ·149     ·0794 square inch.
    Elastic limit - - - 7·2     — tons per square inch.
    Maximum tensile strength - 8·75     7·95 ,, ,, ,,
    Elongation - - - 2·1 per cent.     1·65 per cent.
                           in 1·49″.     in 1·214″.
    Fractures - - - - coarse.     coarse.
    Copper - - - - - - - 1·5
    Zinc - - - - - - - - 8·4
    Silicon - - - - - - - ·51
    Iron - - - - - - - - ·56
    Manganese - - - - - - - ·05
    Nickel - - - - - - - Nil.
    Aluminium - - - - - - - 88·98 (by difference).
                                                      100·00

*Gear Box Arm :—*

    Copper - - - - - - - 1·6
    Zinc - - - - - - - - 10·9
    Silicon - - - - - - - —
    Iron - - - - - - - - —
    Manganese - - - - - - - ·02
    Nickel - - - - - - - Nil.
    Aluminium - - - - - - - 87·48 (by difference).
                                                      100·00

*Gear Box Bearer Arm Stud :—*

    Sectional area of specimen - - - - ·0315 square inch.
    Elastic limit - - - - - - 57·0 tons per square inch.
    Maximum tensile strength - - - - 59·6 ,, ,, ,,
    Elongation - - - - - - - 22·6 per cent.
    Contraction - - - - - - - 59·0 ,, ,,
    Fracture - - - - - - - Fibrous. Slightly serrated.
    Analysis :—
    Carbon - - - - - - - ·42
    Silicon - - - - - - - ·28
    Sulphur - - - - - - - ·030
    Phosphorus - - - - - - - ·067
    Manganese - - - - - - - ·39
    Nickel - - - - - - - 3·00
    Chromium - - - - - - - ·42

**General.**

*Propeller Boss Bolt:—*

| | |
|---|---|
| Sectional area of specimen | ·129 square inch. |
| Elastic limit | 40·5 tons per square inch. |
| Maximum tensile strength | 41·5 ,, ,, ,, |
| Elongation | 21·5 per cent. |
| Contraction | 65·0 ,, ,, |
| Fracture | Fibrous. |
| Brinell | 170 |
| Sclero | 27 |

Analysis:—

| | |
|---|---|
| Carbon | ·18 |
| Silicon | ·24 |
| Sulphur | — |
| Phosphorus | — |
| Manganese | 1·01 |
| Nickel | Nil. |
| Chromium | Nil. |

*Radiator Guy Wire:—*

| | |
|---|---|
| Area of specimen | ·0075 square inch. |
| Elastic limit | 92·2 tons per square inch. |
| Maximum tensile strength | 99·5 ,, ,, ,, |
| Elongation | 7·5 per cent. in 2 inches. |

Analysis:—

| | |
|---|---|
| Carbon | ·61 |
| Silicon | ·17 |
| Sulphur | ·036 |
| Phosphorus | ·047 |
| Manganese | ·92 |
| Nickel | Nil. |
| Chromium | Nil. |

*Oil Pipes:—*

Analysis:—

| | Long. ¾ inch O.D. | "U" ¾ inch outside. | Short Thin Pipe. ½ inch outside. |
|---|---|---|---|
| Carbon | ·33 | ·31 | ·34 |
| Silicon | ·15 | ·16 | ·30 |
| Sulphur | — | — | — |
| Phosphorus | — | — | — |
| Manganese | ·83 | ·94 | ·46 |
| Nickel | Nil. | Nil. | 4·10 |
| Chromium | Nil. | Nil. | ·96 |

*4-Way Connection on Oil Pipe.*

| | |
|---|---|
| Copper | 4·4 |
| Zinc | Nil. |
| Iron | ·28 |
| Silicon | ·29 |
| Manganese | ·51 |
| Nickel | Nil. |
| Manganese | ·67 |
| Aluminium | 93·85 (by difference). |
| | 100·00 |

*Union Nut from " U " Oil Pipe.*

| | |
|---|---|
| Carbon | ·06 |
| Silicon | ·08 |
| Sulphur | — |
| Phosphorus | — |
| Manganese | ·32 |
| Nickel | Nil. |
| Chromium | Nil. |

*Screw-down Lubricator Body on Oil Pipes.*

| | |
|---|---|
| Copper | 87·80 |
| Tin | 7·87 |
| Zinc | 2·29 |
| Lead | 1·02 |
| Iron | ·14 |
| Manganese | Nil. |
| Phosphorus | ·08 |
| Nickel | Nil. |
| | 99·20 |

*Wing Propeller Ball-Bearing Standard.*

Analysis :—

| | Head Casting. | Cap. |
|---|---|---|
| Copper | 2·0 | 2·0 |
| Zinc | 10·8 | 10·6 |
| Iron | — | ·70 |
| Silicon | — | ·73 |
| Manganese | ·03 | Trace. |
| Nickel | Nil. | Nil. |
| Aluminium | 87·17 | 85·97 (by difference). |
| | 100·00 | 100·00 |

*Stud from above Component.*

| | |
|---|---|
| Carbon | ·08 |
| Silicon | ·02 |
| Sulphur | — |
| Phosphorus | — |
| Manganese | ·40 |
| Nickel | Nil. |
| Chromium | Nil. |

*Brass Strip Lining from above.*

| | |
|---|---|
| Copper | 63·20 |
| Zinc | 35·95 |
| Tin | Nil. |
| Lead | ·58 |
| Iron | ·21 |
| Manganese | Nil. |
| Phosphorus | Trace. |
| | 99·94 |

*Wing Propeller Shaft.*

The parts used for test were those next to the reversing box, *i.e.*—
- (1) Short flanged piece with internal castellations.
- (2) Short flanged sleeve piece for taking tubular shaft.
- (3) Tubular shaft.

Analysis:—

|  | C. | Si. | S. | P. | Mn. | Ni. | Cr. |
|---|---|---|---|---|---|---|---|
| (1) | ·19 | ·242 | ·044 | ·05 | ·45 | 3·98 | 1·03 |
| (2) | ·17 | ·220 | ·049 | ·058 | ·47 | 3·99 | 1·09 |
| (3) | ·40 | ·171 | ·05 | ·035 | 1·09 | — | — |

|  | Tensile Tons. | Yield Tons (Approx.). | Elongation. Per Cent. |
|---|---|---|---|
| (1) | 54·3 | 43·0 | 20 |
| (2) | 47·6 | 40·0 | 19 |
| (3) | 56·1 | 44·0 | 15 |

The tubular shaft has either been "treated" by quenching and reheating, or simply what the Air Department call "normalised," that is, tempered at a temperature of about 450° after the last pass.

Results have been seen, after "normalising," very much as that given above from a ·42 carbon steel in a thinner tube (11 gauge).

The (1) and (2) samples have been "treated."

*Aluminium from Petrol Tanks.*

| Cu | 0·36. |
|---|---|
| Zn | Trace. |
| Tin | Nil. |
| Fe | 0·50. |
| Ni | Nil. |
| Lead | Trace, if any. |
| Manganese | ,,   ,, |
| Silicon | 0·68. |
| Aluminium | 98·46. |

### Examination of Motor Spirit and Oils from L. 33.

The following results of the examination of motor spirit and lubricating oil from L. 33 has been received from a representative of the Ministry of Munitions:—

*Motor Spirit.*

It cannot be definitely stated whether this is Galician, although the evidence is in that direction. As a spirit, it exhibits no striking peculiarities, although apparently it has, from exposure and heat,* lost some of its lighter constituents. The final distilling point of 90 per cent. under 141° C. is about common practice. Some physical tests are as follows:—

Specific gravity at 60° F. - 0·724

Distillation (Engler method). Degree F.
- First drop — 196
- 10 per cent. distilled below — 196
- 20 ,, ,, ,, — 203
- 30 ,, ,, ,, — 210
- 40 ,, ,, ,, — 217
- 50 ,, ,, ,, — 225
- 60 ,, ,, ,, — 232
- 70 ,, ,, ,, — 244
- 80 ,, ,, ,, — 259
- 90 ,, ,, ,, — 286

* The petrol was found in a drum which had burst and fire burnt itself out for want of air.

## Lubricating Oil.

This is obviously a pure Galician oil and is a standard type which Galicia used to export to England before the war. Some of its physical tests are as follows:—

| | |
|---|---|
| Specific gravity at 60° F. | 0·918 |

| | Degree F. |
|---|---|
| Closed flash point | 418 |
| Open „ „ | 432 |
| Fire test | 445 |

| | Seconds. |
|---|---|
| Viscosity (Redwood) at 70° F. | 1,090 |
| „ „ „ 140° F. | 122 |

*Grease found in Outboard Shaft Bracket Cap.*

| | | |
|---|---|---|
| Moisture | 1·8 | |
| Oil, fatty | 17·9 | } 86·0 |
| „ mineral | 68·1 | |
| Fats as soap | 3·4 | |
| Silica | 1·3 | |
| Alum. and iron oxide | ·5 | } 8·8 |
| Lime | 1·2 | |
| Alkalies | 5·8 | |
| | 100·0 | |

A yellow grease, distinctly alkaline to litmus, containing minute particles of mica. Melts at 67° C. As the grease was dirty the figures are slightly dubious, but are probably fairly accurate.

### Examination of small Pieces removed from Crankshaft Web of LZ 85 (Salonica).

The following results have been obtained:—

*Analysis:—*

| | Per Cent. |
|---|---|
| Carbon | ·46 |
| Manganese | ·45 |
| Phosphorus | ·022 |
| Sulphur | ·019 |
| Silicon | ·354 |
| Nickel | 3·06 |
| Chromium | ·59 |

*Brinell Tests.*—These tests indicated a hardness of 286 (3·6 m/m. at 3,000 kgs. load). This is equivalent to a tensile strength of about 63 tons.

*Micro-Examination.*—This showed the steel to be made up of martensitic sorbite with very little free ferrite. Judging from the structure it is thought that the steel was in excellent condition, and would be expected to give a high yield ratio.

*Impact Test.*—There was sufficient metal to give a single notch standard Izod Test, which, on breaking, absorbed 27 ft.-lbs.

*Conclusions.*—From the small sample of material examined, it has been ascertained that the crankshaft was of nickel chromium steel with distinctly higher carbon than we care to use here. It had been hardened and tempered to give a good structure and the results were:—

| | |
|---|---|
| Brinell test | 286 |
| Max. stress (estimated from Brin.) | 60 to 66 tons sq. in. |
| Yield ratio (deduced from structure and composition) | about 80 per cent. |
| Impact test (actual) | 27 ft.-lbs. |

### Report on two Valves from Engine of LZ 85 (Salonica).

Two valves were submitted, one with a longer stem marked "Bohler Rapid"— evidently an exhaust valve; while the other, with a shorter stem, appeared to be an inlet valve.

The material of the two valves has been submitted to chemical analysis and microscopic examination. Chemical analysis gave the following results:—

|  | Exhaust Valve. "Bohler Rapid." | Inlet Valve. |
|---|---|---|
|  | Per cent. | Per cent. |
| Carbon | 0·676 | 0·184 |
| Silicon | 0·235 | 0·222 |
| Sulphur | 0·045 | 0·031 |
| Phosphorus | 0·010 | 0·032 |
| Manganese | 0·080 | 0·425 |
| Nickel | 0·180 | 4·600 |
| Chromium | 4·210 | 1·102 |
| Tungsten | 13·200 | Nil. |
| Vanadium | 0·370 | Nil. |

It will be seen that the chemical composition of the exhaust valve corresponds with its mark "Bohler Rapid" in that it is typical of a tungsten high-speed tool steel of good quality. The inlet valve is a nickel-chrome steel of very moderate hardness.

*Microscopic Examination.*—Microscopic examination of the material of the exhaust valve under a magnification of 1,000 diameters shows the typical structure of a steel containing carbide (tungsten carbide). This is characteristic of a high-speed tool steel in the unhardened condition.

The structure of the inlet valve, under a magnification of 1,000 diameters, is typical of pearlitic nickel-chrome steel which has been quenched and tempered. In the valve in question this treatment has not been carried out in the best possible manner, the time and temperature of re-heating having been rather greater than is desirable. It seems, however, that, for the purpose intended, the material in this condition is adequate.

### Report on Mechanical Tests of Inlet and Exhaust Valves from LZ 85 (Salonica).

The two valves were marked respectively—

(a) 419 C2 E.R.  (Length 8 inches.)
(b) Bohler Rapid.  (Length 4 inches.)

Valve (b) was so hard that a tensile test specimen could not be prepared from it.

A tensile test on a test piece machined from (a) gave the following result:—

| | |
|---|---|
| Test No. | 1099 |
| Marks on valve | 419 C2 E.R. |
| Cross sectional area, square inches | 0·0491 |
| "Primitive" elastic limit, tons per square inch | 20·5 |
| Yield stress, tons per square inch | 32·1 |
| Ultimate stress, tons per square inch | 47·1 |
| Per cent. extension on 1 inch | 25·5 |
| Per cent. reduction of area | 70·9 |
| Modulus of elasticity, lbs. per square inch | $28·9 \times 10^6$ |

Brinell hardness tests on both valves (a) and (b) gave the following results:—

Diameter of ball - 10 mm.
Load - 1,000 kg.

$$\text{Brinell hardness No.} = \frac{\text{Load in kilogrammes}}{\text{Spherical area of indentation in square mm.}}$$

| Test No. | Marks on Valve. | Brinell Hardness No. |
|---|---|---|
| 1107 | Bohler Rapid | 387 to 55 |
| 1108 | 419 C2 E.R. | 194 |

### Six-Hour Trial of Maybach Engine No. 699 (from the Starboard Car of L 33), carried out at Kingsnorth.

The average speed during this test was *1,418 r.p.m.*, and the average B.H.P. was *202*.

The engine ran very steadily and was economical in petrol, only consuming *81·3 gallons* during the six hours, which gives a rate of *·537 pints per B.H.P. hour*. After running for four hours the engine stopped for 5 mins. 15 secs. owing to the accidental stoppage of the petrol supply. The petrol consumption was measured by taking the mean of the quarter-hourly readings of the rate of flow.

The oil tank was filled to $\frac{1}{8}$-inch of top, and a total quantity of $10\frac{3}{4}$ gallons of oil was added at regular intervals. This works out at *·071 pints* of oil per B.H.P. hour.

The air blast through crank case was at a pressure of 15 mms. of water.

Mean temperature of cooling water—Inlet - 48·5° C. ⎱ Diff. 22·5° C.
    „     „     „     „     —Outlet - 71° C. ⎰
    „     „     „ oil (to engine) - - 40° C.
    „     „     „ crank case top - - 65° C.
    „ oil pressure - - - - - - 1·05 kgs. per sq. cm.

After this trial the engine was run for 15 minutes at 450 r.p.m. at no load, and was then opened up to 200 B.H.P. at 1,400 r.p.m. There was no misfiring and the exhaust was quite clean while opening up. The load was then gradually removed until the automatic cut-off came into action at 1,675 r.p.m.

*Exhaust Gas Analysis.*—The exhaust gas analysis taken during the six-hour run gave the following result:—

|  | Per cent. |
|---|---|
| $CO_2$ | 12·2 |
| $O_2$ | ·1 |
| $CO$ | 3·3 |
| $CH_4$ | ·4 |
| $H_2$ | ·3 |
| $N_2$ | 83·7 |
|  | 100·0 |

## APPENDIX C.—FURTHER STATISTICAL DETAILS* OF MACHINERY, &c.

### Reduction Gear for Stern Drive.

*Driving Shaft* :—
- Marks on shaft — z.28377. No. 10.
- Sclero — 55
- Brinell — 341 to 375

    Analysis :—
- Carbon — ·16
- Silicon — ·25
- Sulphur — ·038
- Phosphorus — ·031
- Manganese — ·39
- Nickel — 4·15
- Chromium — 1·07

*Driven Shaft* :—
- Marking — z.17192. No. 10.
- Sclero — 55
- Brinell — 325 to 375.

    Analysis :—
- Carbon — ·13
- Silicon — ·16
- Sulphur — ·030
- Phosphorus — ·039
- Manganese — ·46
- Nickel — 4·76
- Chromium — ·73

*Large Gearwheel* :—
- Brinell (ours) — 514 on side of teeth.
-   „  (German) — 514   „  „
- Sclero — 75–80.

    Analysis :—
- Carbon — ·15
- Silicon — ·16
- Sulphur — ·041
- Phosphorus — ·037
- Manganese — ·38
- Nickel — 4·68
- Chromium — 1·01

*Driving Pinion* :—
- Mark — "4"

| Hardness. | Brinell. | Sclero. |
|---|---|---|
| Side of teeth | 555 | 80 |
| Top of teeth | 601, 534 | 85 |

    Analysis :—
- Carbon — ·13
- Silicon — ·25
- Sulphur — ·033
- Phosphorus — ·038
- Manganese — ·28
- Nickel — 4·84
- Chromium — ·65

*Propeller Boss* :—
- Mark — R.1. 428.
- Brinell — 156
- Sclero — 25

---
* Received too late for inclusion in Appendix B.

*Tensile Test:—*

| | |
|---|---|
| Test piece | ·46-inch by ·217-inch. |
| Elastic limit | 24·6 tons per square inch. |
| Maximum tensile strength | 34·1 ,, ,, |
| Elongation | 24·6 per cent. in 1·26 ins. |
| Fracture | Fibrous. |
| Brinell | 163 |
| Sclero | 25 |

Analysis:—

| | |
|---|---|
| Carbon | ·40 |
| Silicon | ·24 |
| Sulphur | ·038 |
| Phosphorus | ·047 |
| Manganese | ·77 |
| Nickel | ·97 |
| Chromium | Trace. |

### Main Shaft from Burned Reverse Box.

| | |
|---|---|
| Mark | z.26555 N. 10. |
| Brinell (at 6 points) | 321 |
| Sclero | 50 |

Analysis:—

| | |
|---|---|
| Carbon | ·29 |
| Silicon | ·17 |
| Sulphur | ·044 |
| Phosphorus | ·066 |
| Manganese | ·40 |
| Nickel | 3·66 |
| Chromium | ·71 |

### Lay Shaft from Burned Reverse Box.

| | |
|---|---|
| Mark | z.21432. |
| Brinell | 229 ⎫ Tempered by heat. |
| Sclero | 30–35 ⎭ |

Analysis:—

| | |
|---|---|
| Carbon | ·16 |
| Silicon | ·19 |
| Sulphur | ·041 |
| Phosphorus | ·038 |
| Manganese | ·46 |
| Nickel | 4·84 |
| Chromium | ·76 |

### Bevel Wheels from Burned Reverse Box.

(1) Mark on back — Co.X1. 13–12–15. C.29746. π7. 35z. NC.4. 2A.

    Sclero — 60–65.

Analysis:—

| | |
|---|---|
| Carbon | ·14 |
| Silicon | ·26 |
| Sulphur | ·044 |
| Phosphorus | ·041 |
| Manganese | ·28 |
| Nickel | 3·01 |
| Chromium | ·68 |

(2) Mark on back — Co.X1. 29–11–15. C.29746. π7. 35z. NC.4. 2.

| | |
|---|---|
| Carbon | ·14 |
| Silicon | ·25 |
| Sulphur | ·050 |
| Phosphorus | ·043 |
| Manganese | ·30 |
| Nickel | 3·17 |
| Chromium | ·67 |

### Bevel Pinion from Burned Reverse Box.

| | |
|---|---|
| Mark - | N.C.29019.  26z.  NC.4. |
| | zN. 9368.  π7.  30-8-15. |
| Sclero - | 60–65. |

Analysis:—

| | |
|---|---|
| Carbon | ·15 |
| Silicon | ·19 |
| Sulphur | ·044 |
| Phosphorus | ·052 |
| Manganese | ·40 |
| Nickel | 4·15 |
| Chromium | 1·01 |

### Components ex Wing Propeller Box.

| | Propeller Box Pinion. | Pinion Shaft. | Propeller Shaft. |
|---|---|---|---|
| Carbon | ·18 | ·18 | ·33 |
| Silicon | ·24 | ·25 | ·24 |
| Sulphur | ·244 | ·041 | ·030 |
| Phosphorus | ·066 | ·063 | ·054 |
| Manganese | ·30 | ·44 | 1·03 |
| Nickel | 4·47 | 4·40 | Nil |
| Chromium | 1·19 | 1·15 | Nil |

# GENERAL INDEX.

### A.

**Aerial :—**
    Arrangement of, in L 33 - - - - 60
    ,, Alternative - - - 60
    Winding Gear in L 33 - - - 61
**Alternators :—**
    L 33 - - - - - - - 58
    Forward Car - - - - - 58
    Starboard, Midship, and After Car - 59
**Ammunition.** *See under* Armament.
**Aneroid** - - - - - - - 84
**Antwerp,** Observation Car of Airships used at the Bombardment - - - - 32
**Armament :—**
    Ammunition :—
        Armour-Piercing Bullets - - 70
        Explosive Bullets - - - 69
        Machine Guns and Bombs (L 33) - 15, 68
        ,, ,, (SL 11) - - 68
        ,, Gun Sights - - - 75
    Maxim-Nordenfeldt Guns - - - 68
    Parabellum Machine Guns - - 68
**Army Airship Service :—**
    Battalions, Organisation of Airship - 10
    Landing Party - - - - 12
    Maintenance Party - - - 12
    Training and Personnel :—
        (*a*) Before War - - - 10
        (*b*) Since Outbreak of War - 11
**Army Bill (German), 1913** - - - 11

### B.

**Badische Process** (Hydrogen Manufacture) - 94
**Ballast,** Jettisoning of - - - 81
**Bamberg,** Carl - - - - 82
**Battalions,** Organisation of Airship - 10
**Bavaria,** Engineer Corps at Munich - 11
**Bearing Plate** - - - - - 83
**Berlin,** Army Airship Station - - 11
**Berlin-Johannisthal** (Naval Airship Station) - 6
**Billericay** - - - - - 13, 37, 74
**Bitterfeld,** Hydrogen Installation - - 94
**Blaavands Huk,** Weather Station - - 90
**Blendbombe** - - - - - 65
**Boecker,** Lieut.-Commander - - - 78
**Bombs :—**
    Armament, Standard, of Naval Airship - 62
    Fuze - - - - - - 63
    General Appearance - - - 63
    H.E. Bomb, Large, New - - 63
    ,, Spherical Bomb - - 63
    High Explosive - - - - 62
    Incendiary Bomb :—
        Dimensions, &c. - - - 64
        Fuze, Effect - - - 65
        Release Gear - - - 55
    Pear-Shaped Bombs - - - 62, 63
    Pieces - - - - - 63
    Weight of Bombs carried - - 62, 63
**Bomb-Dropping Gear and Electrical Installation :—**
    Comparison with Earlier Dropping Gear - 55
    Positions - - - - - 52
    Releases - - - - - 52
    ,, Electrical Installation - 54
    Release Gear for Incendiary Bombs - 55
**Bomb Sights.** *See under* "Sights."
**Bordelichte** - - - - - 57
**Borkum,** Directional Station - - 79, 80
**Bruges,** Directional Station - - 79

### C.

**Cellon** - - - - - - 50, 80
**Cellenstoffabrik,** G.m.b.H. - - - 46

**Cologne :—**
    Army Airship Station - - - 11
    Hydrogen Installation - - - 94
**Communications, &c. :—**
    L 33 - - - - - - 29
    ,, Telephones - - - 56
**Compasses** - - - - - 77, 82
**Construction** - - - - - 13–23
    *See also under* L 33, Z 4, L 3, LZ 85, &c.
**Crews (Army Airship Service) :—**
    Age (Average) of LZ 85 - - 12
    Formation and Training before War - 10
    ,, ,, ,, after Outbreak of War - 11
    Landing Party - - - - 12
    Maintenance Party - - - 12
    Number and Disposition - - 12
**Crews (Naval Airship Service) :—**
    Ages, Average (L 7, L 15, L 33) - 7
    Formation of, since Outbreak of War - 7
    Landing Party - - - - 10
    Maintenance Party - - - 9
    Number, Disposition, and Size - 8, 76
    Training - - - - - 8
    "War Stations" - - - - 9
**Cuffley,** Schütte-Lanz brought down - 18, 37, 68

### D.

**Darmstadt,** Army Airship Station - - 11
**Deck Insulator** - - - - 61
**Delag Co.** (Deutsche Luftschifffahrt Aktien-Gesellschaft) - - - - 6, 7
**Dresden,** Naval and Army Airship Station - 7, 11
**Drift Indicator** - - - - 78, 85
**Düsseldorf,** Army Airship Station - 11

### E.

**Electrical Installation :—**
    Bomb Release - - - - 54
    Electric Lighting (L 33) - - 57
    Telephones (L 33) - - - 56
    Thermometer - - - - 86
**Elevators.** *See* Rudders.
**Engine Revolution Indicator** - - 88
**Engineer Corps** (at Munich) - - 11

### F.

**Fabric and Gas Valves :—**
    Components - - - - 46
    Covers (Outer), Arrangement on L 33 - 48
    Lacing - - - - - 50
    Material, Weight and Strength - 49
    Proofing and Dope - - - 50
    Sundry (L 33, &c.) - - - 51
    Water Ballast Bags - - - 50
**Fire Extinguisher** - - - - 88
**Flares, Illuminating :—**
    Action, Appearance, and Dimensions - 65
    Fuze - - - - - - 65
    (*See also* Flares, Water.)
**Flares, Water** - - - - - 65
    Patt. No. 1—Action, Appearance, and Dimensions - - - - 66
    Patt. No. 2—Action, Appearance, and Dimensions - - - - 67
    Patt. No. 3—Action, Appearance, and Dimensions - - - - 67
    Patt. No. 4—Action, Appearance, and Dimensions - - - - 68
**Frankfort-on-Main :—**
    Naval Airship Station - - - 6
    Hydrogen Supply - - - 94

**Friedrichshafen:—**
　Army Airship Station　— — — 11, 13, 81
　Hydrogen Installation　— — — 94

## G.

**Galvanometer**　— — — — — 87
**Garnisondienstfähige**　— — — — 12
**Garnisontruppe**　— — — — 12
**Gasbags** (*see also* Fabric):—
　Arrangement (General)　— — — 46
　Filling　— — — — — 46
　Gas, Ventilation and Leakage of　— 48
　Materials　— — — — — 46
　Repairs　— — — — — 48
　Seams　— — — — — 46
　Shear Wire and Central Wire Ropes,
　　Special Fittings for　— — — 48
　Supports and Protection　— — — 47
　Temperature and Condensation of Moisture　48
　Valves:—
　　Type 1　— — — — — 51
　　　" 2　— — — — — 52
**Graticule**　— — — — — 76
**Graudenz**, Army Airship Station　— — 11
**Griesheim Electric Co.**　— — — — 94

## H.

**Hamburg-Fuhlsbüttel**; Naval Airship Station　— 6
**Hamburg**, Meteorological Observatory　— — 79
**Hanover**, Army Airship Station　— — 11
**"Hansa"**　— — — — — 6, 7
**Heligoland**　— — — — — 7
**Hydrogen**, Production in Germany:—
　i. By Bottles　— — — — 94
　ii. By Plant　— — — — 94
　iii. By-product　— — — — 94
　iv. By Silicon Process　— — — 94
　v. Oil Cracking　— — — — 94

## I.

**Instruments and Fittings:—**
　Aneroid　— — — — — 84
　Bearing Plate and Pelorus　— — 77, 83
　Compasses　— — — — 77, 82
　Drift Indicator, Assembly and Use　— 78, 83
　Engine Revolution Indicator　— — 88
　Fire Extinguisher　— — — — 88
　Lifebelt　— — — — — 89
　Petrol Gauge of Storage Petrol Tanks　— 88
　Rate of Rise Indicator　— — — 85
　Sextant, Observing　— — — 77
　Statascope　— — — — — 85
　Station Pointer　— — — — 77
　Thermometer, Electric　— — — 86

## K.

**Königsberg:—**
　Army Airship Station　— — — 11
　Hydrogen Installation　— — — 94

## L.

L 1　— — — — — — 6, 7
L 2　— — — — — — 4, 7
L 3　— — — — — — 7, 16, 62
L 4　— — — — — — 7, 62
L 7　— — — — — — 7, 8, 9
L 11　— — — — — — 8
L 13　— — — — — — 8
L 14　— — — — — — 8, 79
L 15　— — 7, 23, 46, 49, 50, 51, 56, 74, 85, 87
L 16　— — — — — — 8
L 17　— — — — — — 8
L 18　— — — — — — 8
L 20　— — — — — — 15
L 21　— — — — — — 8
L 30　— — — — — 4, 8, 62, 77
L 31　— — — 8, 13, 37, 72, 74, 75, 83

**L 32:—**
　Petrol　— — — — 8, 13, 30, 31
　Speeds　— — — — — 30
　Propellers　— — — — — 35, 37
　Bomb Sights　— — — — 74, 75
　Heights attained　— — — — 80
　Compasses　— — — — — 83
　Rate of Rise Indicator　— — — 85

**L 33:—**
　Access Hatches　— — — — 22
　Ammunition:—
　　Armour Piercing　— — — 70
　　Explosive Bullets　— — — 69
　Armament (*see* Guns)　— — — 15
　Bomb Release:—
　　Electrical Installation　— — 54, 55
　　Indicator Lamp　— — — 55
　　Switches, Use of different　— 54
　　For Incendiary Bomb　— — 55
　Bomb Sights　— — — — 74, 75
　　" Stowage　— — — — 21
　Cars, Dimensions, &c.　— — — 14, 25
　Communications　— — — — 29
　Comparison with previous Classes　— 15
　Control, Rudder, Elevator　— — 19, 22
　Control, Minor　— — — — 23
　Covers, Outer:—
　　Arrangement　— — — — 48
　　Fabric　— — — — — 49
　　Fabrics, Sundry　— — — 51
　　Lacing　— — — — — 50
　　Proofing and Dope　— — 50
　　Water Ballast Bags　— — 50
　Crew, Age and Training of　— — 7
　Crew Space　— — — — — 22
　Dimensions　— — — — — 13
　End—Fall in Essex　— — — 13
　Fabrics, Sundry　— — — — 51
　Fins, Rudders and Elevators　— — 18
　Gasbags:—
　　Arrangement of　— — — 46
　　Support of　— — — — 47
　　Valves, Type 1　— — — 51
　　　　　" 2　— — — 52
　Guns, Machine　— — — — 68
　Gun Platform (After)　— — — 25
　　　" (Forward)　— — — 24
　Hull Wiring (compared with SL 11)　— 33
　Instruments, &c.　— — — — 82–89
　(*See under* main heading **Instruments** for details.)
　Keel　— — — — — — 20
　Lighting, Electric　— — — — 57
　Machinery and Propellers:—
　　Analysis of Parts (*see under* **Machinery** for details)　— — — — 95
　　Carburettors　— — — — 39
　　Connecting Rod Cooling Arrangement　— 38
　　Engine Seatings　— — — 43
　　General Remarks—Dimensions and
　　　Weights of Parts of Engines, &c.　— 42, 95
　　Governor Control　— — — 41
　　Lubricating System　— — — 40
　　Magnetos　— — — — 42, 95
　　Motor Spirit and Oils (Analysis of)　105
　　Petrol Filling Arrangements　— 45
　　　" Pipe　"　— — 45
　　　" Stowage　"　— — 44
　　　" Supply System　— — 21, 40
　　　" Tank Nos., compared with L 32　— 31
　　　" Tanks, Aluminium in　— 105
　Propellers:—
　　Diameters and Pitch　— — 34
　　Hub Fitting　— — — — 35
　　Leading and Trailing Edges　— 35
　　Plywood Panels in way of　— 27
　　Veneer of　— — — — 35
　　Weight of　— — — — 35
　　Woods composing Laminations　— 36, 37
　Propeller Shaft Supports. "A" Brackets
　　for Wing Propellers　— — — 27
　Starting Arrangements　— — — 42
　Transmission Gear　— — — — 43
　Water Cooling System　— — — 38
　Weights of Parts　— — — — 95

| L 33—continued. | PAGE |
|---|---|
| Materials:— | |
|   Chemical Analysis of | 28 |
|   „ „ Remarks on | 29 |
| Maybach Engine No. 699 | 108 |
| Mooring Point, &c. | 23 |
| Navigation—Crew, Landing, &c. | 76–81 |
| (See under main heading **Navigation** for details.) | |
| Origin | 13 |
| Petrol Tanks and Piping | 21, 105 |
| (See also under Machinery above.) | |
| Propellers. (See under Machinery above.) | |
| Repairs during Flight | 48 |
| Shear Ropes | 18 |
| Statascope | 85 |
| Stranded Wires | 29 |
| Structure | 16 |
| Suspension Strops | 24 |
| Tail Structure | 25 |
| Telegraphs | 65 |
| Telephones | 56 |
| Tensile Tests | 29 |
| Thermometers, Use of Electrical | 48, 86 |
| Ventilation | 22 |
| Walking Way | 20 |
| Water Ballast Bag Arrangements | 21 |
|   „ „ Bags (Emergency) | 48 |
| Weight Carried | 13 |
| Wireless Telegraphy Apparatus | 58, 61 |
| (See under main heading **Wireless Telegraphy** for details.) | |
| Wires and Ropes | 17, 29 |
| Workmanship, General | 27 |
| **L 34** | 8, 13 |
| **L 35** | 8 |
| **LZ 72** | 12 |
| **LZ 77** | 14, 49, 51 |
| **LZ 85** | 12, 15, 16, 18, 23, 33, 72, 74, 75, 106, 107 |
| Landing:— | |
| German Rigid Airships: | |
|   (a) By day | 80 |
|   (b) By night | 81 |
| Mooring on Water | 81 |
| Landing Party:— | |
| Army Airship Service | 12 |
| Naval Airship Service | 10 |
| **Lahr**: Army Airship Station | 11 |
| **Leipzig**: Naval Airship Station | 8 |
| **Leuchtbombe** | 65 |
| **Liebern's Quadrant** | 77 |
| **Life Belt** | 89 |
| **Linde Frank, Caro Process.** (Installation of Hydrogen) | 94 |
| **Little Wigborough** | 13, 37, 74 |
| **Lübbert, Kapitän z. S.** | 6 |
| **Ludolf of Bremerhaven** | 83 |
| **Luftschiffbau Zeppelin Aktien Gesellschaft** | 13 |
| **Luftschifftrupp** | 12 |

## M.

| "M" type of Airship | 10, 11 |
|---|---|
| **Machine Guns.** See under Armament; for L 33, under **L 33—Guns.** | |
| **Machinery** | 34–46 |
| (See under L 33 for details.) | |
| Maintenance Party:— | |
|   (Army Airship Service) Duties | 12 |
|   (Naval Airship Service) Duties | 9 |
| **Mannheim**:— | |
| Army Airship Station | 11 |
| Hydrogen Installation | 94 |
| **Machinery in L 33**:— | |
| Weights of Engine, Oil, Petrol, Water | 95 |
| Analysis and Mechanical Tests of Parts:— | |
|   Clutch | 97, 98, 99, 100 |
|   Engine | 95, 96, 97 |
|   Motor Spirit and Oils | 105, 106 |
|   Oil Pipes | 103 |
|   Petrol Tanks—Aluminium from | 105 |
|   Propeller Boss Bolt | 103 |
|   Radiator Guy Wire | 103 |
|   Reversing Box | 100, 101, 102, 110, 111 |

| Machinery in L 33—continued. | |
|---|---|
| Analysis and Mechanical Tests of Parts—continued. | PAGE |
|   Wing Propeller Ball Bearing Standard | 104 |
|   Wing Propeller Box | 102, 109, 110, 111 |
|   Wing Propeller Shaft | 105 |
| (See also under **L 33.**) | |
| **Machinery in LZ 85**:—Analysis of Valves | 106, 107 |
| **Materials (L 33)**: Chemical Analysis of | 28, 29 |
| **Maxim-Nordenfeldts** | 68 |
| **Maybach Engine, No. 699** | 108 |
| **Messerschmidt System** for production of Hydrogen | 94 |
| **Meteorological Conditions during Zeppelin Raids**:— | |
| Analysis of Weather conditions during raids up to September 1915 | 89 |
| Analysis (Table) of Weather conditions during raids between September 1915 and November 1916 | 90, 92 |
| Graphs and Details showing main conditions for Zeppelin raids | 90 |
| Information obtained by Enemy Raiders from Neutrals | 91 |
| **Metz**: Army Airship Station | 11 |
| **MG. Type**, Wireless Telegraphy | 58 |
| **Munich** (Bavaria), Station of Engineer Corps | 11 |

## N.

| Naval Airship Service:— | |
|---|---|
| Training and Personnel: | |
|   (a) Before War | 6 |
|   (b) Since outbreak of War | 7 |
| **Naval Bills (German)** 1913–14 | 6 |
| **Navigation:** | |
| Calculating ground speed and Fixing Positions | 78 |
| Control Position | 76 |
| Controls | 77 |
| Crew: Officers and Men | 76 |
| Directional Wireless | 78 |
| General: | |
|   Airship Lights | 80 |
|   Course during Bomb-dropping | 80 |
|   Course for Attack on Midlands | 80 |
|   Courses of Rigid Airships | 79 |
|   Cruising and Full Speed | 79 |
|   Directional Stations | 78 |
|   Heights attained | 80 |
|   Housing | 81 |
|   Landfall, making a | 79 |
|   Landing: | |
|     By day | 80 |
|     By night | 81 |
|   Lightships | 80 |
|   Mooring on Water | 81 |
|   Position at Night | 80 |
|   Restarting, Preparations for | 81 |
|   Weather Reports | 79 |
| Instruments: | |
|   General | 76–79 |
| (See also under main heading **Instruments.**) | |
| **Niedersgörsdorf** | 11, 12 |
| **Nordholz**: Directional Station | 79 |

## O.

| Observation Car of German Rigid Airships | 31 |
|---|---|
| Ostend Weather Station | 79 |

## P.

| Parabellum Machine Guns | 68 |
|---|---|
| Pelorus | 77, 83 |
| Personnel: See Training. | |
| Petrol: | |
|   Consumption of in L 32 | 30 |
|   Comparison of Tank Numbers of L 32 and L 33 | 31 |

| | PAGE |
|---|---|
| Petrol Gauge of Storage Petrol Tanks | 88 |
| (See also under L 33, Machinery.) | |
| Potsdam, Naval Airship Station | 7 |
| Potters Bar | 13, 37, 74 |
| Propellers | 34–37 |
| (See under L 33, Machinery and Propellers, for details.) | |
| Chauvière shape | 34 |
| Comparisons of L 31, L 32, L 33, and SL 11 | 37 |
| Prussia, Complement to Army Airship Battalions under Army Act, 1913 | 11 |
| Putzig, Naval Airship Station | 6 |

### R.

| | |
|---|---|
| Rate of Rise Indicator | 85 |
| Reinickendorf, Training Centre for Army Airship Service | 11 |
| Revigny | 14 |
| Rudders :— | |
| Balanced Monoplane | 4 |
| Box | 4 |

### S.

| | |
|---|---|
| "Sachsen" | 7, 8 |
| Salonika, Destruction of LZ 85 | 15, 72, 74, 75 |
| Saxony, Complement to Army Airship Battalions under Army Act, 1913 | 11 |
| Schiffspflege-gruppe | 12 |
| Schneidemühl, Army Airship Station | 11, 94 |
| Schulkommando | 11 |
| Schütte-Lanz Airships :— | |
| Airship down at Cuffley | 18 |
| Armament | 33 |
| Cars | 33 |
| Comparison with Zeppelins | 5 |
| " " L 33 | 14, 15 |
| Fins, Rudders, and Elevators | 33 |
| Gasbags, Arrangement of | 46 |
| " Fabric of | 47 |
| " Support of | 47 |
| Hull Wiring | 33 |
| Keel | 33 |
| Observation Car, Type of | 32 |
| Structure and Form | 32 |
| SL 7–9 | 32 |
| SL 11 | 33, 34, 35, 37, 48, 49, 68, 71, 83 |
| "Seewehr" | 7 |
| Seifersdorf (School for Tailoring, Naval Airship Service) | 8 |
| Sextant, Observing | 77 |
| Sights :— | |
| Attack, Method of | 75 |
| Bomb Sight and Ground Speed Indicator— | |
| (a) Description | 72 |
| (b) Instructions for Use | 74 |
| Bomb Sights (Variations in L 31, L 32, L 33, LZ 85) | 74, 75 |
| Machine Gun Sights, Angles of Elevation and Ranges | 76 |
| Silicon Process (Hydrogen Production) | 94 |
| Statascope | 85 |
| Stations (Army and Navy). | |
| (See with Training Centres.) | |
| Statistical Details | 95–111 |
| (For details see under Machinery, LZ 85, Crankshaft Web and Engine Valves.) | |
| Stations, Army and Naval. | |
| (See under Training Centres.) | |
| Stavanger, Destruction of L 20 | 15 |
| Steuerleute | 6 |
| Sylt, Directional Station | 79 |

### T.

| | PAGE |
|---|---|
| Tegel : Hydrogen Installation | 94 |
| Telefunken Installations | 58 |
| Telephones (L 33) | 56 |
| Thermometers, Electric | 86 |
| Training and Personnel :— | |
| Army Airship Service | 10–12 |
| Naval Airship Service | 6–10 |
| Training Centres and Stations—Past and Present :— | |
| Army Airship Service :— | |
| Allenstein | 11 |
| Berlin | 11 |
| Cologne | 11 |
| Darmstadt | 11 |
| Dresden | 11 |
| Düsseldorf | 11 |
| Friedrichshafen | 11 |
| Gotha | 11 |
| Graudenz | 11 |
| Hanover | 11 |
| Königsberg | 11 |
| Lahr | 11 |
| Liegnitz | 11 |
| Mannheim | 11 |
| Metz | 11 |
| Niedersgörsdorf | 11 |
| Potsdam | 11 |
| Reinickendorf | 11 |
| Schneidemühl | 11 |
| Naval Airship Service :— | |
| Berlin–Johannisthal | 6, 7, 11 |
| Dresden | 7 |
| Frankfort-on-Main | 6, 7 |
| Hamburg-Fuhlsbüttel | 6, 7, 8 |
| Leipzig | 8 |
| Posen | 11 |
| Potsdam | 7 |
| Putzig | 6 |
| Seifersdorf (School for Tailoring) | 8 |
| Tegel | 11 |

### V.

| | |
|---|---|
| Viktoria–Luise | 6, 7, 8 |

### W.

| | |
|---|---|
| Wattmeter Principle | 87 |
| Wheatstone Bridge Principle | 87 |
| Wireless Telegraphy Apparatus :— | |
| Aerial Gear | 60 |
| Aerial Winding Gear | 61 |
| Alternator, Forward Car | 58 |
| Alternators, Starboard, Midship and After Car | 59 |
| Arrangement, General | 61 |
| Deck Insulator | 61 |
| Earth Connection | 61 |
| Receiver W/T | 60 |
| Results of Tests | 59 |
| Wireless Cabin | 61 |
| Würtemberg, Complement to Army Airship Battalions under Army Act, 1913 | 11 |

### Z.

| | |
|---|---|
| Z 1 (First Army Airship) | 10 |
| Z 2 | 10 |
| Z 2, Ersatz | 10 |
| Z 3 | 10 |
| Z 4 | 12, 16 |
| Zeppelin Company | 6 |
| Zeppelin, Count, Early Designs | 10 |
| Zeppelin Raids, Weather during. | |
| (See under Meteorological Conditions.) | |
| Zeppelins, German Rigid :— | |
| Armament | 5 |
| Development, Table showing | 5 |
| Types (four) | 4 |

## INDEX TO PLATES.*

**Plate 1 :** L 31 with Landing Party.

**Plate 2 :—**
Z IV.—General View.
Z IV.—Fore Car.
Z IV.—After Car.

**Plate 3 :—**
Sachsen.—General View.
L 3.—General View.
Z VIII.—Landing.

**Plate 4 :—**
L 2.—General View.
L 2.—Stern View.
L 2.—Fore Car and Fore Engine Car.
L 13.—General View.

**Plate 5 :—**
L 30.—General View.
L 31.— ,, ,,
L 31.—Fore Car and Crew.

**Plate 6 :—**
SL 1.—General View.
SL 2.—General Views.

**Plate 7 :—**
SL 1.—Hull on Stocks.
SL 1.—Fore Car.
SL 1.—Elevator and Rudder.

**Plate 8 :** L 33.—Views of Wreck.

**Plate 9 :—**
LZ 85.—Double-braced Triangular Girder.
LZ 85.—Single-braced Triangular Girder.
LZ 85.—"W" Girder.
LZ 85.—Fin Girders.

**Plate 10 :—**
L 33.—Joints of Main Longitudinal and Intermediate Transverse Frames.
L 33.—Joint of Main Longitudinal and Main Transverse.

**Plate 11 :** L 33.—Joints of Main Transverse Frames.

**Plate 12 :—**
L 33.—Bow—Interior View.
L 33.—Main Transverse Frame Truss.
L 33.—Casting for Handling Rope Strop.

**Plate 13 :—**
L 33.—Diagonal Bracing Wires.
L 33.—Bow supporting Wires for Outer Cover.
L 33.—Wiring forward.

**Plate 14 :—**
L 33.—Upper Rudder and Fin.
L 33.—Triangular Boom and After edge of Fin.
L 33.—Tail, showing Fins, Upper Rudder, &c.
L 33.—Fin showing Junction of Rectangular Boom with outer edge Girder.

**Plate 15 :—**
L 33.—Top Joint of Keel.
L 33.—Rudder
LZ 85.—Rudder.

**Plate 16 :** L 33.—Guide Sheaves for Rudder and Elevator Controls, Keel Control Leads, Gas Valve Controls and Water Ballast Controls.

**Plate 17 :** L 33.—Keel Structure, showing Water Ballast and Bomb Positions ; Walking-way, W.C. and Handling Rope Tub.

**Plate 18 :—**
L 32.—Stern Cap and Flagpost.
L 33.—W.C.
LZ 85.—Keel Structure.

**Plate 19 :—**
L 33.—Athwartship Gangway to Wing Cars.
L 33.—Structure behind Plywood in way of Wing Propellers.
L 33.—Wing Propeller Bracket.

**Plate 20 :—**
L 33.—Wing Propeller and Brackets.
L 33.—Swivelling Joint for Propeller Brackets and Car Struts.

**Plate 21 :** L 33.—Water Ballast Valves, Water Ballast Bag Attachments and Buckets.

**Plate 22 :** L 33.—Petrol Tank Stowage, Guide Girders and Slip Bolt.

**Plate 23 :—**
L 33.—Oil and Water Tanks for Wing Car.
L 33.—Keel Ventilator Shafts.

**Plate 24 :** LZ 85.—Mooring Point, with Double Strop to Slip Bolt.

**Plate 25 :—**
L 33.—Mooring Point.
LZ 85.—Girders at Mooring Position.

**Plate 26 :—**
LZ 85.—Tubs for Trail Ropes.
L 33.—Handling Rope Tub Aft.
L 33.—Wing Car (Interior View).

**Plate 27 :** L 33.—Forward Gun Platform, showing Structure and Tripod.

**Plate 28 :—**
L 33.—Forward Gun Platform and Gun Mounting.
L 33.—After Gun Platform, showing Gun Mounting in place.

**Plate 29 :—**
L 32.—Bottom of Wing Car.
L 33.—Side View of Wing Car.

**Plate 30 :** Engine and Rudder Control Orders.

**Plate 31 :** Observation Car.—End and Side Views.

**Plate 32 :** Observation Car—Winch and Cable.

**Plate 33 :** L 33.—Diagram of Typical Structure.

**Plate 34 :** L 33.—Sections of Girders.

**Plate 35 :** L 33.—Mooring Arrangements (Section at Frame 5).

**Plate 36 :** L 33.—Frame 25.

**Plate 37 :** L 33.—Central Fitting for Diametrical Wires in Main Frame.

**Plate 38 :** L 33.—Arrangement of Rudder and Elevator Control Cables.

**Plate 39 :** L 33.—Section at Frame 18.

**Plate 40 :** L 33.—Incendiary Bomb Frame and Shutter Arrangement.

**Plate 41 :** L 33.—Rudder Elevator Controls (Sprocket Gears, &c.).

**Plate 42 :** L 33.—General Arrangement of Gas and Water Ballast Valve Control Gear Box in Control Car.

**Plate 43 :** L 33.—Arrangement of Forward Gun Platform.

**Plate 44 :** Plan of Wing Cars.

**Plate 45 :** Elevation of Wing Cars.

**Plate 46 :** L 33.—Wing Car—Section through Middle.

**Plate 47 :** L 33.—Details of Fore and Aft Girder for Wing Car.

**Plate 48 :** L 33.—Diagrammatic Sketch of Control Car.

**Plate 49 :** L 33.—Brackets supporting Wing Propellers.

**Plate 50 :** L 33.—Protective Ply Wood and Girder Supports in way of Wing Propellers.

**Plate 51 :** Schütte-Lanz Airship—Typical Arrangement.

**Plate 52 :** Schütte-Lanz Airship—Sections of Girders.

**Plate 53 :** Schütte-Lanz Airship—Sections of Box Girders.

\* **Plate 1** faces page 3 ; **Plates 2 to 106** are at end of book ; **Plates A to M** are in separate packet (C.B. 1265A).

Plate 54 : L 33.—Wing and direct-drive Propellers.
Plate 55 : L 33.—Views of Engines.
Plate 56 : L 33.—Views of Cylinder, Piston, Crank Case Ventilator Outlet, and Connecting Rod.
Plate 57 : L 33.—Views of Crank Shaft, Oil Tank, Oil Filters and Silencer.
Plate 58 : L 33.—Views of Bevel Box, Reverse Box, Wheels and Pinion of Wing Propeller Box, and Wheels and Pinion of Reverse Box.
Plate 59 : L 33.—Wing Propeller.
Plate 60 : Blades of L 33, L 32, and L 31.
Plate 61 : L 33.—Starboard Wing Car, showing arrangement of Circulating Water Pipes.
Plate 62 : L 33.—Telegraph Dial in Starboard Wing Car and Oil Sump under Engine Crank Case.
Plate 63 : L 33.—Engine Governor.
Plate 64 : L 33.—Arrangement of Transmission Gear for Non-Reversible Propeller.
Plate 65 : L 33.—Arrangement of Machinery in After Car.
Plate 66 : L 33.—Engine Girder of Starboard Wing Car.
Plate 67 : L 33.—Starboard Wing Car (Sectional Elevation, After end).
Plate 68 : L 33.—Starboard Wing Car (Sectional Elevation, Forward end).
Plate 69 : L 33.—Wing Cars (Fore and Aft Engine Brackets).
Plate 70 : L 33.—Petrol Tank and Sling.
Plate 71 : L 33.—Petrol Tank Fittings.
Plate 72 : L 33.—Arrangement of Stowage and Method of Slipping. Petrol Tanks.
Plate 73 : L 33.—Petrol Tank Release.
Plate 74 : L 33.—Diagrammatic Sketch of Petrol Arrangement.
Plate 75 :—
L 33.—Lower Rudder with Fabric.
L 33.—Emergency Water Ballast Bag.
Plate 76 : L 33.—Fragments of Outer Cover and Lacing, &c.
Plate 77 : Manœuvring Valves of L 33 and LZ 85.
Plate 78 : L 33.—Pad Piece and Turnbuckle for Shear Wire.
Plate 79 : L 33.—Outer Cover on Parallel Portion.
Plate 80 : Automatic Gas Valve (Zeppelin Type).

Plate 81 : L 33.—Sectional Elevation of Manœuvre Gas Valve.
Plate 82 : L 33.—Bomb Releases.
Plate 83 : Bomb Releases from L 33 and LZ 85.
Plate 84 : LZ 85.—Distribution Board.
Plate 85 : LZ 85.—Main Switchboard with two Electro Magnetic Switches.
Plate 86 : LZ 85.—Distribution Board and Main Switchboard, with three Electro Magnetic Switches.
Plate 87 : LZ 77.—Pilot's Switchboard.
Plate 88 : L 15.—Mechanical Bomb Release Gear.
Plate 89 : LZ 77.—Diagram of Pilot's Switchboard and Release Gears.
Plate 90 : LZ 77.—Diagrams of Release Gear.
Plate 91 : L 33.—Telephone Instruments.
Plate 92 : L 33.—D.C. Motor (or Generator) and Alternator from Forward Car.
Plate 93 : L 33.—Parts of Valve Receiver.
Plate 94 : L 32.—Arrangement of Aerial Winding Gear and W/T Instrument Cabinet.
Plate 95 : L 33.—Leading-in Insulator (Porcelain).
Plate 96 : Water and Illuminating Flares.
Plate 97 : L 33.—Maxim-Nordenfeldt G.A. Machine Gun and Tripod Gun Mounting from top platform.
Plate 98 : Bomb Sight from L 31 and Ground Speed Stop-Watch from LZ 85.
Plate 99 : Machine Gun Sight.
Plate 100 : Tracing showing Frisian Islands in Line 273° (Mag.)—North Sea to the Wash.
Plate 101 :—
L 33.—Aneroid.
L 15.—Fire Extinguisher.
L 32.—Statascope.
Plate 102 : Zeppelin Engine Revolution Indicator, Life Belt, Petrol Gauge, and Electric Thermometer.
Plate 103 : Diagrammatic Sketch of Drift Indicator.
Plate 104 : Chart showing Height of Barometer, &c. for Raids, September 1915—February 1916.
Plate 105 : Chart showing Height of Barometer, &c. for Raids, March 1916—August 1916.
Plate 106 : Chart showing Height of Barometer, &c. for Raids, September 1916—November 1916.

# ZEPPELIN AIRSHIPS.
## 1st TYPE.

Z IV. (completed 1913).

Z IV. Fore Car.

Z IV.--Rear Car.

## 1ST TYPE (continued).

"SACHSEN" (completed 1913).

L 3 (completed 1914).

Z VIII. about to land at Friedrichshafen (completed 1914).

## 2ND TYPE.

L 2.—General View.

L 2.—Stern View.

L 2.—Fore Car & Fore Engine Car.

## 3RD TYPE.

L 13.

C.B. 1265.
**Plate 5.**
See Chapt. I.

4TH TYPE.

L 30.

L 31.

Fore Car of L 31.

*Ordnance Survey, February, 1917.*

# SCHÜTTE-LANZ AIRSHIPS.

C.B. 1265.
**Plate 6.**
See Chapt. 1.

## 1ST TYPE.

S L. 1.

## 2ND TYPE.

S.L. 2.

## 3RD TYPE.

Similar in every way to the 2nd Type except that the two foremost cars have been joined together and form one car.

Ordnance Survey, February, 1917.

# SCHÜTTE-LANZ AIRSHIPS.
## 1st Type.

S.L. 1.

S.L. 1. Fore Car.        Rudder and Elevator of S.L. 1.

C.B. 1265.
**Plate 8**
See Chapt. III.

Figs. 31 and 32.—Views of the wreck of L 33 at Little Wigborough.

Ordnance Survey, February, 1917

C.B. 1265.
**Plate 9.**
*See* Chap. III.

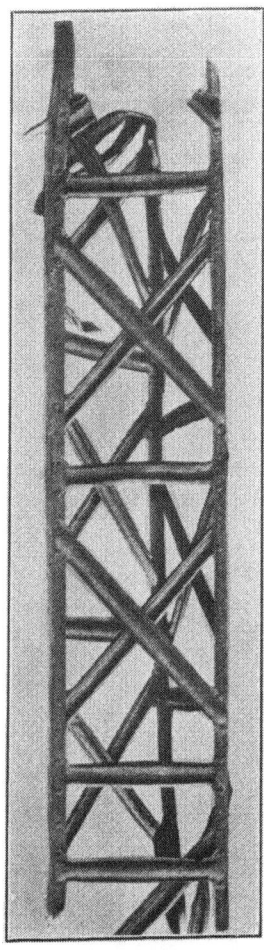

Fig. 33.—Double-braced triangular girder from LZ 85.

Fig. 34.—Single braced triangular girder from LZ 85.

Fig. 35.—Top view of "W" girder from LZ 85.

Fig. 36.—End view of "W" girder from LZ 85.

Fig. 37.—Fin girders from LZ 85.

O   AS 2318

C.B. 1265.
**Plate 10.**
See Chapt. III.

Fig. 39.—Joint of Main Longitudinal and Intermediate Transverse Frame of L 33.

Fig. 41.—Joint of Main Longitudinal and Intermediate Transverse Frame amidships (L 33).

Fig. 38.—Joint of Main Longitudinal and Intermediate Transverse near ends of ship (L 33).

Fig. 40.—Joint of Main Longitudinal and Main Transverse Frame (No. 3) near Bow (L 33).

Ordnance Survey, February, 1917.

C.B. 1265.
**Plate 11.**
See Chapt. III.

Fig. 43.—Joint of Main Transverse Frame Kingpost with Frame and with Intermediate Longitudinal (L 33).

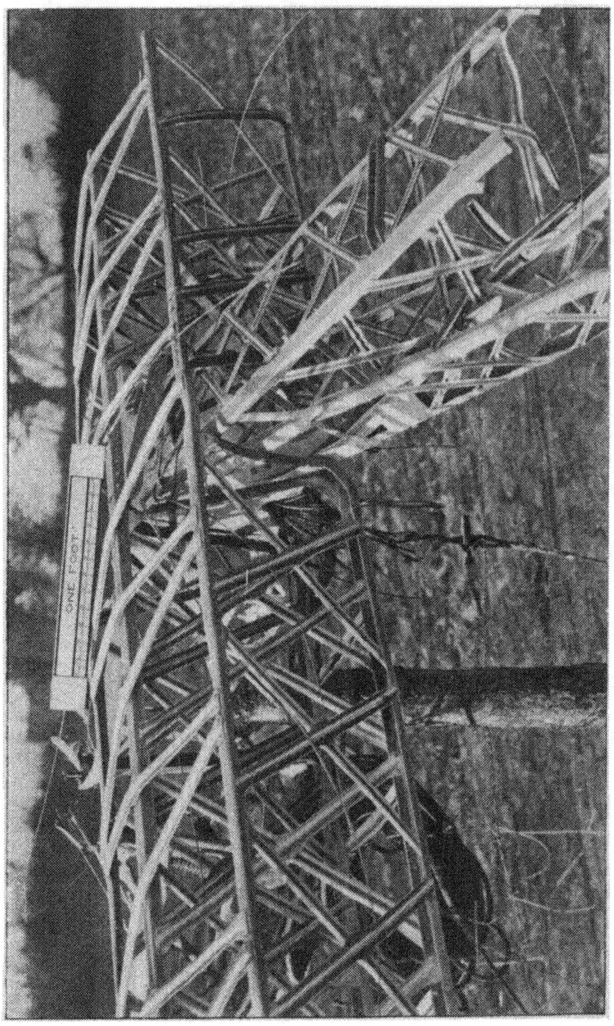

Fig. 45.—Joint of Main Transverse Frame with "A" Longitudinal (L 33), with coil of shear rope and gastight pad fittings.

Fig. 42.—Joint of Main Longitudinal and Main Transverse Frame (L 33).

Fig. 44.—Joint of Main Transverse Frame with "A" Longitudinal (L 33).

Fig. 46.—Bow of L 33. View from inside.

Fig. 47.—Main Transverse Frame Truss (L 33).

Fig. 48.—Joint of "G" Girder with Main Transverse Frame, showing casting for Handling Rope Strop (L 33).

Plate 13.
See Chap. III.

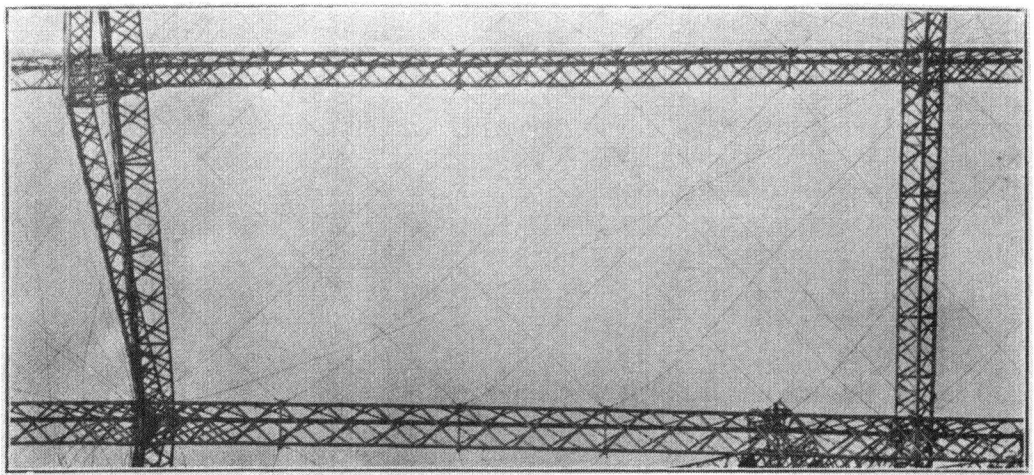

Fig. 49.—Diagonal bracing wires in L 33.

Fig. 50.—Wiring near bow of L 33 showing extra wires for supporting outer cover.

Fig. 51.—Wiring forward of frame 1 in L 33.

C.B. 1265.
Plate 14.
See Chap. III.

Fig. 52.—Upper rudder and fin of L 33.

Fig. 53.—Triangular boom and after edge of fin of L 33.

Fig. 54.—Tail of L 33 showing fins, upper rudder, &c.
Fig. 55 (Inset).—Fin of L 33 showing junction of rectangular boom with outer edge girder.

C.B. 1265.
**Plate 15.**
*See* Chap. III.

Fig. 56.—Top joint of keel of L 33.

Fig. 57.—Rudder of L 33.

Fig. 58.—Rudder of LZ 85.

C.B. 1265.
**Plate 16.**
See Chapt. III.

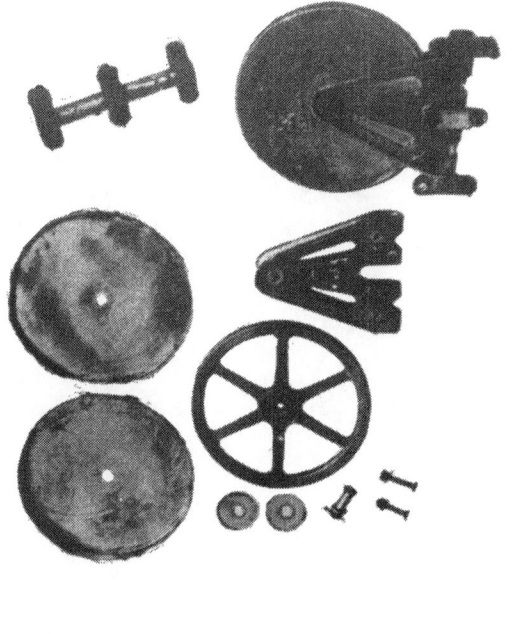

Fig. 59.—Guide Sheaves in Corridor (L 33) for Rudder and Elevator Controls (A and B) and for minor controls (C).
NOTE.—B and C have hardwood rollers.

Fig. 60.—7½″ Guide Sheave for Rudder and Elevator Controls (L 33).

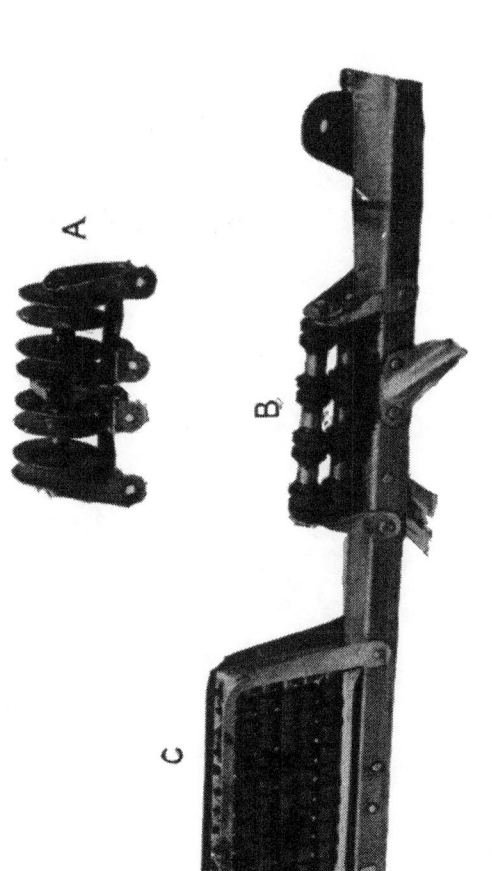

Fig. 61.—Nest of 1¾″ Guide Sheaves for Control Leads fitted in Keel over Control Car (L 33).

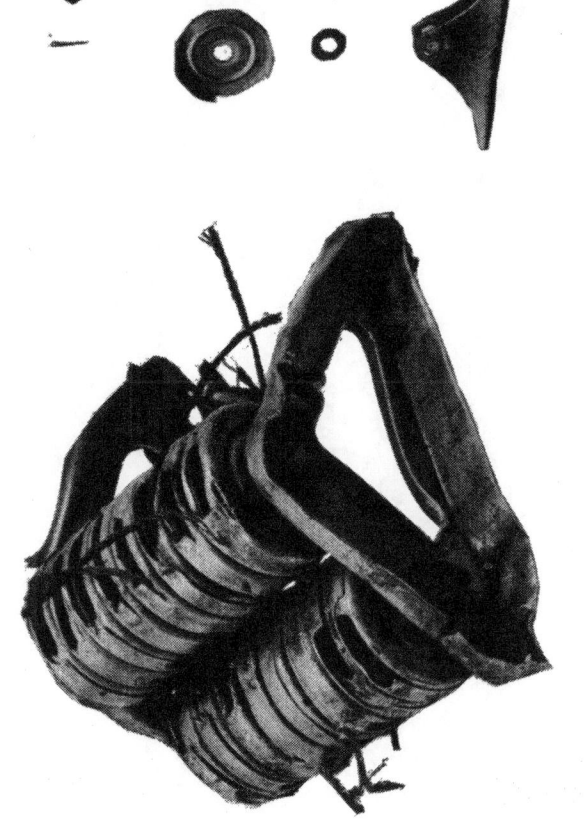

Fig. 62.—1¾″ Guide Sheaves for Gas Valve Controls, etc. (L 33).

Fig. 63.—2″ Sheave for Water Ballast Controls, etc. (L 33).

*Ordnance Survey, February, 1917.*

Fig. 64.—Keel structure L 33. Water ballast position on right; petrol tank position on left.

Fig. 65.—Keel structure L 33 at 100 kg. Bomb position one side partly removed. Part of ventilator shaft at top.

Fig. 66.—Side view of keel structure of L 33.

Fig. 67.—Base view of keel structure, showing bomb-shutter frame in place.

Fig. 68.—Keel of L 33 aft, showing walking way, W.C., and handling rope tub.

Fig. 69.—Keel structure of L 33 near stern.

Fig. 70.—Stern Cap and Flagpost of L 32.

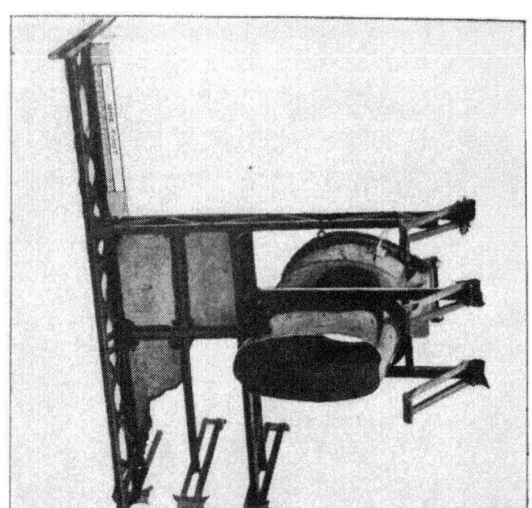

Figs. 71 & 72.—W.C. from L 33 as seen from underneath and in position.

Fig. 73.—Portion of Keel structure of LZ 85 (Salonica).

Fig. 74.—Athwartship Gangway to Wing Cars in L 33; outboard end on left of Picture. (Walking way has been removed between longitudinals G and F).

Fig. 75.—Athwartship Gangway to Wing Cars in L 33. View looking inboard.

Fig. 76.—Structure behind plywood in way of Wing Propellers in L 33.

Fig. 77.—Wing Propeller Bracket in L 33, and remains of structure behind plywood.

C.B. 1265.
**Plate 20.**
See Chapt. III.

Figs. 78 & 79.—View from Aft of Wing Propeller and Brackets of L 33. Inset in top left-hand corner is Fig. 42c showing type of swivelling joint used for Propeller Brackets and Car Struts.

Figs. 80 & 81.—Wing Propeller and Brackets of L 33 as seen from aft. Inset in top right-hand corner is Fig. 42b showing a side view of the Wing Propeller and Brackets.

*Ordnance Survey, February, 1917.*

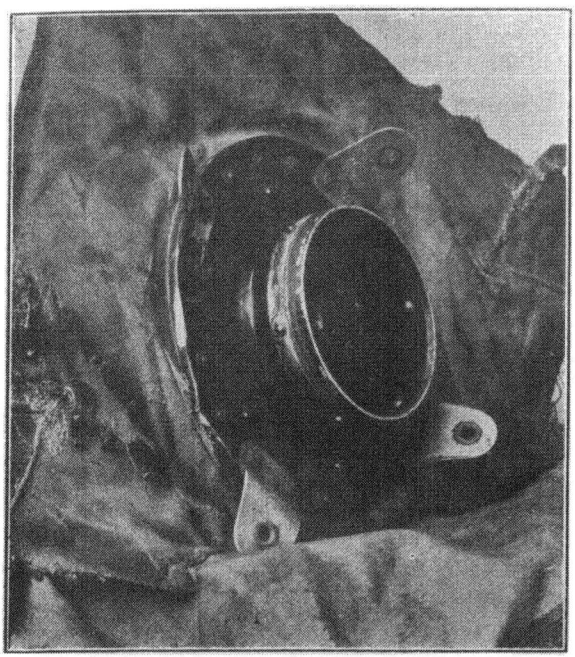
Fig. 82.—Water ballast valve from L 33, pipe removed. View from below.

Fig. 83.—Water ballast valve from L 33.

Fig. 84.—Water-ballast bag attachment to girder above on L 33.

Fig. 85.—Emergency water-ballast buckets aft on outside of No. 9 Longitudinal in L 33.

Fig. 86.—Emergency water-ballast bucket forward in L 33.

C.B. 1265.
**Plate 22.**
*See* Chap. III.

Fig. 87.—Petrol tank stowage in L 33. "Fixed" positions.

Fig. 88.—Petrol tank guide girders in L 33 showing links for suspension of "fixed" tanks.

Fig. 89.—Petrol tank in L 33 with slip-bolt.

O AS 2318

Fig. 90.—Oil and water tanks for wing car of L 33. (The water tank is the upper one.)

Fig. 91.—Oil and water tanks for wing car of L 33.

Fig. 92.—Smaller keel ventilator shaft of L 33.

Fig. 93.—Keel ventilator shaft of L 33 with part of wooden hood.

Fig. 94.—Keel ventilator shaft of L 33, attachment of upper end.

C.B. 1265.
**Plate 24.**
See Chapt. III.

Fig. 95.—Mooring point, with double strop attached to slip bolt from LZ 85.

Fig. 96.—Mooring point, slip bolt partly withdrawn, and thimble of double strop removed. From LZ 85.

*Ordnance Survey, February, 1917*

C.B. 1265.
**Plate 25.**
See Chapt. III.

Fig. 97.—Mooring Point of L 33.

Fig. 98.—Arrangement of girders at mooring position (the forward part is to the left) of LZ 85.

*Ordnance Survey, February, 1917.*

C.B. 1265.
**Plate 26.**
See Chapt. III.

Fig. 99.—Tubs for Trail Ropes from LZ 85.

Fig. 100.—Handling Rope Tub Aft. (at side of walking way) from L 33.

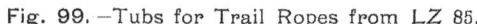

Fig. 101.—View inside Wing Car of L 33.

*Ordnance Survey, February, 1917.*

C.B. 1265.
**Plate 27.**
See Chapt. III.

Fig. 102.—Forward Gun Platform of L 33. Transverse structure at Frame 7.

Fig. 103.
Forward Gun Platform of L 33. showing Tripod in place.

Fig. 104.
Forward Gun Platform of L 33, showing view of structure below.

*Ordnance Survey. February, 1917.*

C.B. 1265.
**Plate 28.**
See Chapt. III.

Fig. 105.—Part of structure of Forward Gun Platform in L 33.

Fig. 106.—View of underside of After Gun Platform in L 33.

Fig. 107.—After Gun Platform in L 33, with Gun Mounting in place on girders (view looking aft.)

Fig. 108.—Part of Forward Gun Platform in L 33.

*Ordnance Survey, February, 1917*

C.B. 1265.
**Plate 29.**
See Chapt. III.

Fig. 109.—View of bottom of Wing Car from L 32.

Fig. 110.—Side view of Wing Car from L 33.

*Ordnance Survey, February, 1917.*

C.B. 1265.
**Plate 30.**
See Chapt. III.

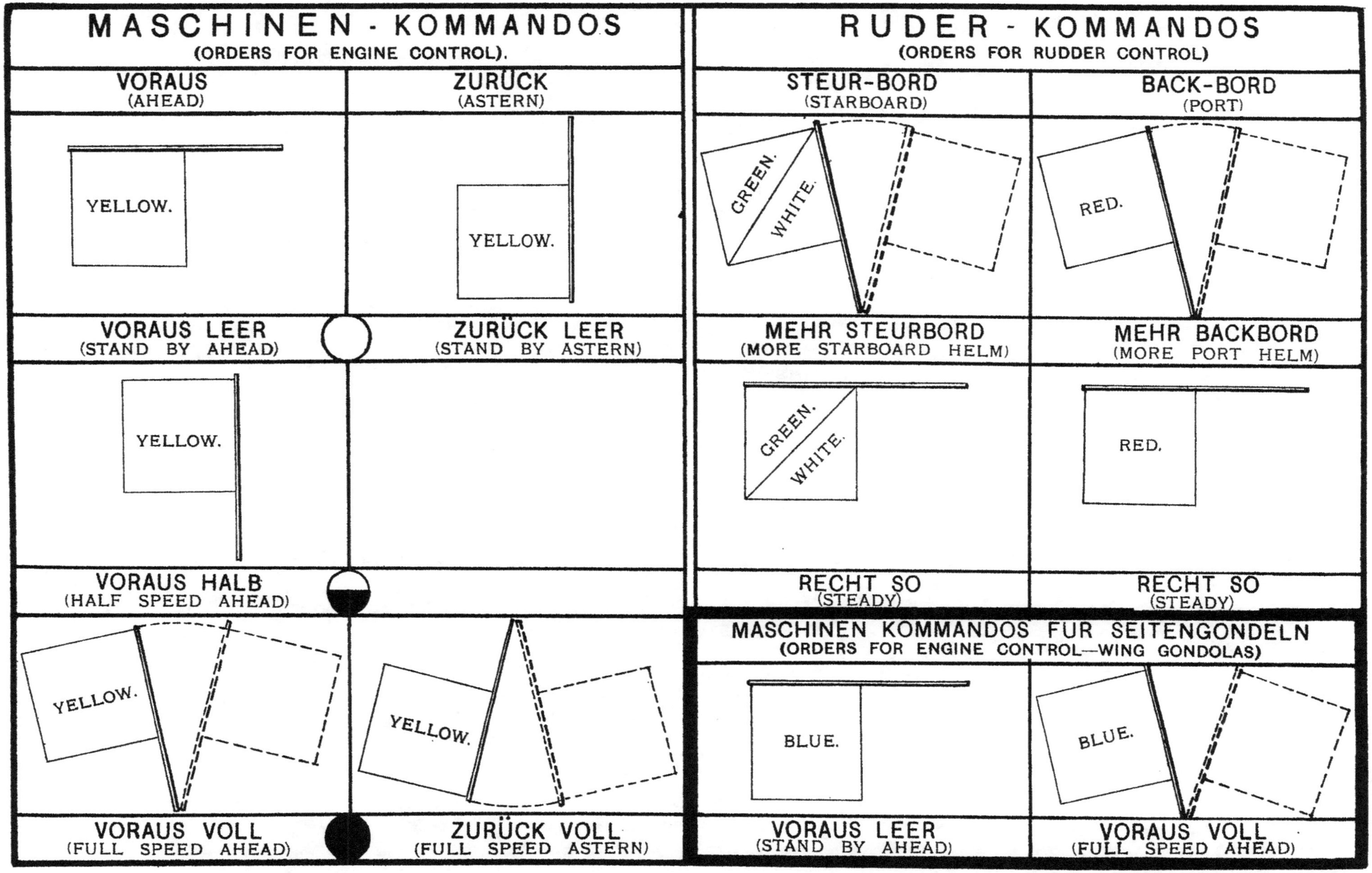

# RUDER - KOMMANDOS
(ORDERS FOR RUDDER CONTROL)

| STEUR-BORD (STARBOARD) | BACK-BORD (PORT) |
|---|---|
| GREEN/WHITE flag | RED flag |
| MEHR STEURBORD (MORE STARBOARD HELM) | MEHR BACKBORD (MORE PORT HELM) |
| GREEN/WHITE | RED |
| RECHT SO (STEADY) | RECHT SO (STEADY) |

# MASCHINEN KOMMANDOS FUR SEITENGONDELN
(ORDERS FOR ENGINE CONTROL—WING GONDOLAS)

| VORAUS LEER (STAND BY AHEAD) | VORAUS VOLL (FULL SPEED AHEAD) |
|---|---|
| BLUE | BLUE |

# MASCHINEN - KOMMANDOS
(ORDERS FOR ENGINE CONTROL)

| VORAUS (AHEAD) | ZURÜCK (ASTERN) |
|---|---|
| YELLOW | YELLOW |
| VORAUS LEER (STAND BY AHEAD) | ZURÜCK LEER (STAND BY ASTERN) |
| YELLOW | (white circle) |
| VORAUS HALB (HALF SPEED AHEAD) | |
| (half black/white circle) | |
| VORAUS VOLL (FULL SPEED AHEAD) | ZURÜCK VOLL (FULL SPEED ASTERN) |
| YELLOW | YELLOW |
| | (black circle) |

Ordnance Survey, February, 1917.

C.B. 1265.
**Plate 31.**
See Chapt. III.

Fig. 111. End View of Observation Car from a German Rigid Airship.

Fig. 112. Side View of Observation Car from a German Rigid Airship.

*Ordnance Survey, February, 1917*

C.B. 1265.
**Plate 32.**
See Chapt. III.

Fig. 114.—End View of Winch for Observation Car from a German Rigid Airship.

Fig. 113.—Winch and Cable from Observation Car of German Rigid Airship.

*Ordnance Survey, February, 1917.*

ELEVATION.

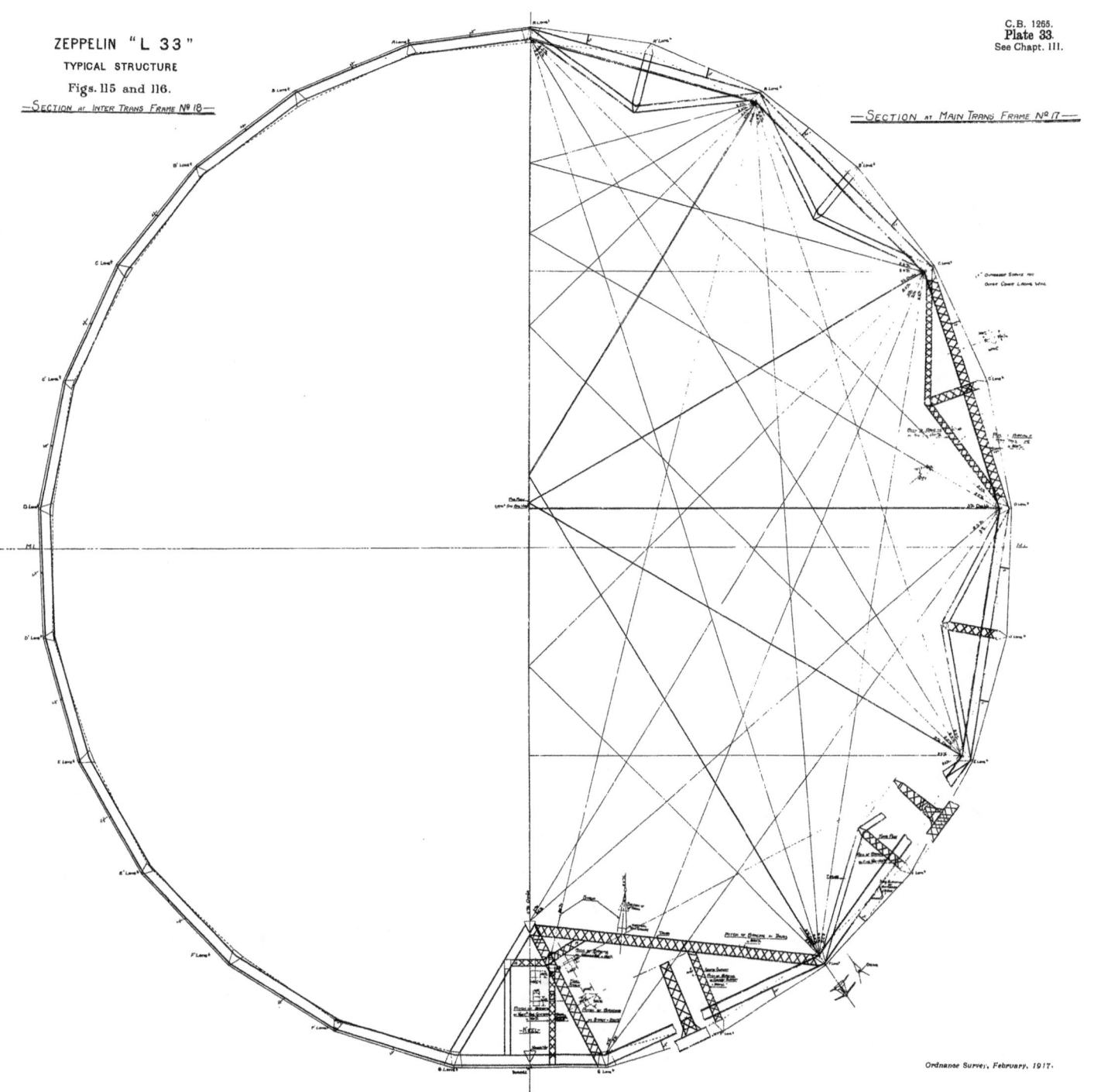

# "L 33."
## SECTIONS OF GIRDERS.

Plate **34.**
C.B. 1265.
See Chapt. III.

Fig. 117.
"A" AND "G" LONGITUDINAL GIRDERS.

Fig. 118.
MAIN TRANSVERSE GIRDER.

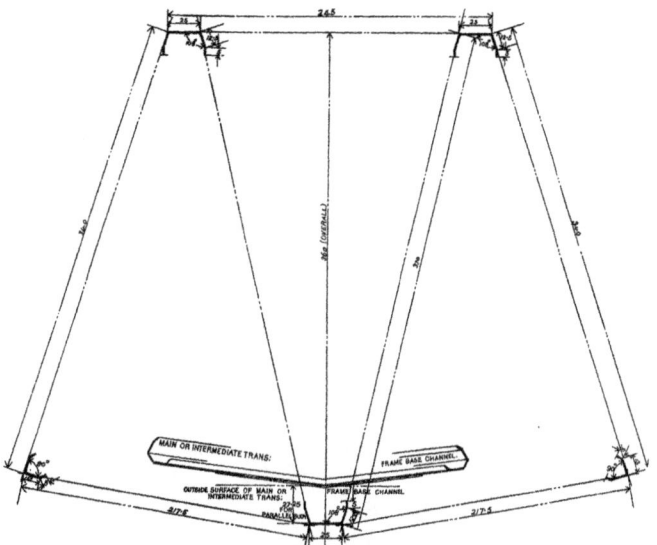

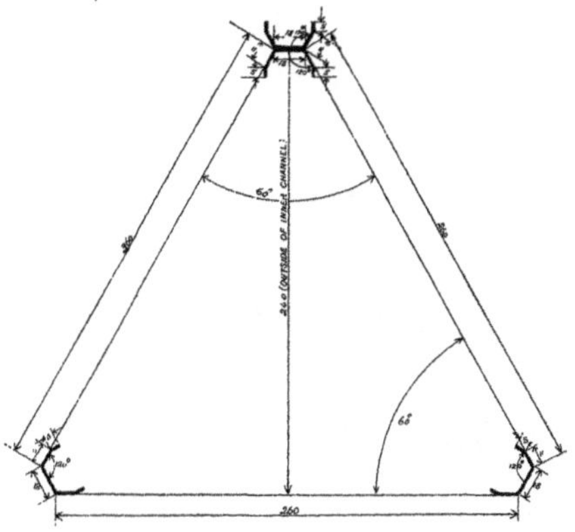

Fig. 119.
MAIN LONGITUDINAL.

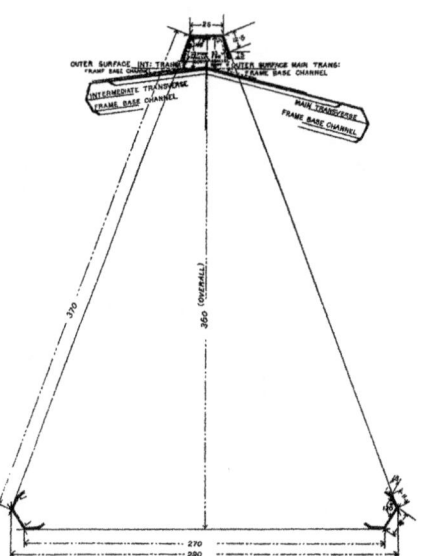

Fig. 120.
INTERMEDIATE LONGITUDINAL.

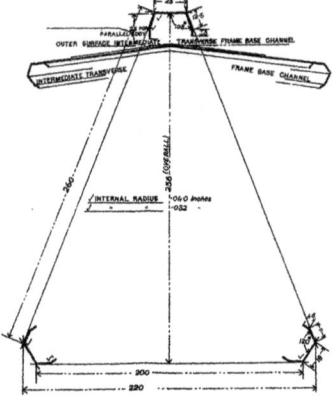

Fig. 121.
INTERMEDIATE TRANSVERSE FRAME.

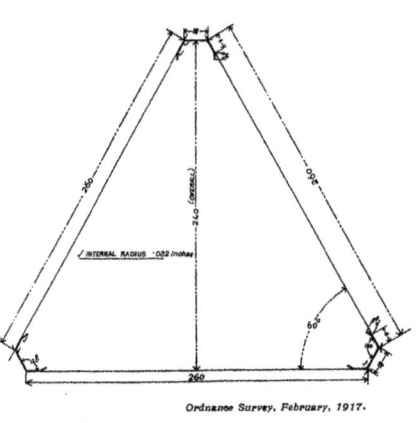

Ordnance Survey, February, 1917.

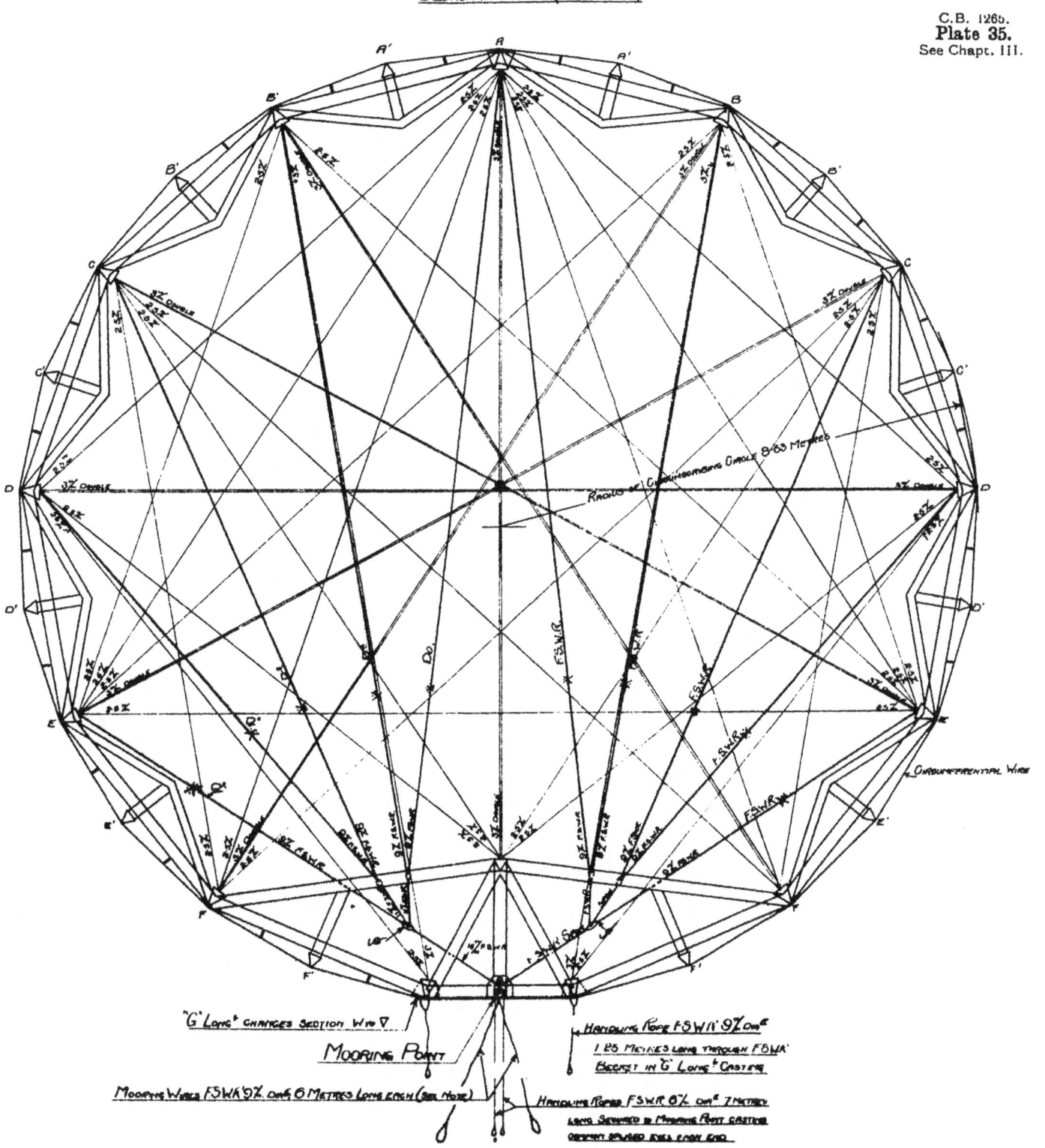

## ZEPPELIN "L 33."
### MOORING ARRANGEMENTS.

NOTE.—The ends of the F.S.W. Ropes in Frame 5 at Longitudinals A to E inclusive, terminated in Common spliced eyes. Also ends marked ↙

The ends of the F.S.W. Ropes Spans in Frame 5 terminate in Steel Thimbles.

*These F.S.W. Ropes distribute the Stresses laterally when Ship is moored.

The 9$^{m.m}/_{m.}$ Mooring Wire is in one length, seized in its bight, thus making a span. This wire secures to a F.S.W.R. Strop, which its turn secures to the Draw Pin Release Gear.

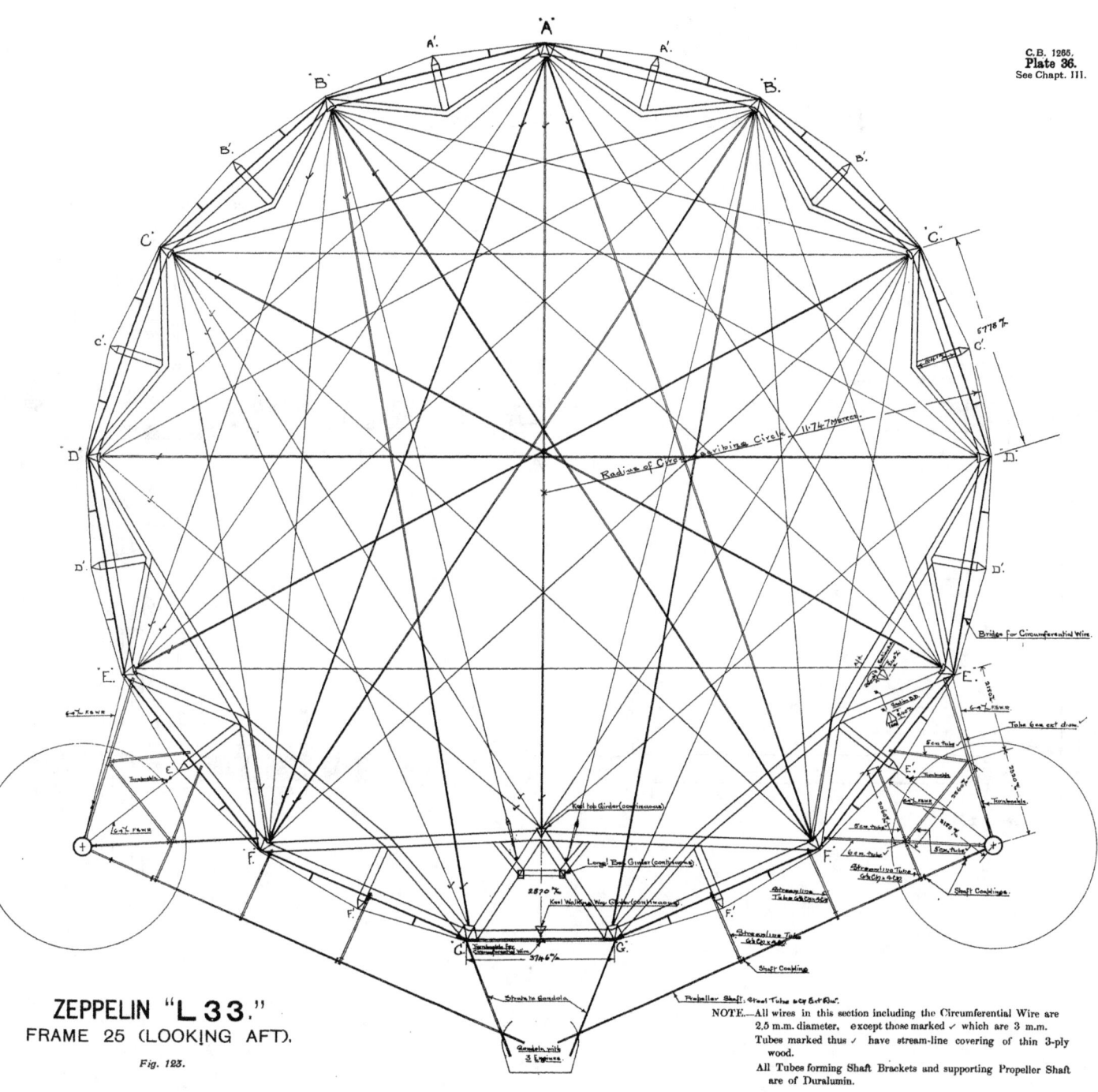

ZEPPELIN "L 33."
FRAME 25 (LOOKING AFT).

Fig. 123.

NOTE.—All wires in this section including the Circumferential Wire are 2.5 m.m. diameter, except those marked ✓ which are 3 m.m.
Tubes marked thus ✓ have stream-line covering of thin 3-ply wood.
All Tubes forming Shaft Brackets and supporting Propeller Shaft are of Duralumin.

# Zeppelin "L33"

Central Fitting for Diametral Wires in Main Frame.
Scale ¾ Full Size

Fig. 124.

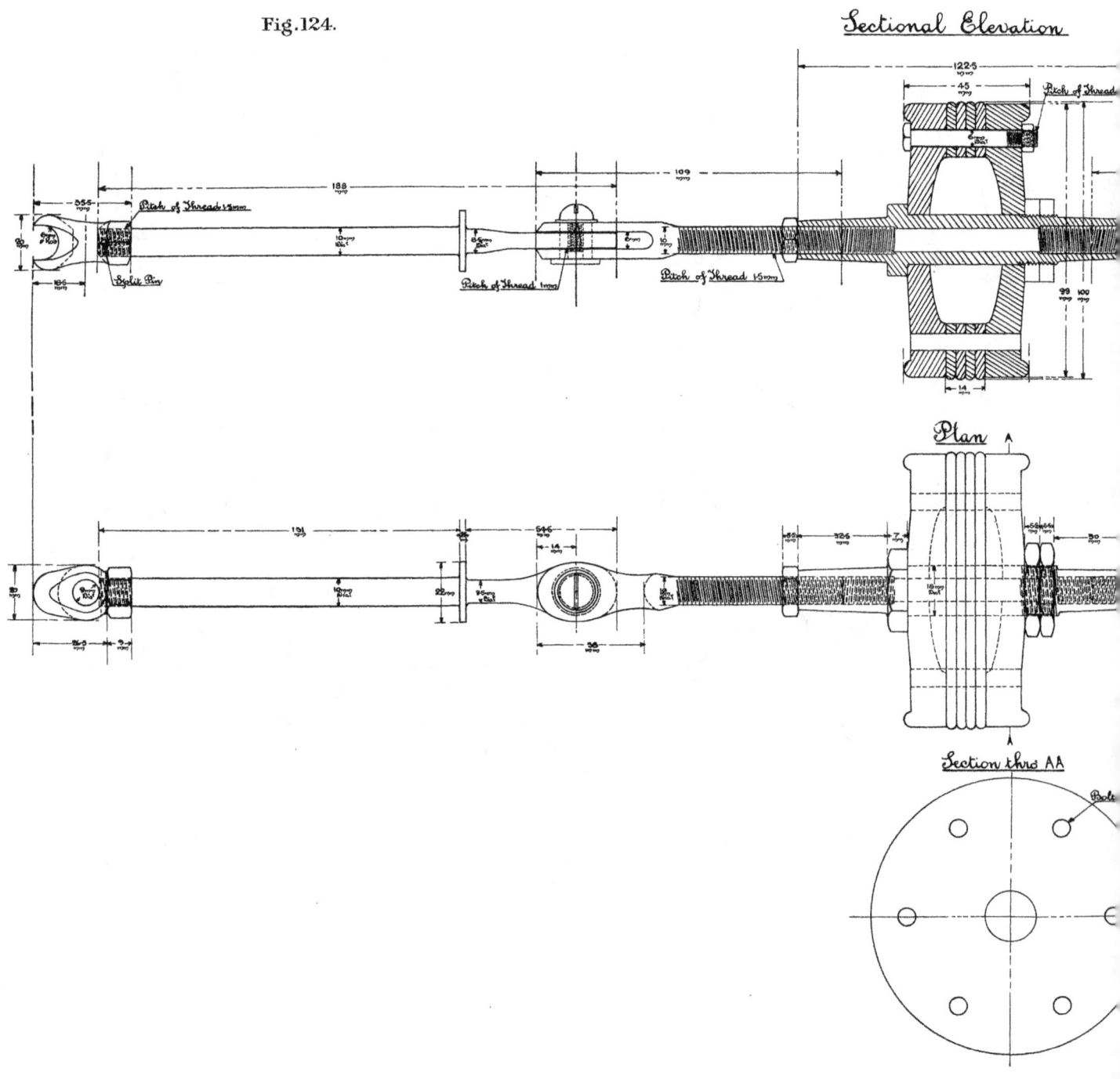

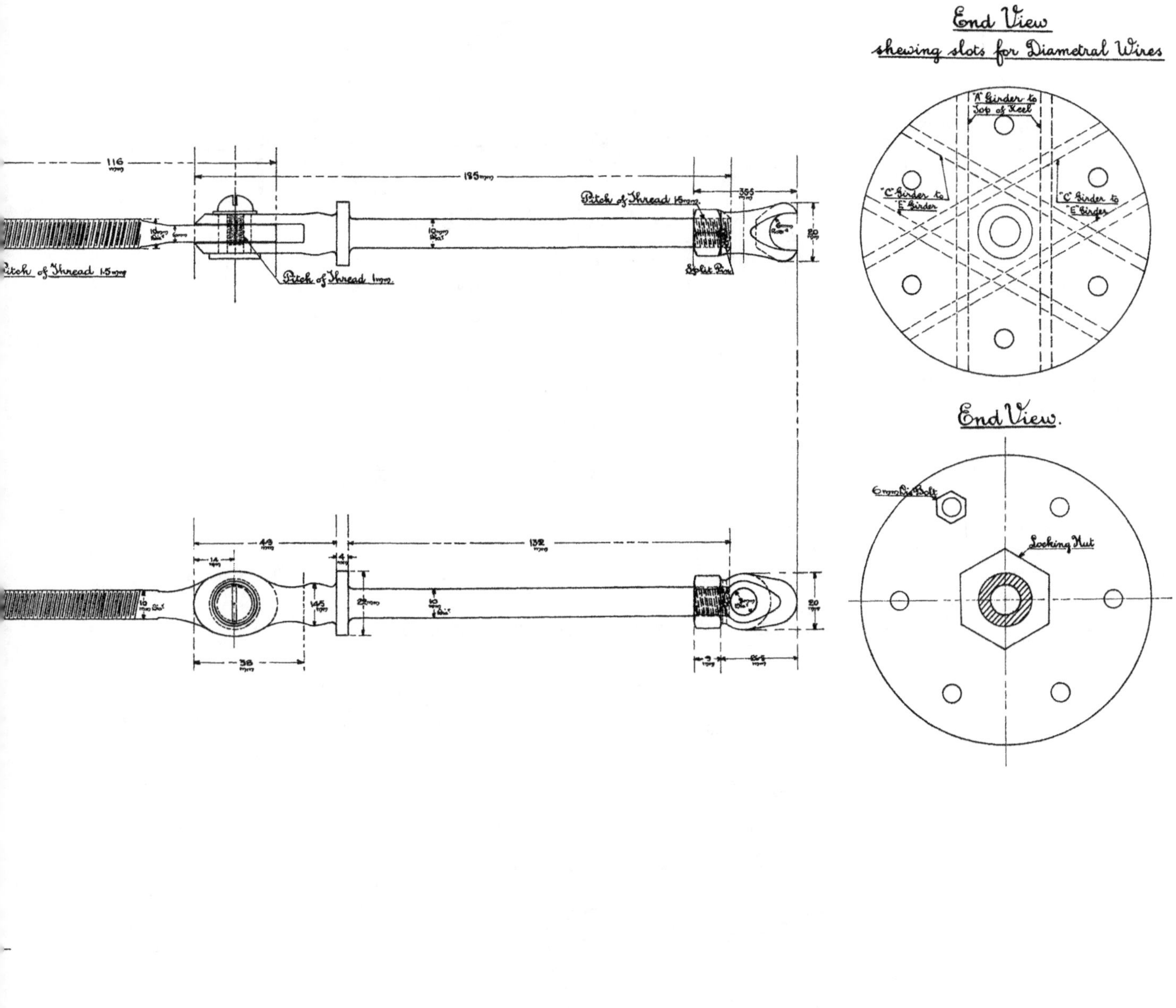

C.B. 1265.
**Plate 37.**
See Chapt. III.

**End View**
shewing slots for Diametral Wires

**End View.**

Ordnance Survey, February, 1917.

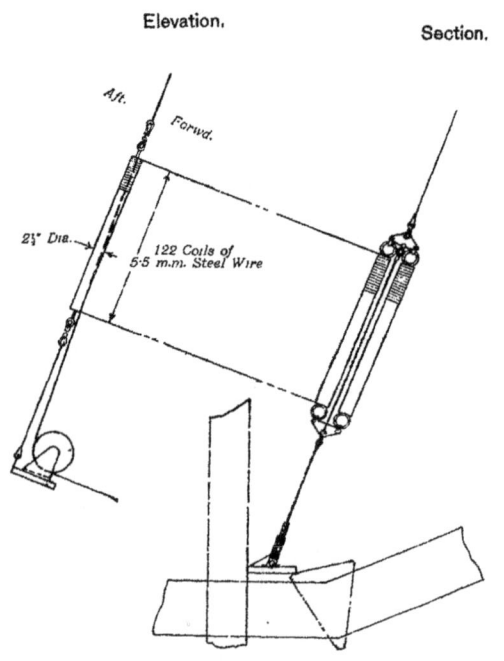

# ZEPPELIN "L 33."

## DIAGRAMMATIC ARRANGEMENT OF RUDDER AND ELEVATOR CONTROL CABLES.

Fig. 125.

Section at 39 Looking aft.

Upper Rudder Starboard Cable
"   "    Port       "
Lower "    Starboard and Port Cable
Starboard Elevator Upper Cable
"       "       Lower "
Port    "    Upper "
"       "    Lower "
Elevator Preventer Ropes

Control Cables 20 m.m. in circumference made up of Flex. Steel Wire F.S.W.R.: Rope of 11 Strands laid up on Hemp Cord, each Strand containing 19 Wires.

Control Cables change from F.S.W.R. to Plain Steel Wire Rope between Frames 37–38.

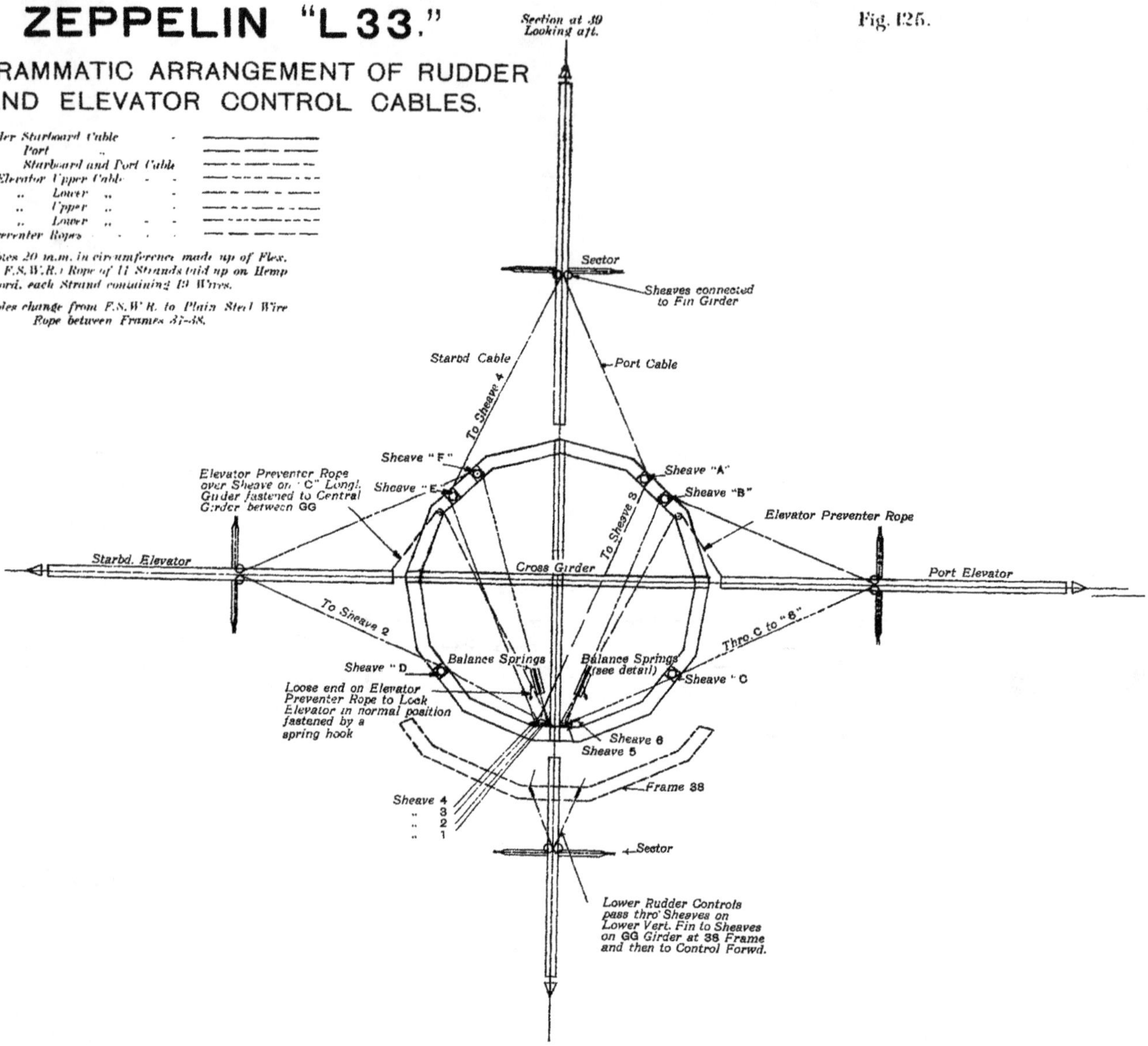

Sector
Sheaves connected to Fin Girder
Starbd Cable
Port Cable
To Sheave 4
Sheave "F"
Sheave "A"
Elevator Preventer Rope over Sheave on "C" Longl. Girder fastened to Central Girder between GG
Sheave "E"
Sheave "B"
Elevator Preventer Rope
Starbd. Elevator
Cross Girder
To Sheave 3
Port Elevator
To Sheave 2
Sheave "D"
Balance Springs
Balance Springs (see detail)
Thro. C to "B"
Sheave "C"
Loose end on Elevator Preventer Rope to Lock Elevator in normal position fastened by a spring hook
Sheave 6
Sheave 5
Sheave 4
"   3
"   2
"   1
Frame 38
Sector
Lower Rudder Controls pass thro' Sheaves on Lower Vert. Fin to Sheaves on GG Girder at 38 Frame and then to Control Forwd.

## DETAILS OF LEAD OF CONTROL CABLE FROM SPRINGS

Elevation.   Section.

Aft.   Forwd.

2¼" Dia.   122 Coils of 5·5 m.m. Steel Wire

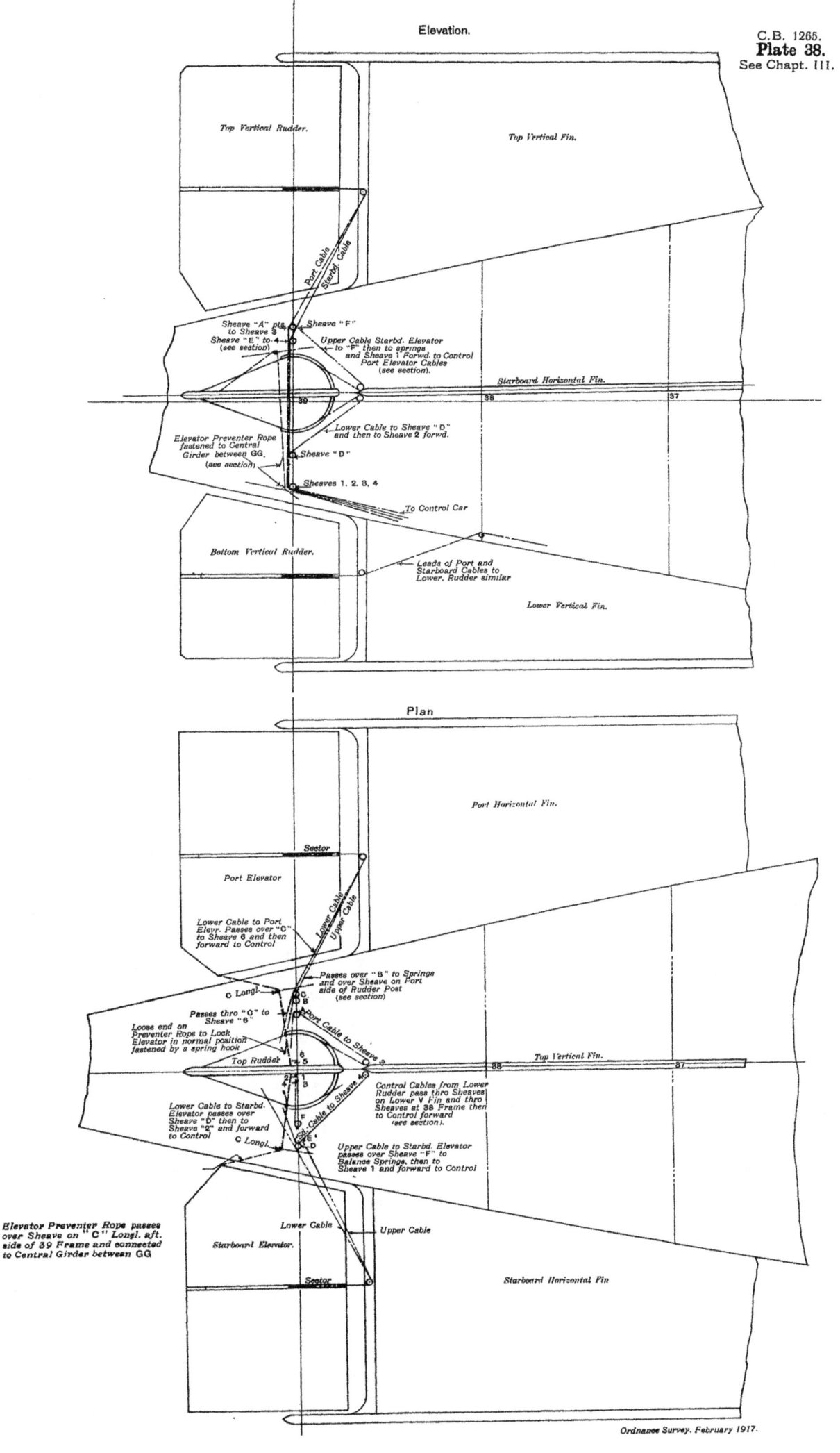

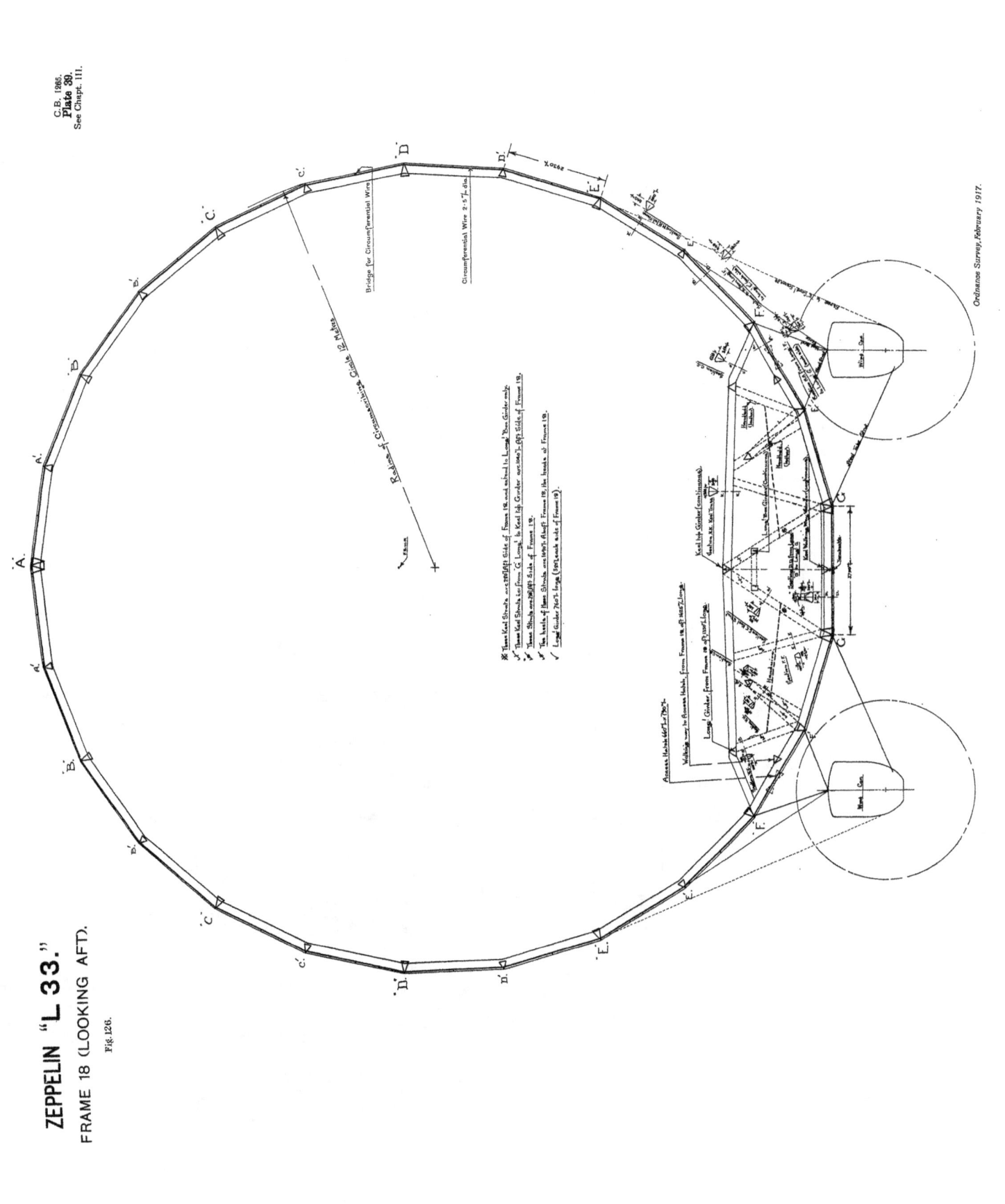

ZEPPELIN "L 33."
FRAME 18 (LOOKING AFT).
Fig. 126.

# ZEPPELIN "L 33."
## INCENDIARY BOMB FRAME AND SHUTTER ARRANGEMENT.
### Fig. 127.

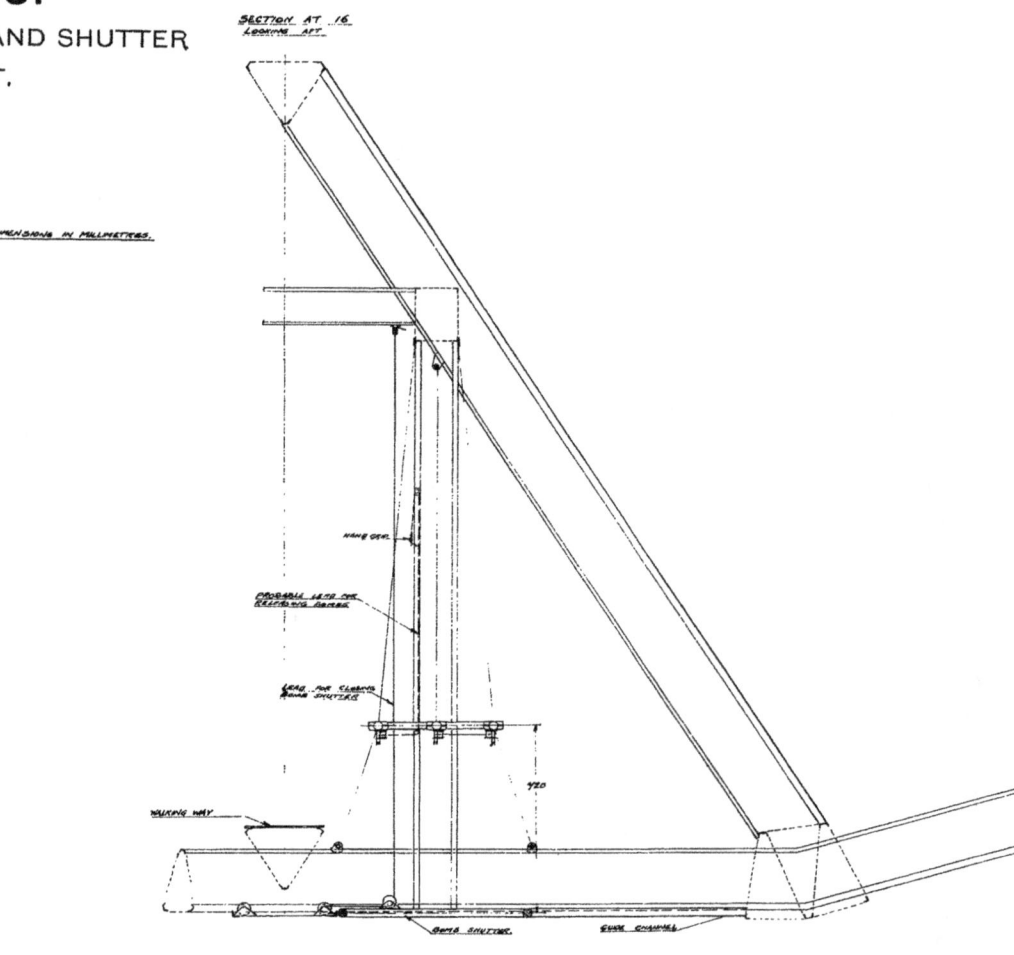

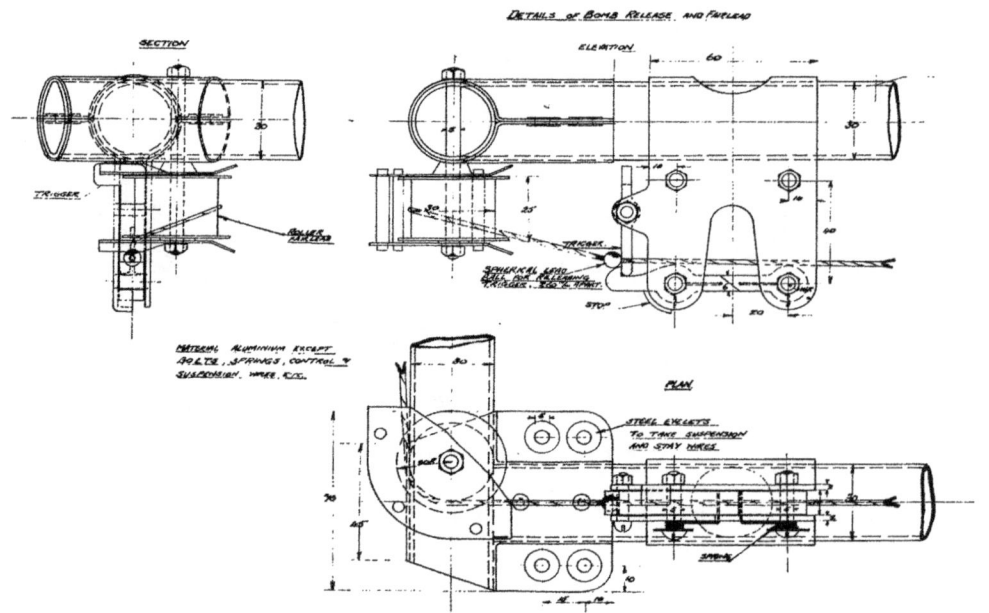

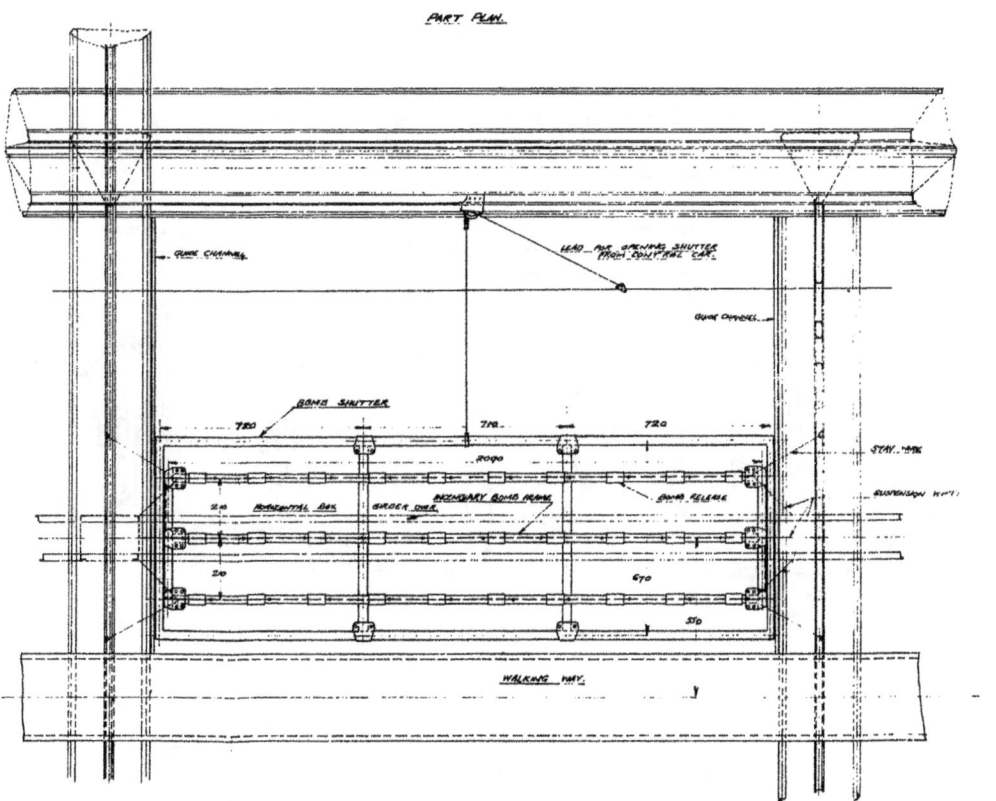

C.B. 1265.
**Plate 40.**
See Chapt. III

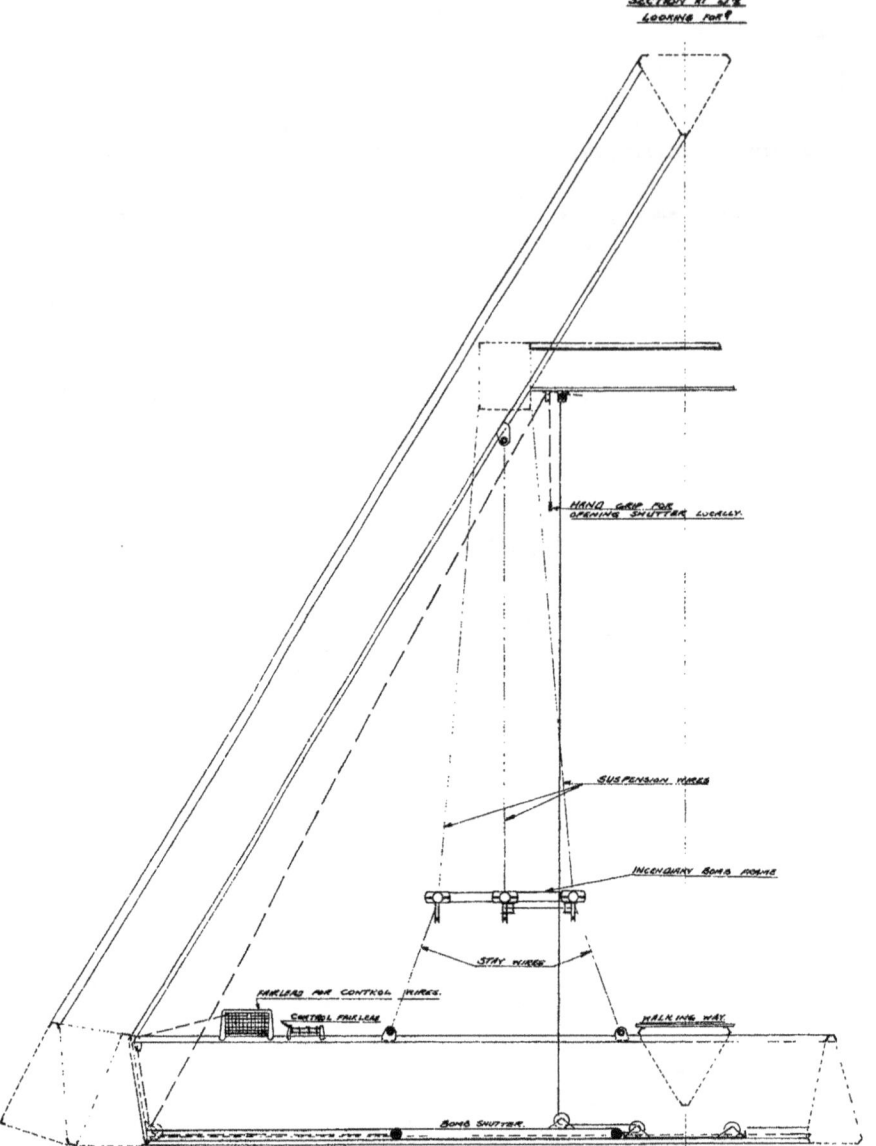

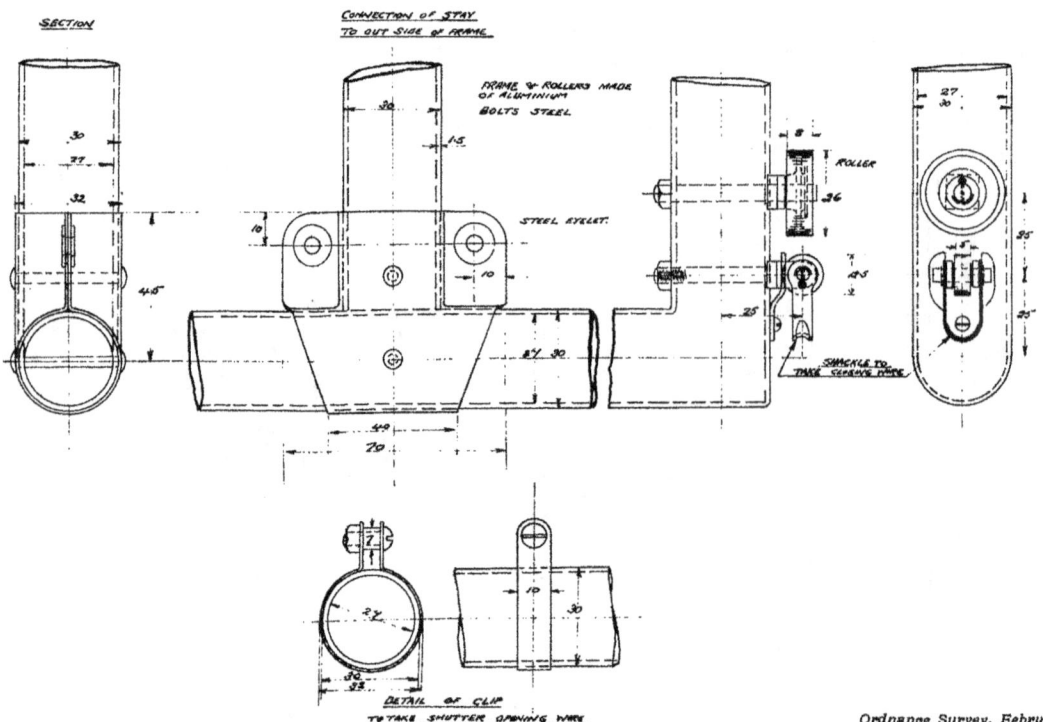

Ordnance Survey, February, 1917.

# "L 33."
## RUDDER & ELEVATOR CONTROLS.
### Transverse Shafts and Sprocket Gears over After Car.
Fig. 128.

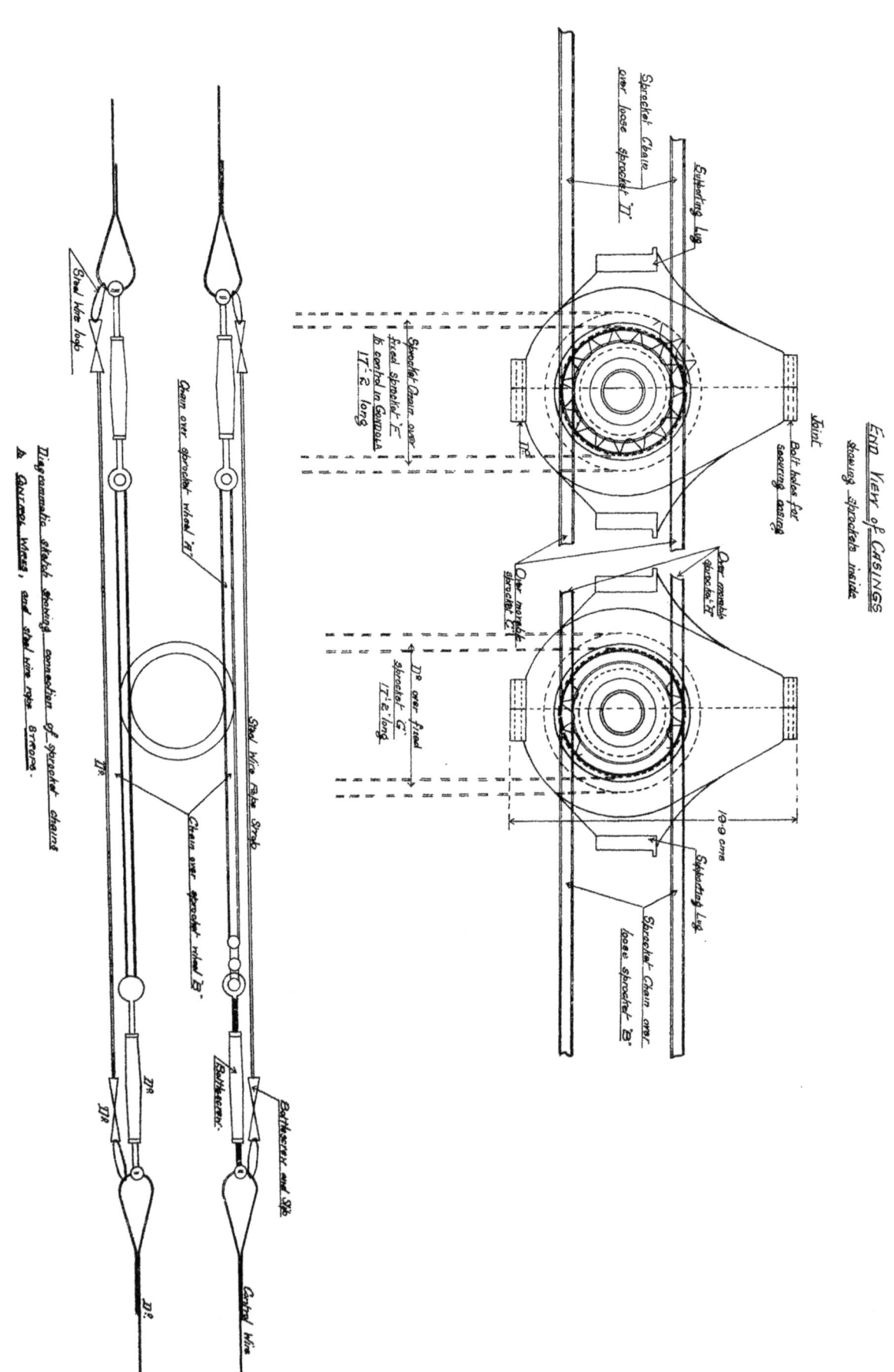

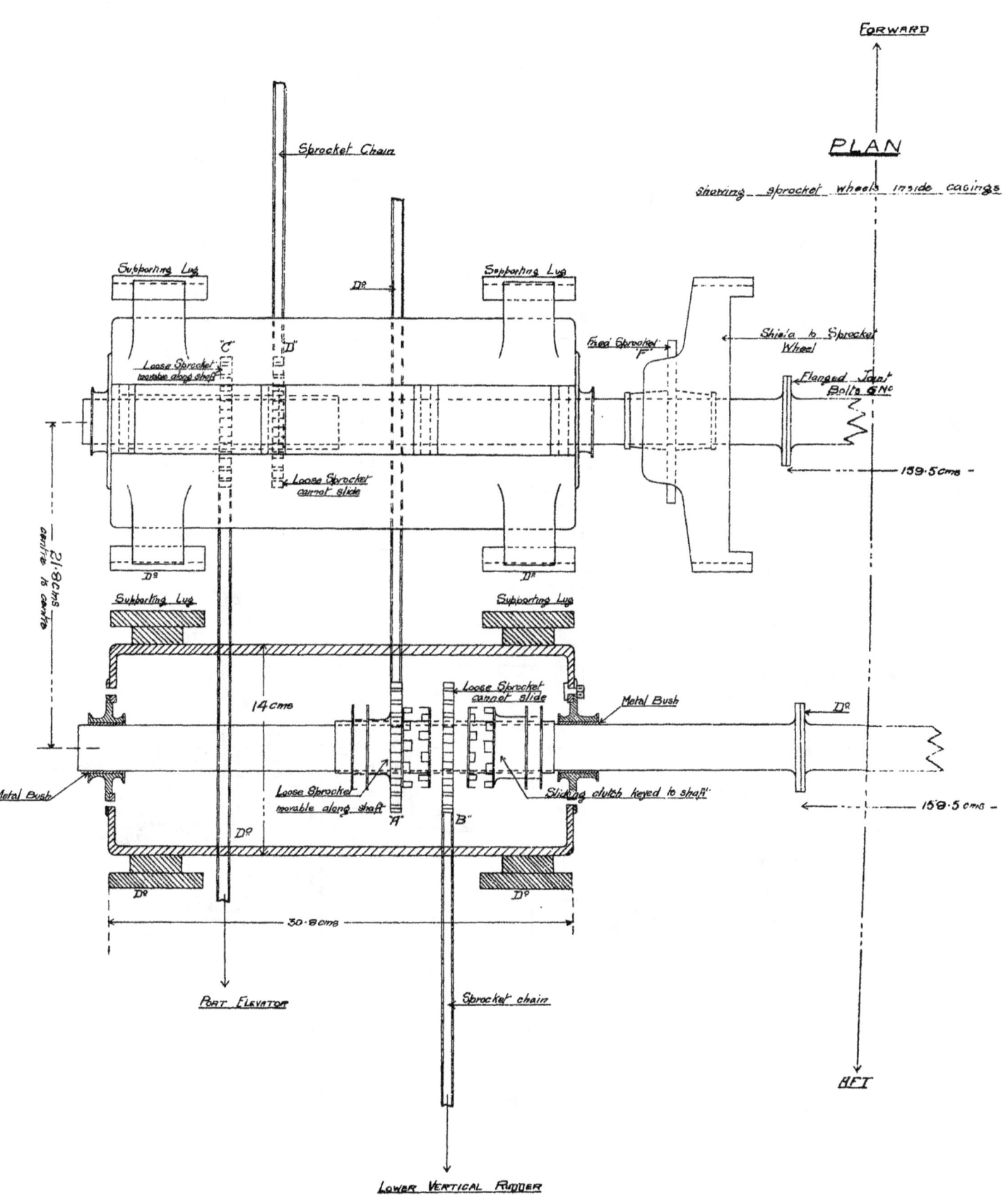

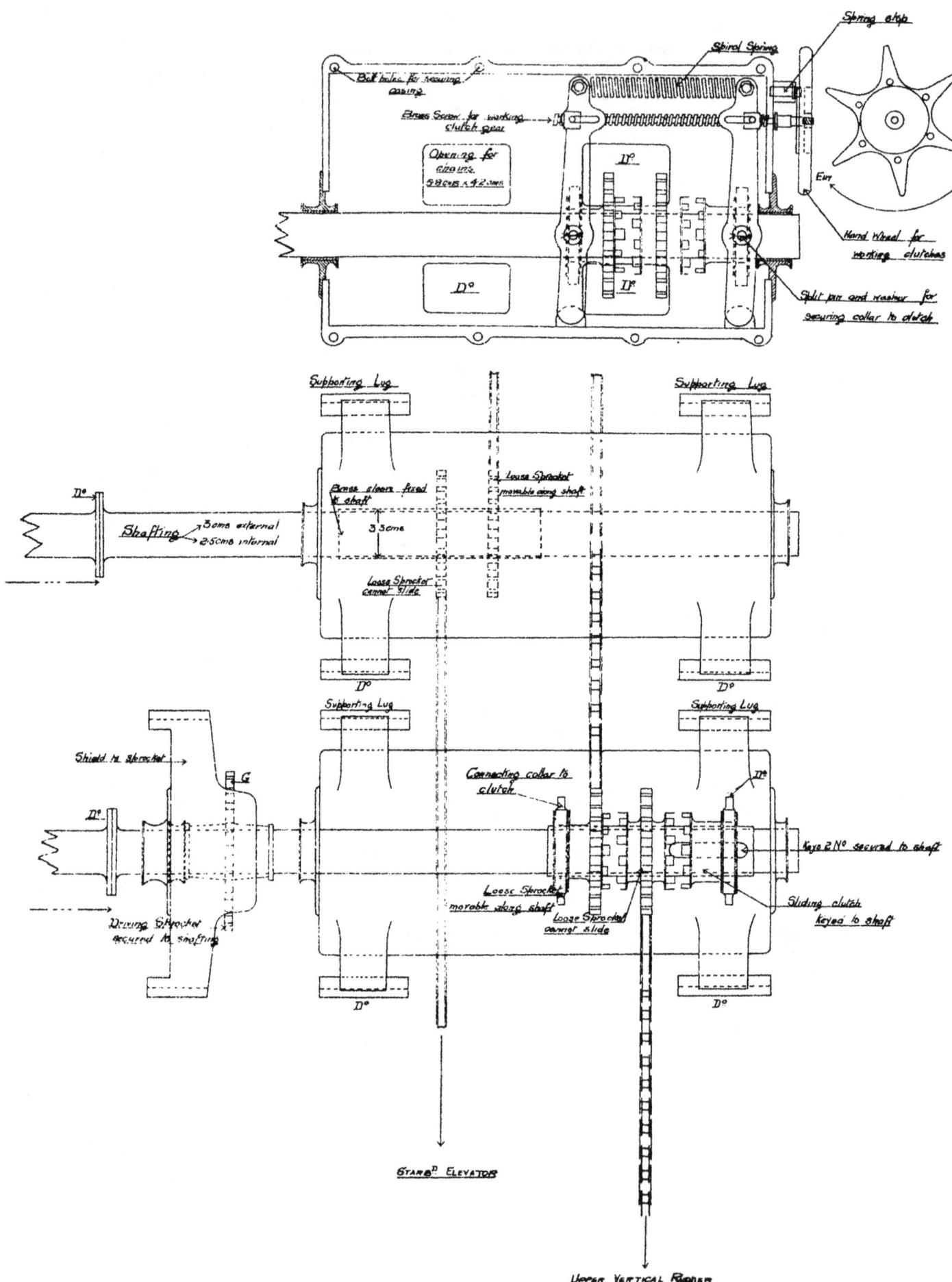

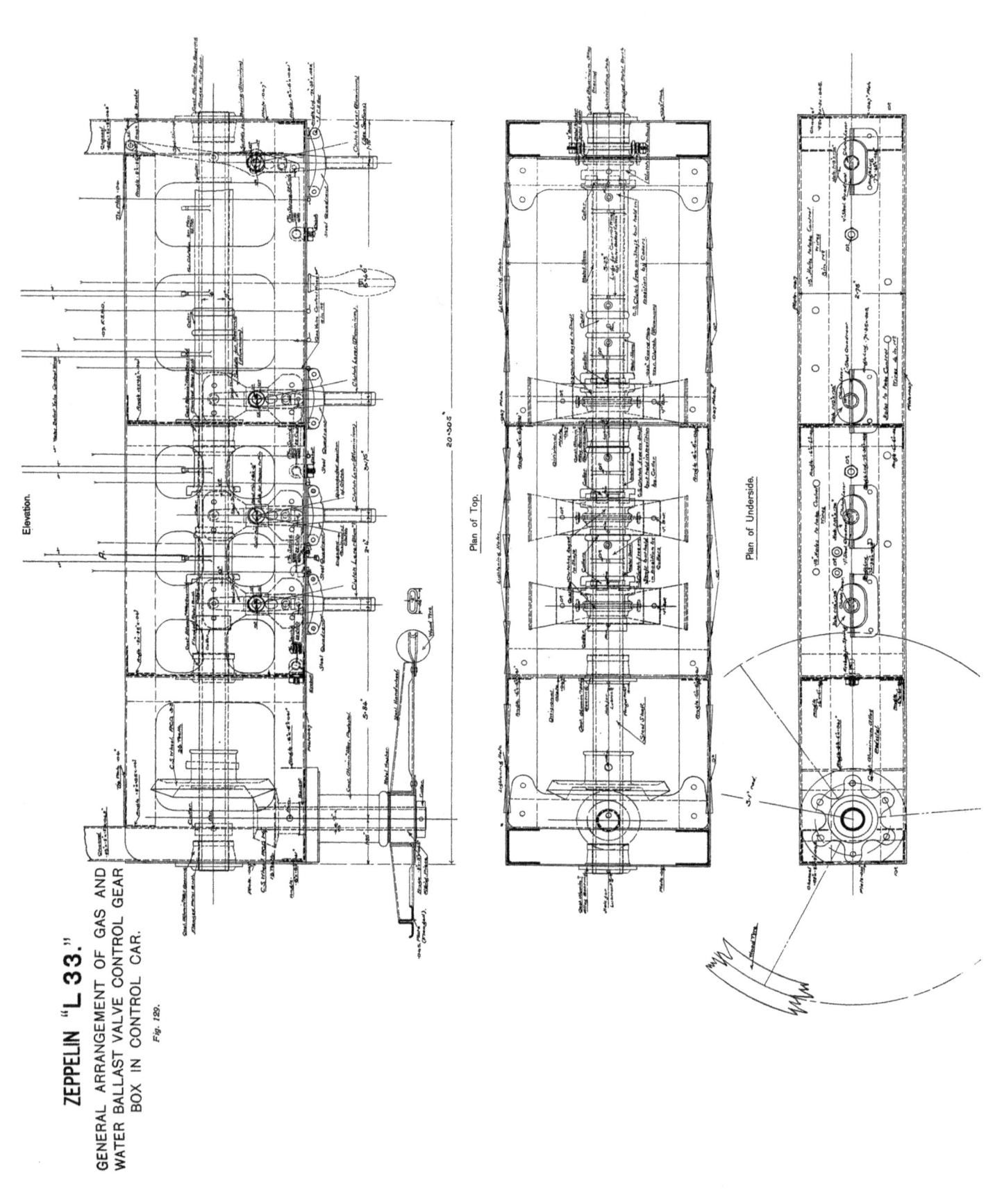

ZEPPELIN "L 33."
GENERAL ARRANGEMENT OF GAS AND WATER BALLAST VALVE CONTROL GEAR BOX IN CONTROL CAR.
Fig. 129.

## Section of End.

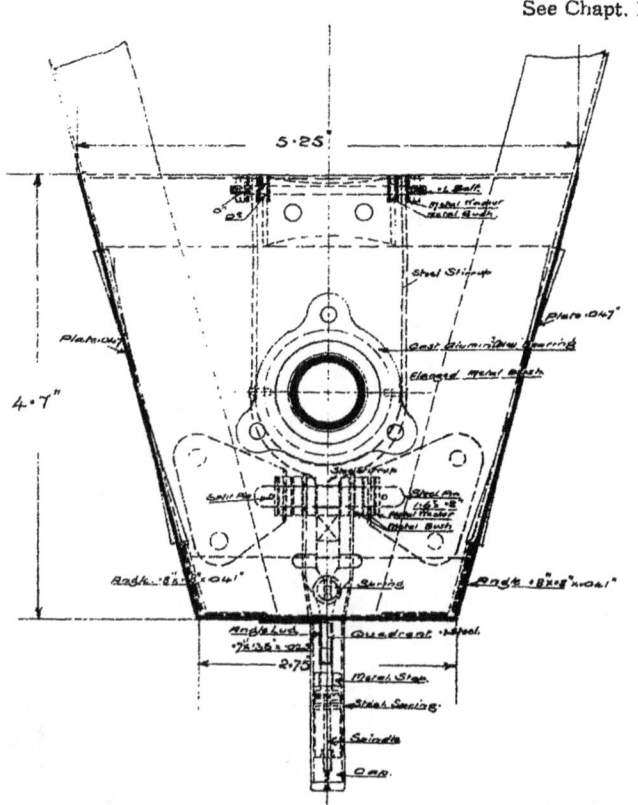

Note: All material Duralumin except where otherwise stated. Gas Valve Control Wires knotted at End presumably to take Wood handle as shewn.

## Section through A.B.

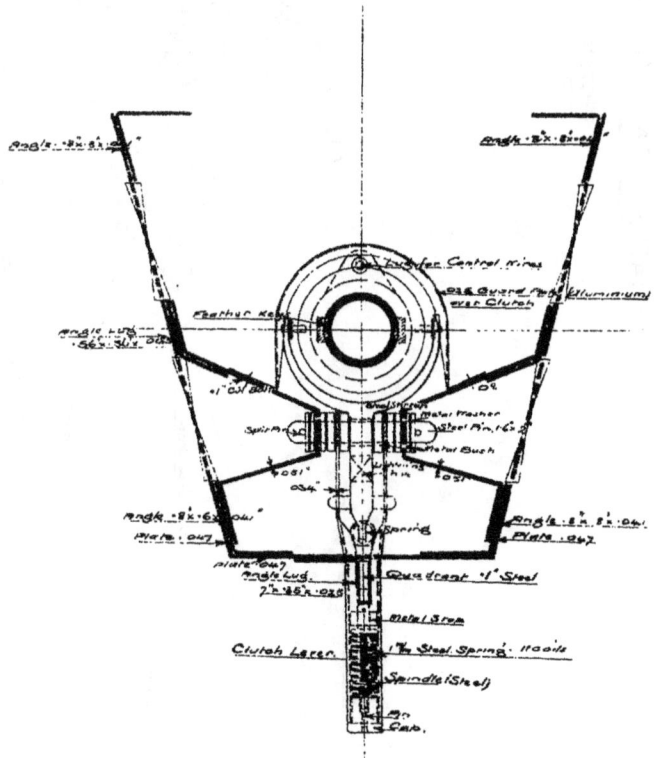

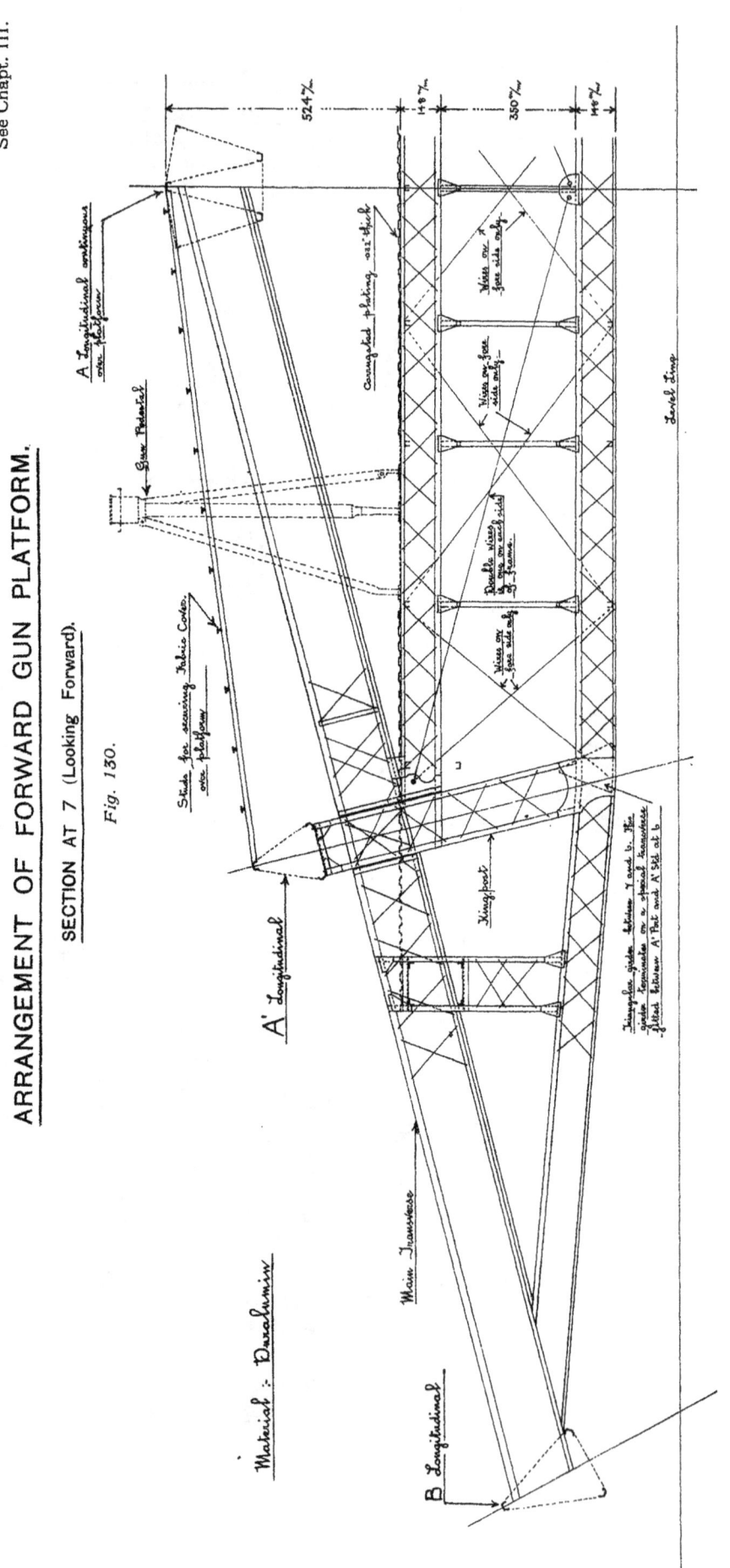

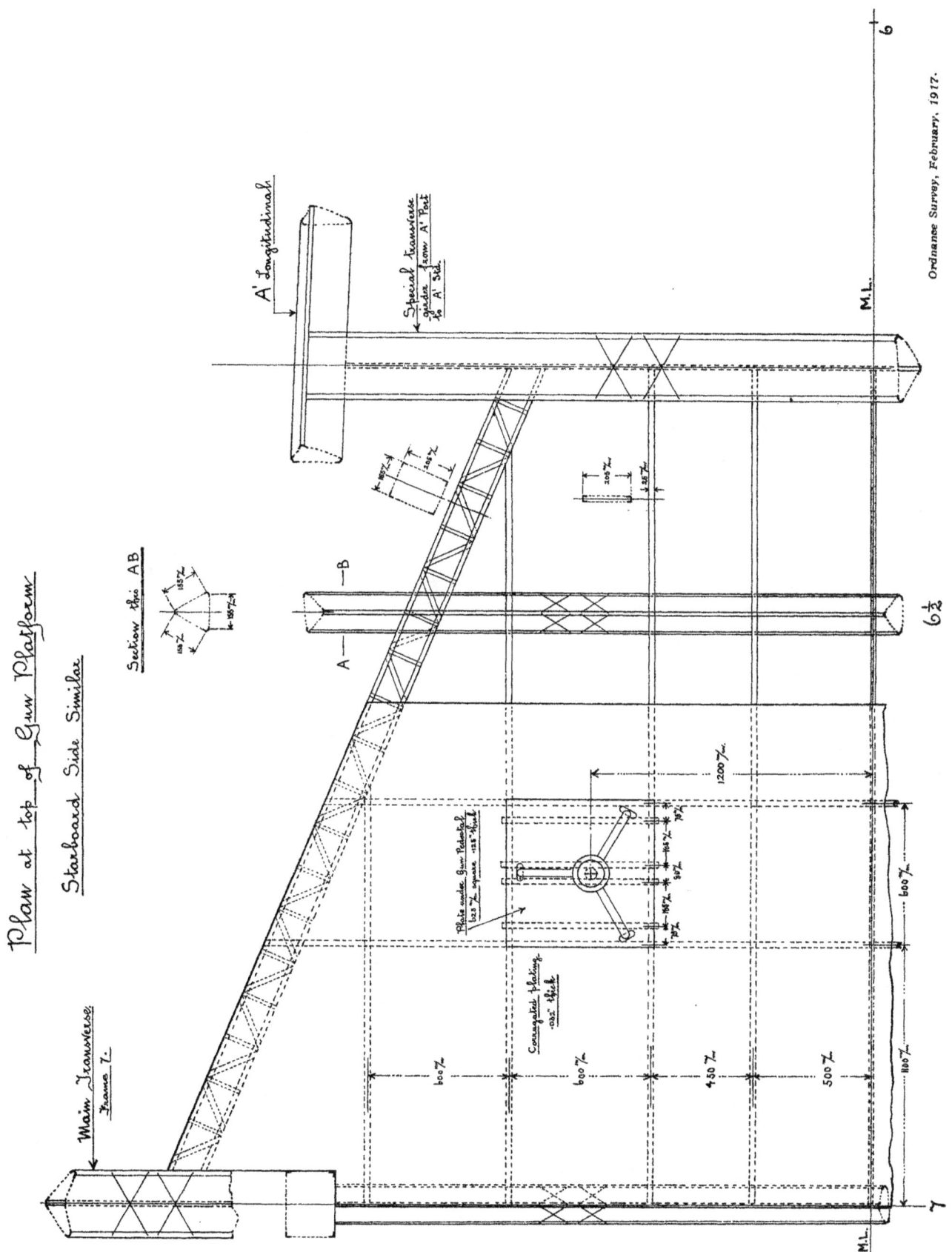

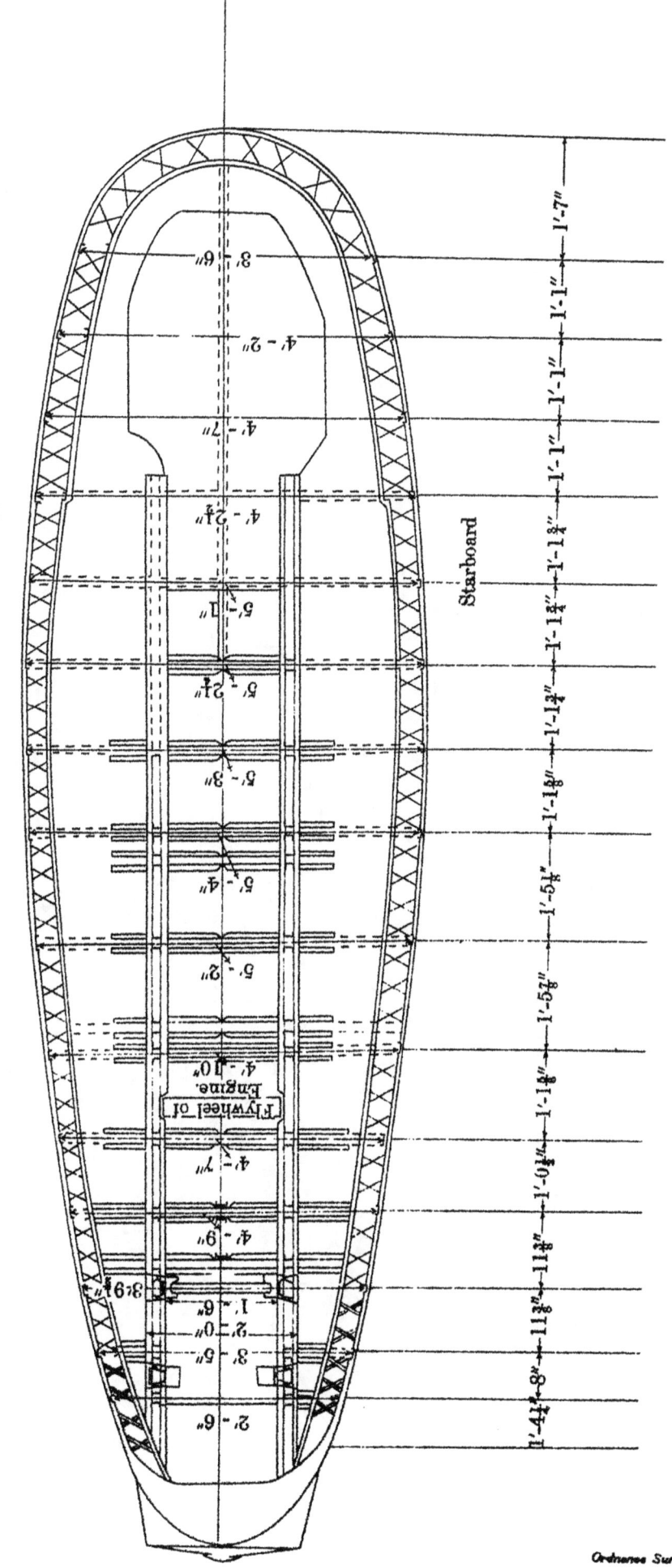

PLAN OF WING CARS.
(WITH FABRIC REMOVED).
Fig. 131.

C.B. 1265.
Plate 44.
See Chapt. III.

Ordnance Survey. February, 1917.

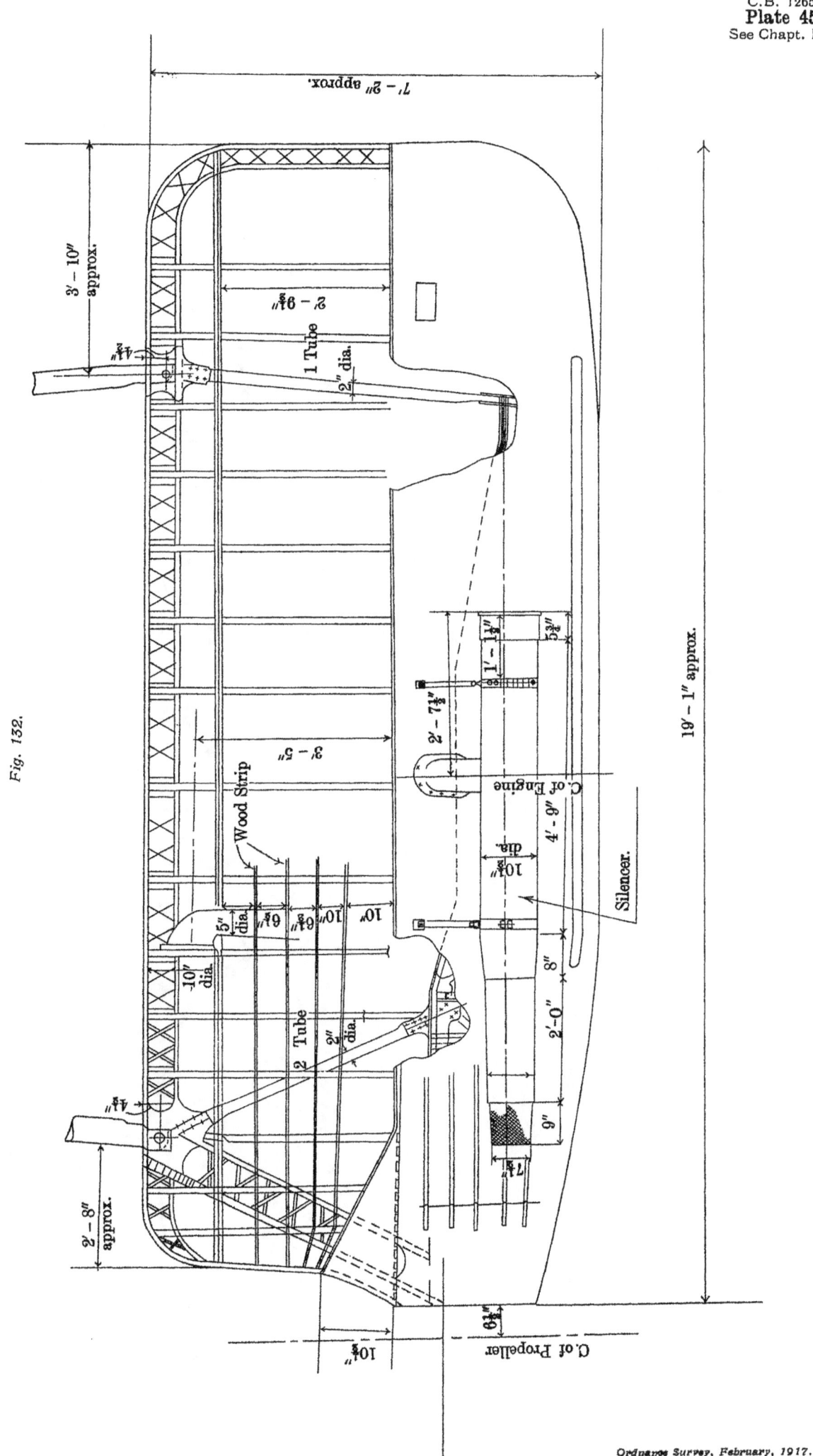

ELEVATION OF WING CARS.
Fig. 132.

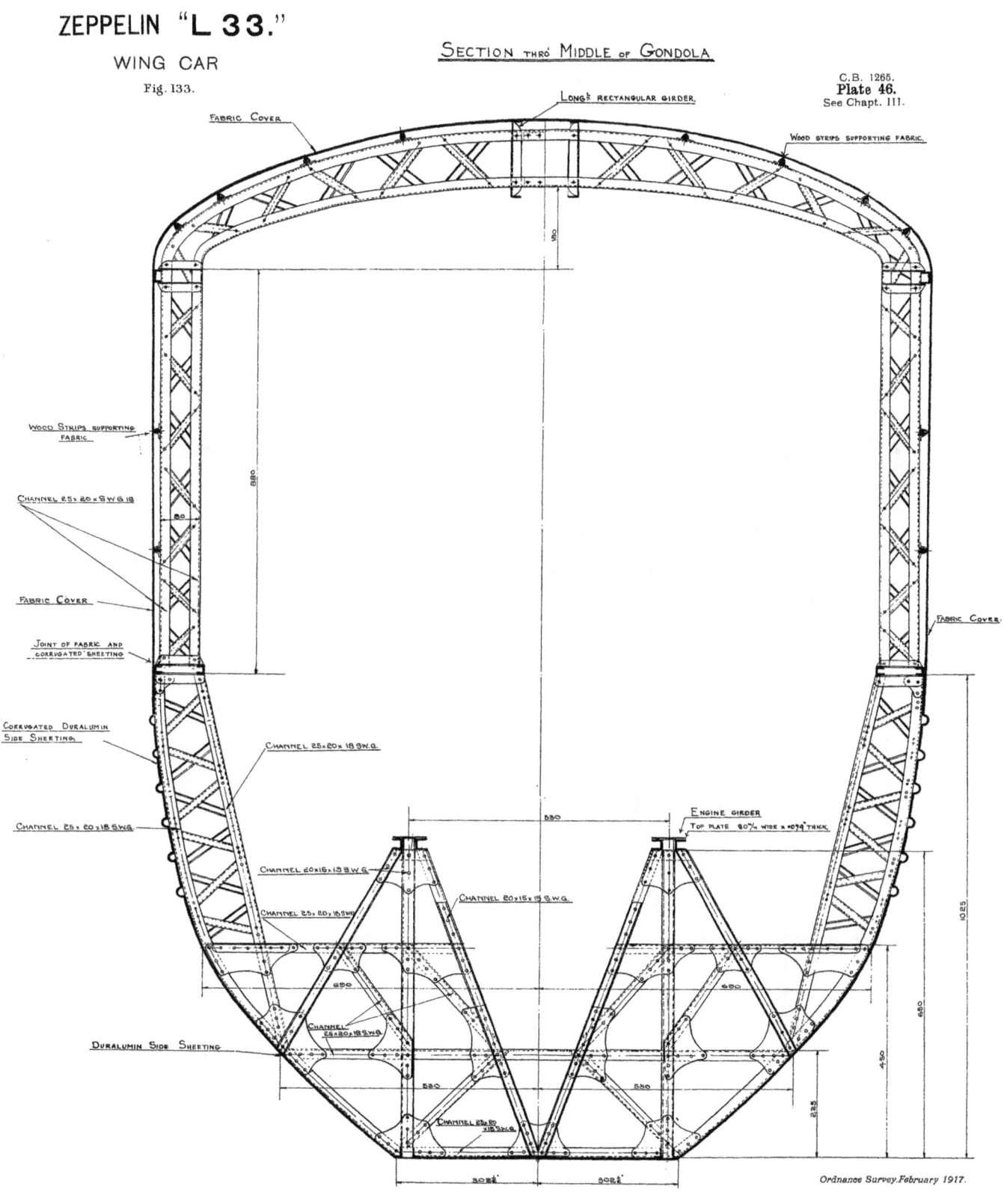

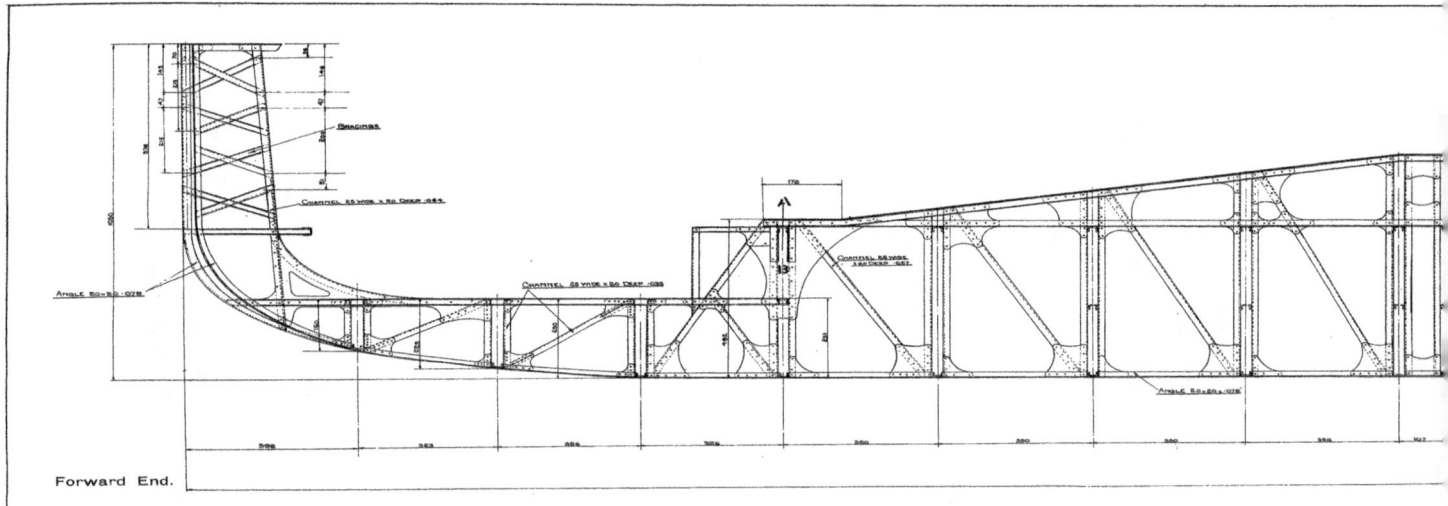

DETAILS OF
FORE AND AFT GIRDER FOR WING CAR OF ZEPPELIN "L 33."
Fig. 134.
DIMENSIONS IN MILLIMETRES.
THICKNESSES IN INCHES.

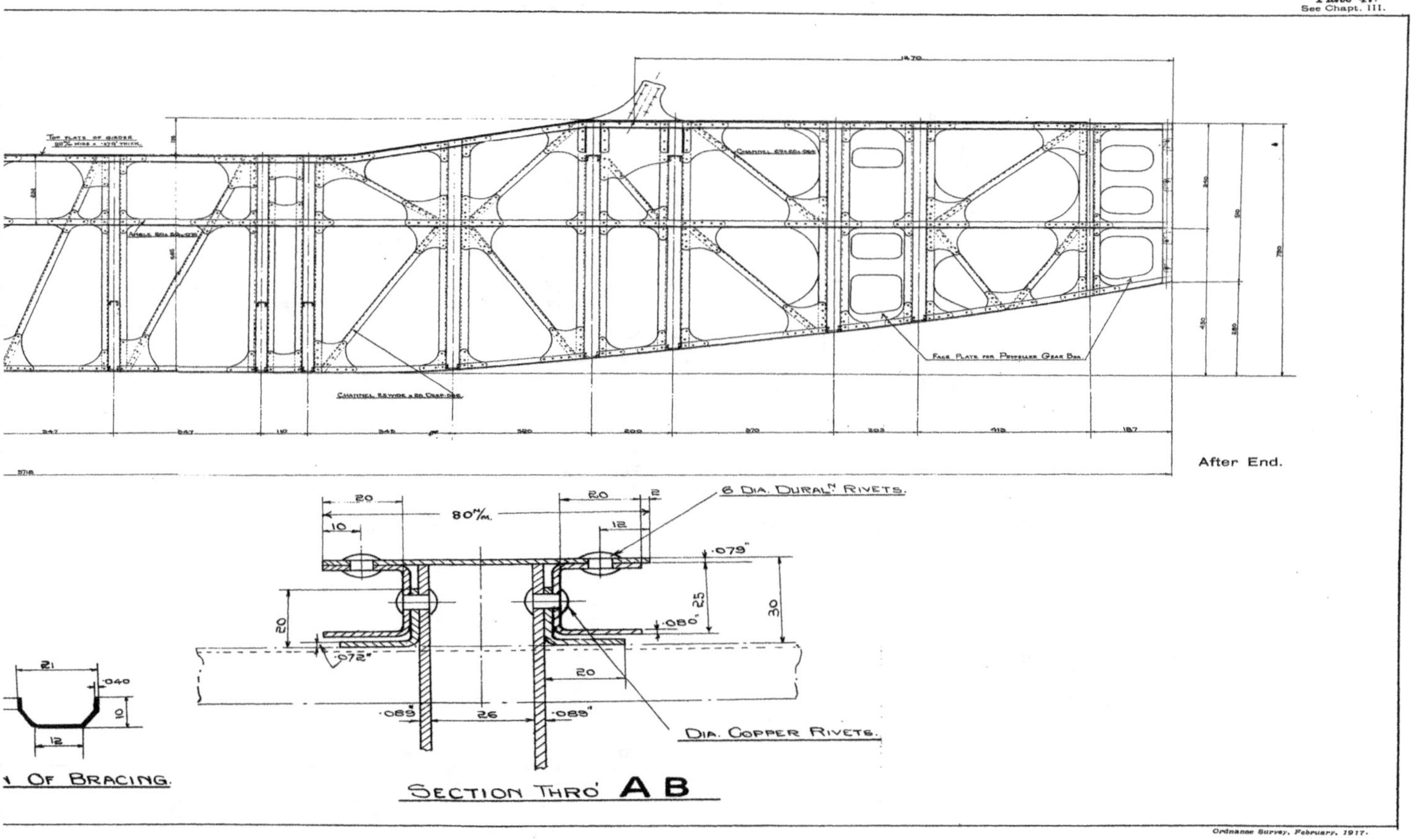

C.B. 1265.
**Plate 47.**
See Chapt. III.

After End.

SECTION THRO' **A B**

OF BRACING.

Ordnance Survey, February, 1917.

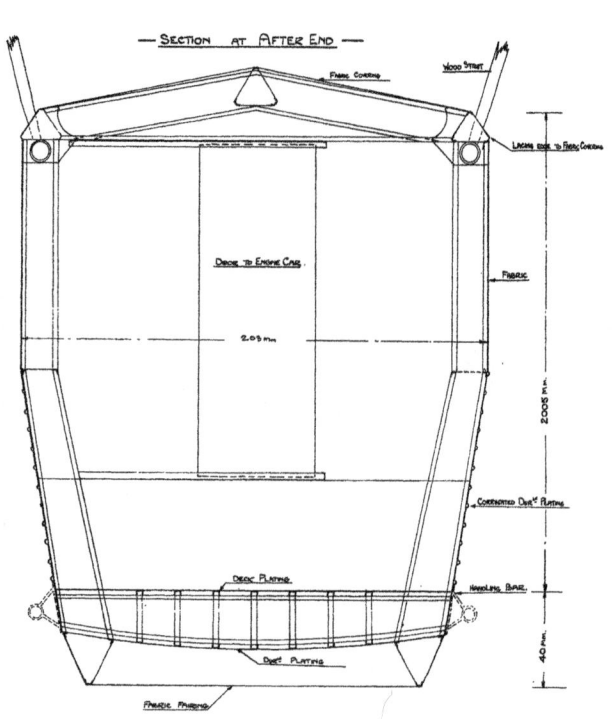
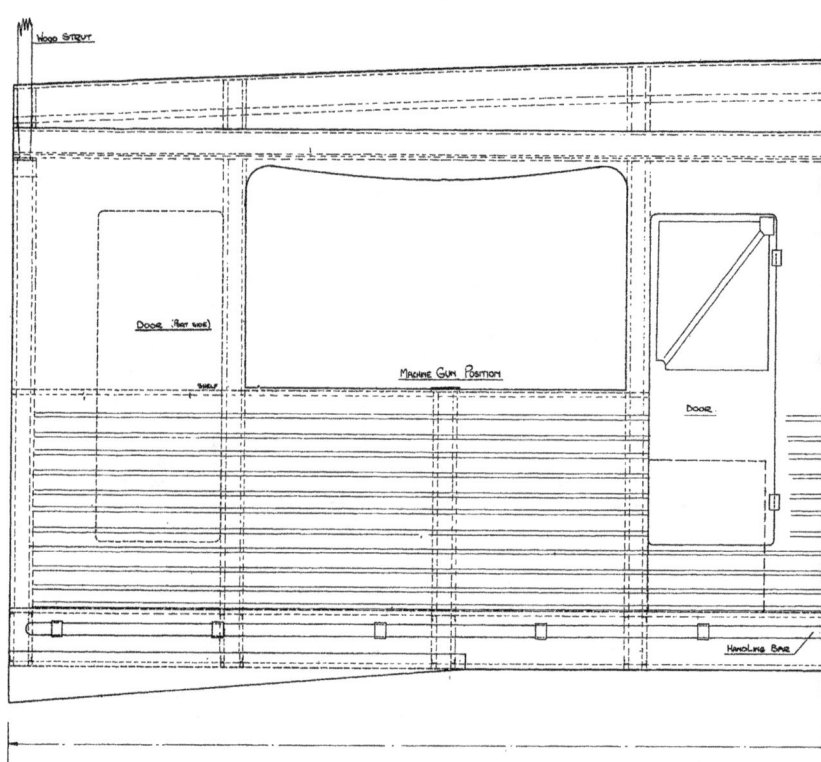
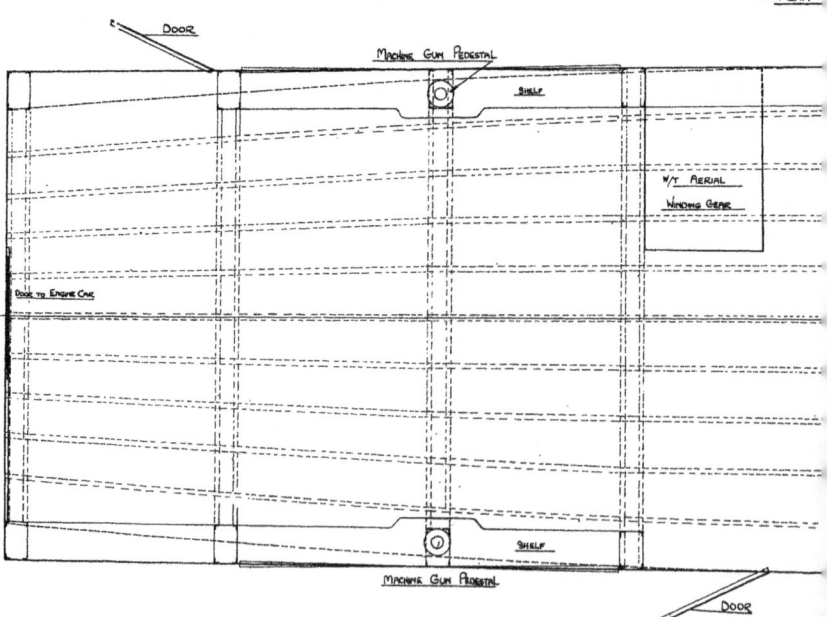

**ZEPPELIN "L 33."**
CONTROL CAR.
*Fig. 135.*

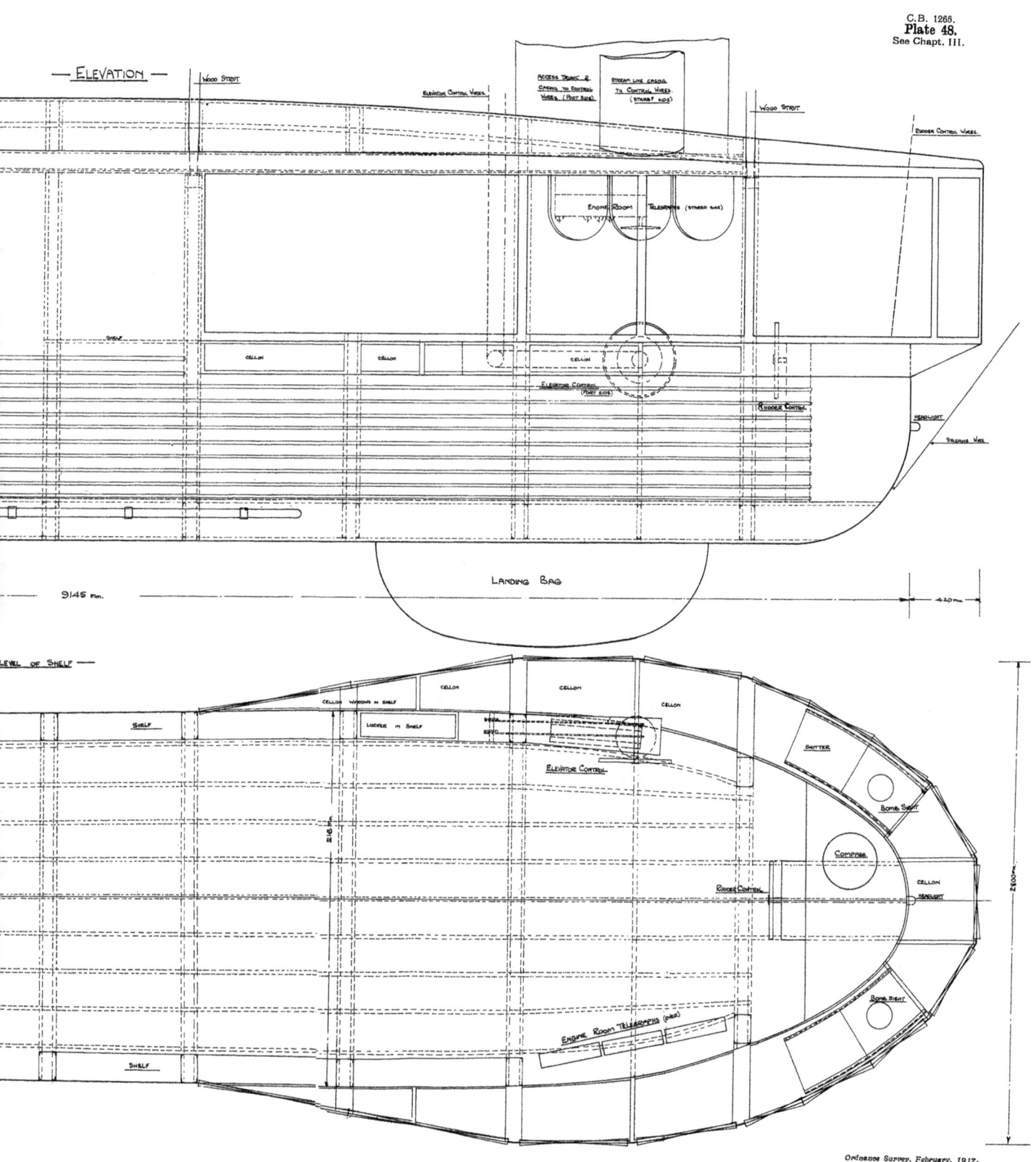

# ZEPPELIN "L 33."
## BRACKETS SUPPORTING WING PROPELLERS.
Fig. 136.

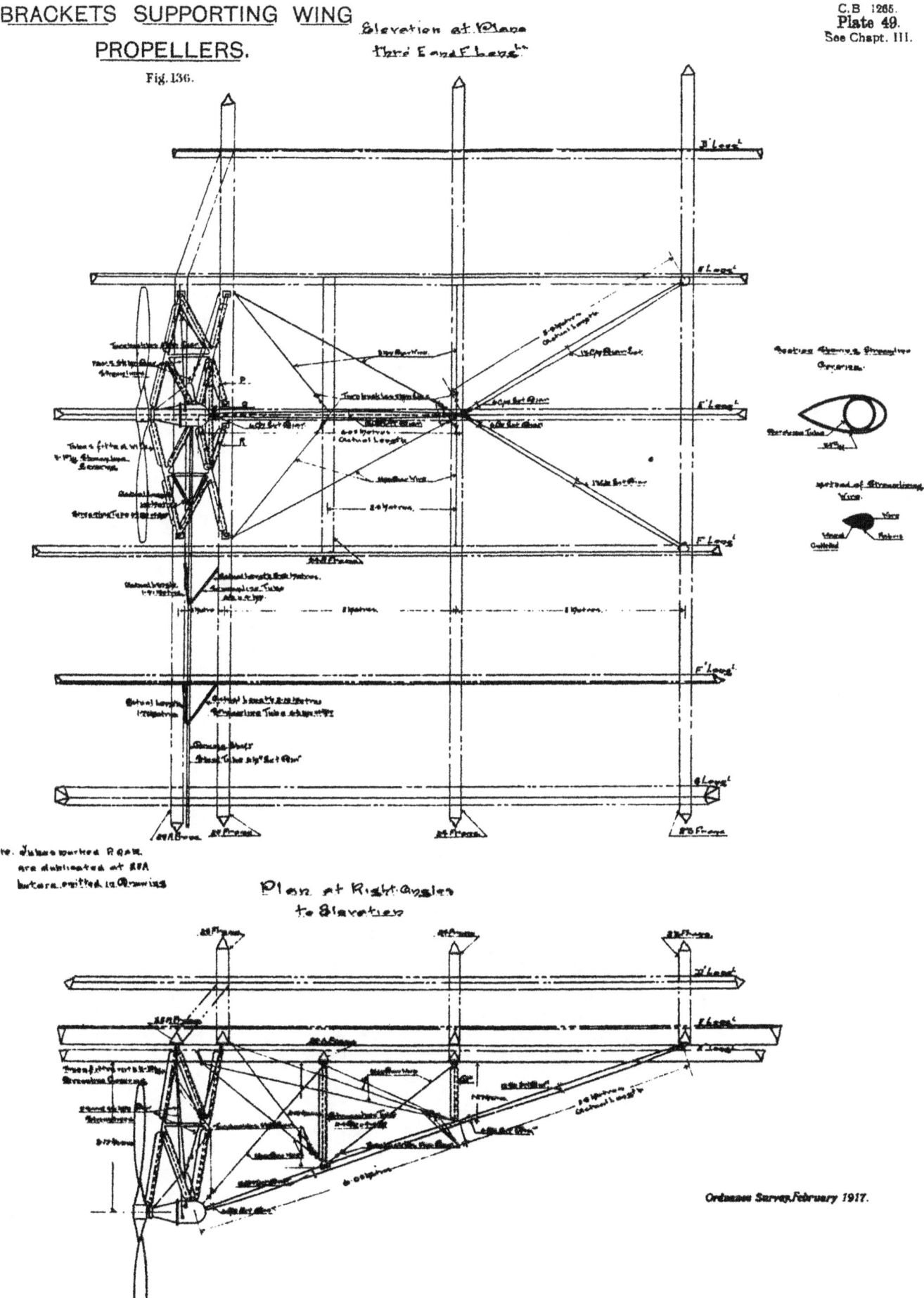

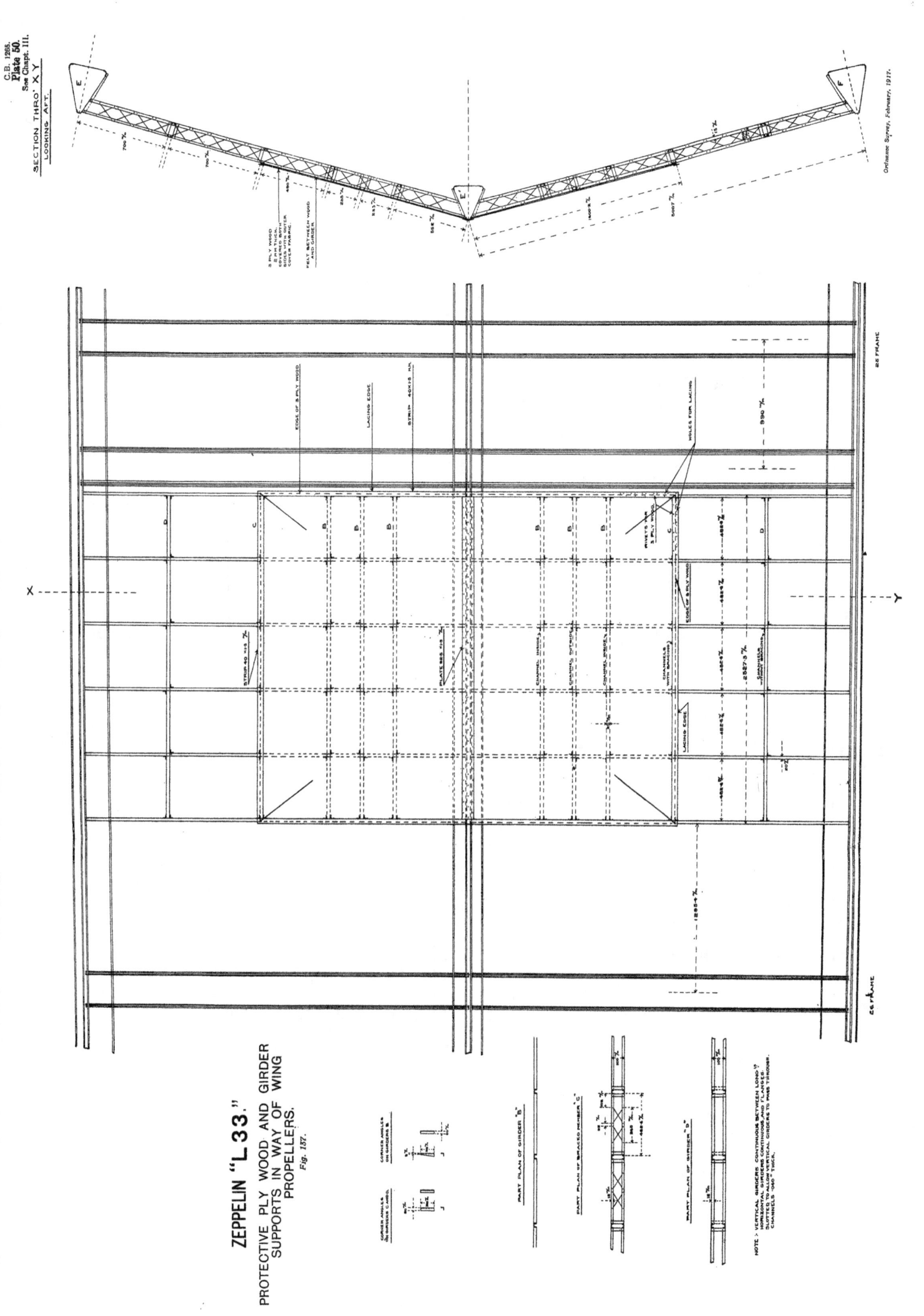

# SCHÜTTE–LANZ AIRSHIP.
## TYPICAL ARRANGEMENT.
Fig. 138

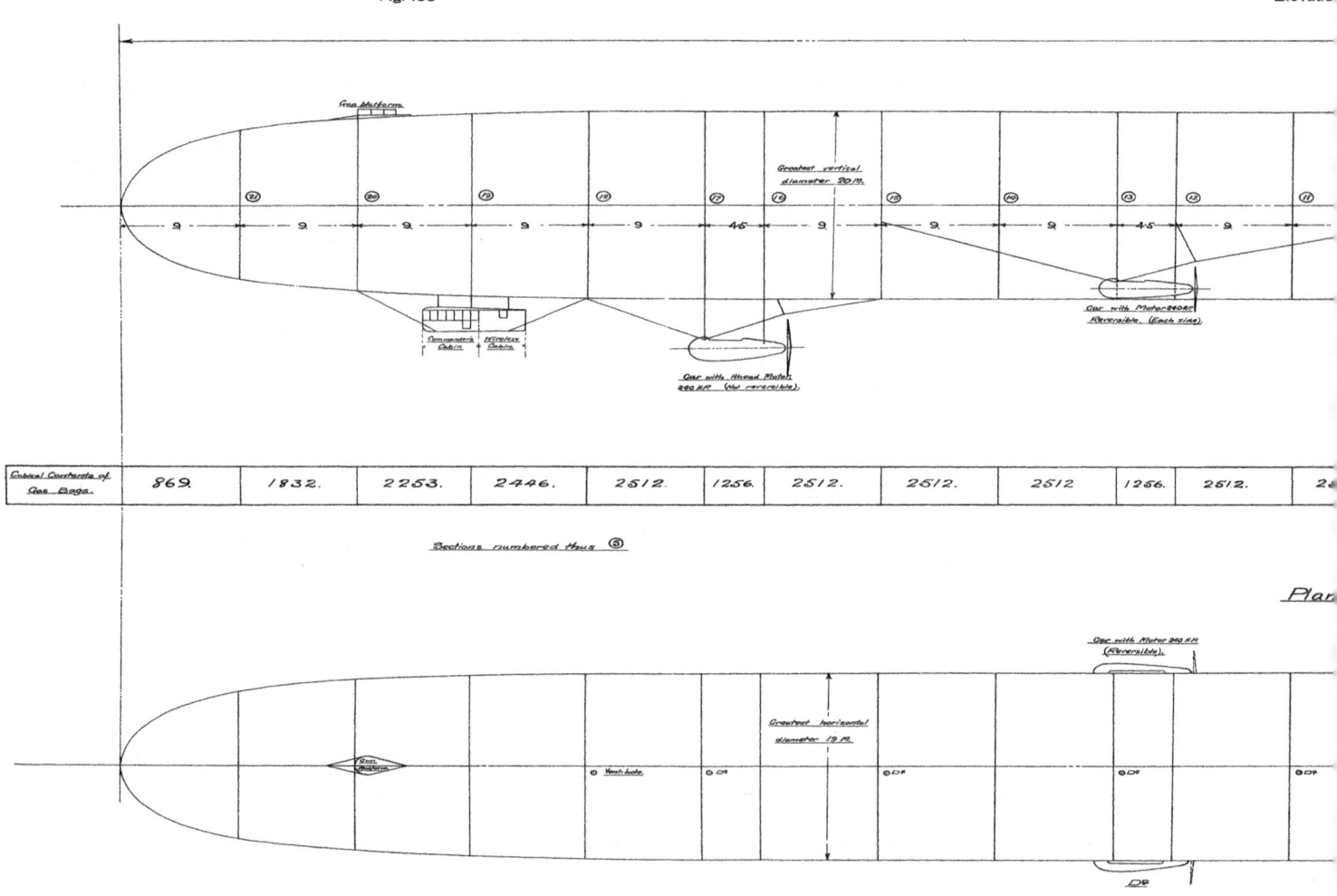

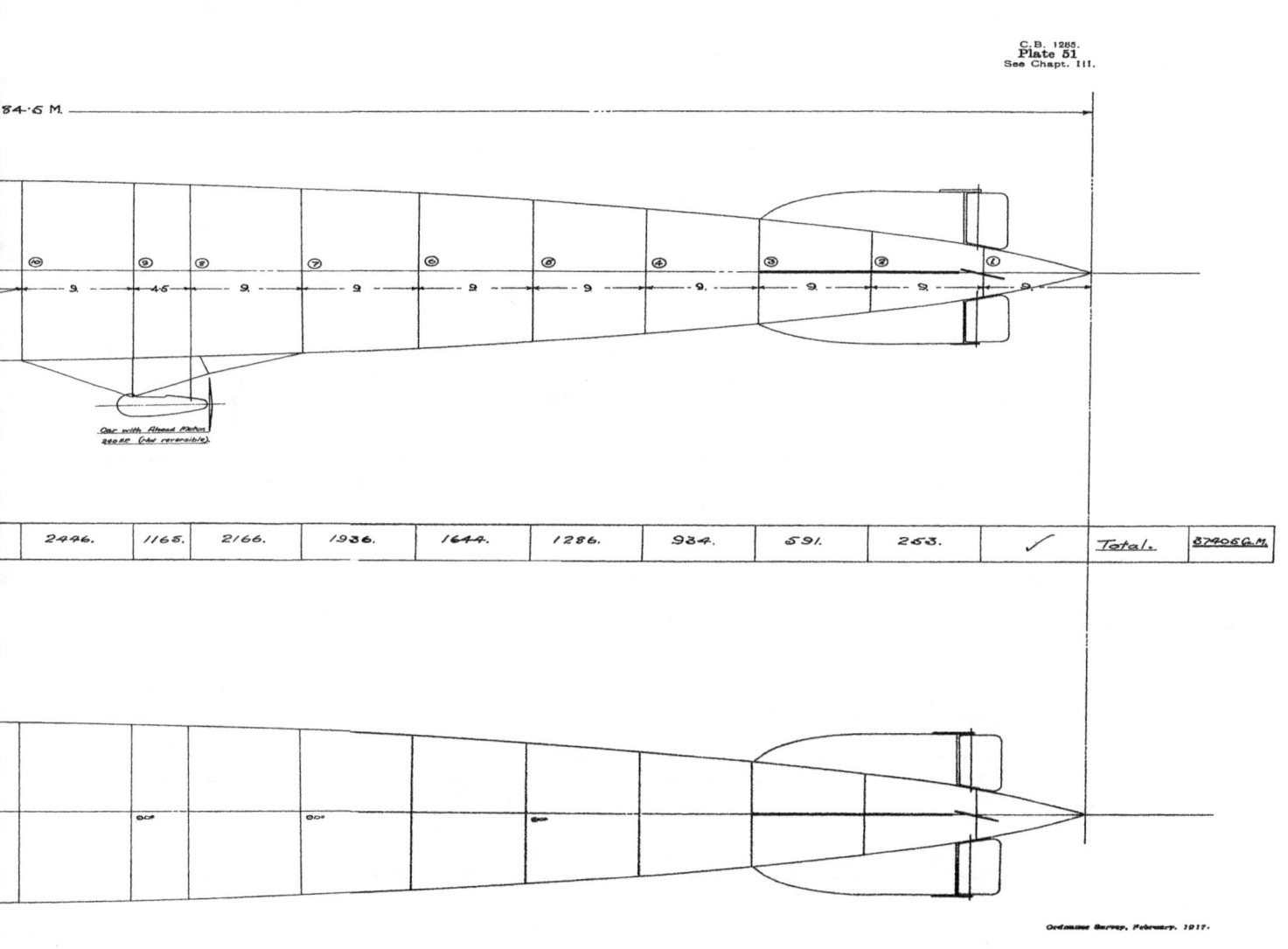

# "SCHÜTTE-LANZ" TYPICAL GIRDERS.

### Fig. 139.

#### LONGITUDINAL GIRDER.

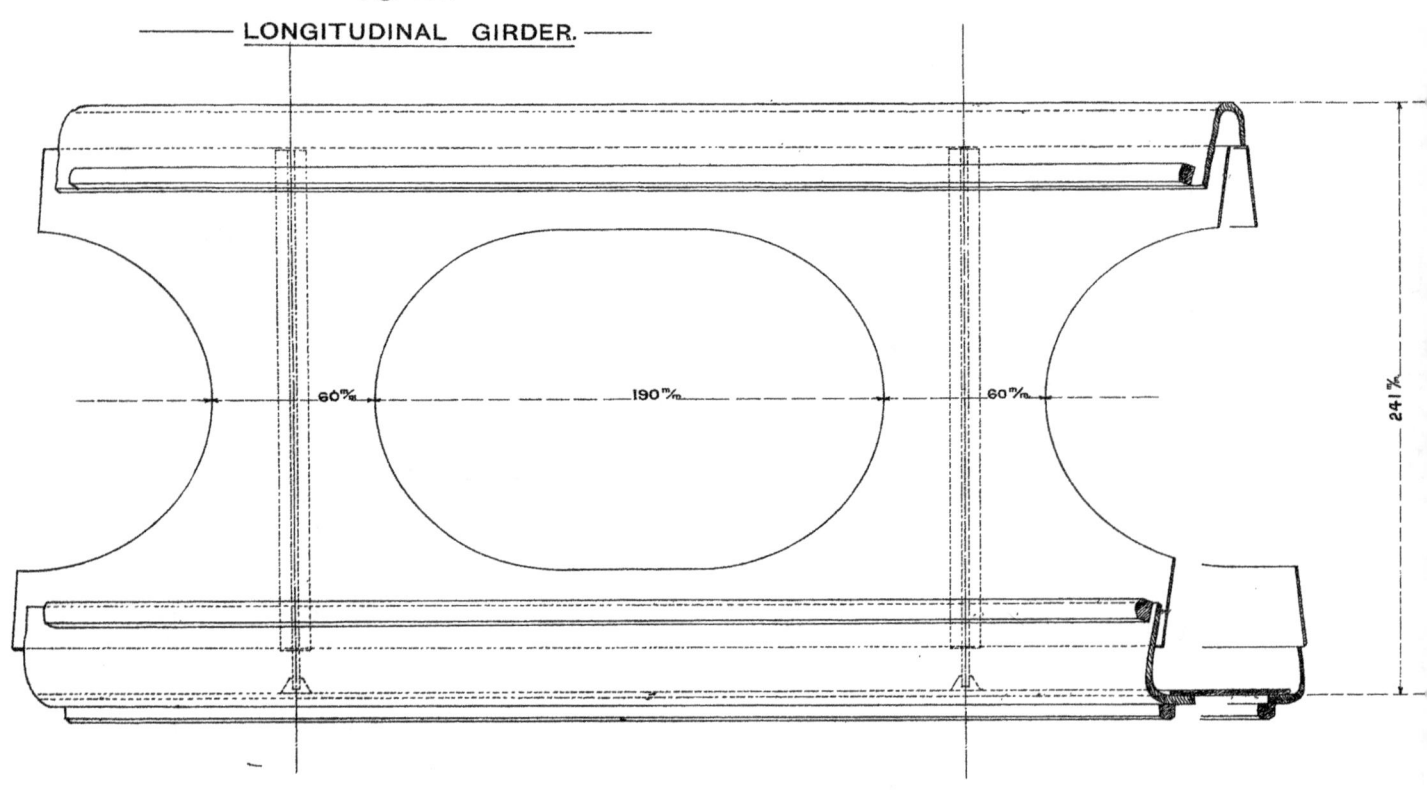

ELEVATION.

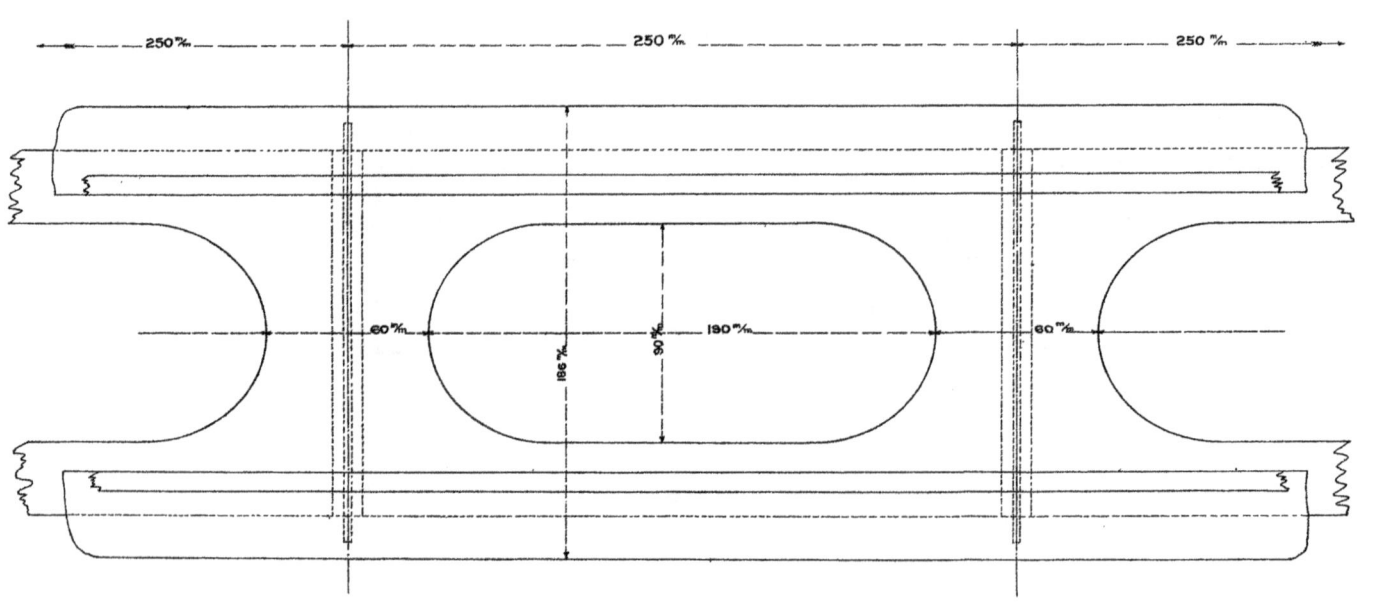

PLAN LOOKING UPWARDS.

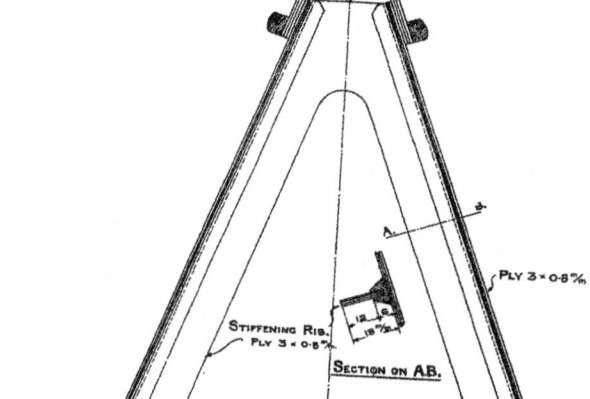

## INTERMEDIATE TRANSVERSE GIRDER.

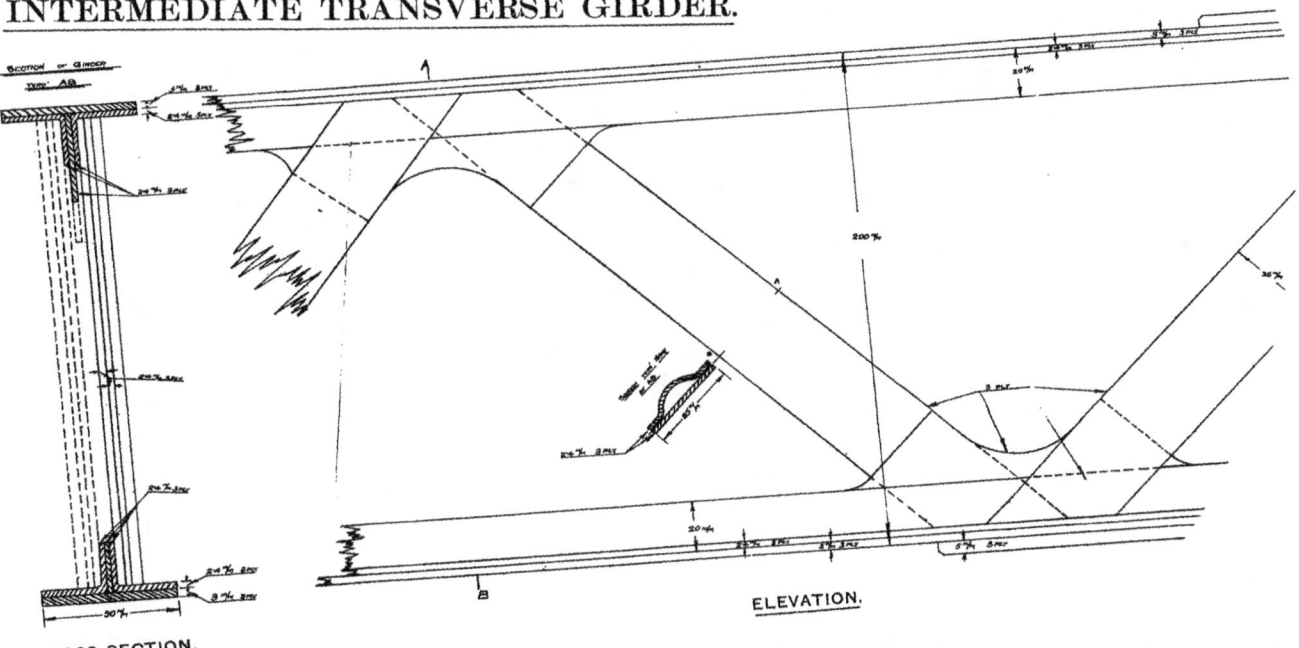

# "SCHÜTTE-LANZ"
## Details of Longitudinal Box Girders.
### Fig. 140.

Plan of Underside (looking in direction of Arrow on Section).

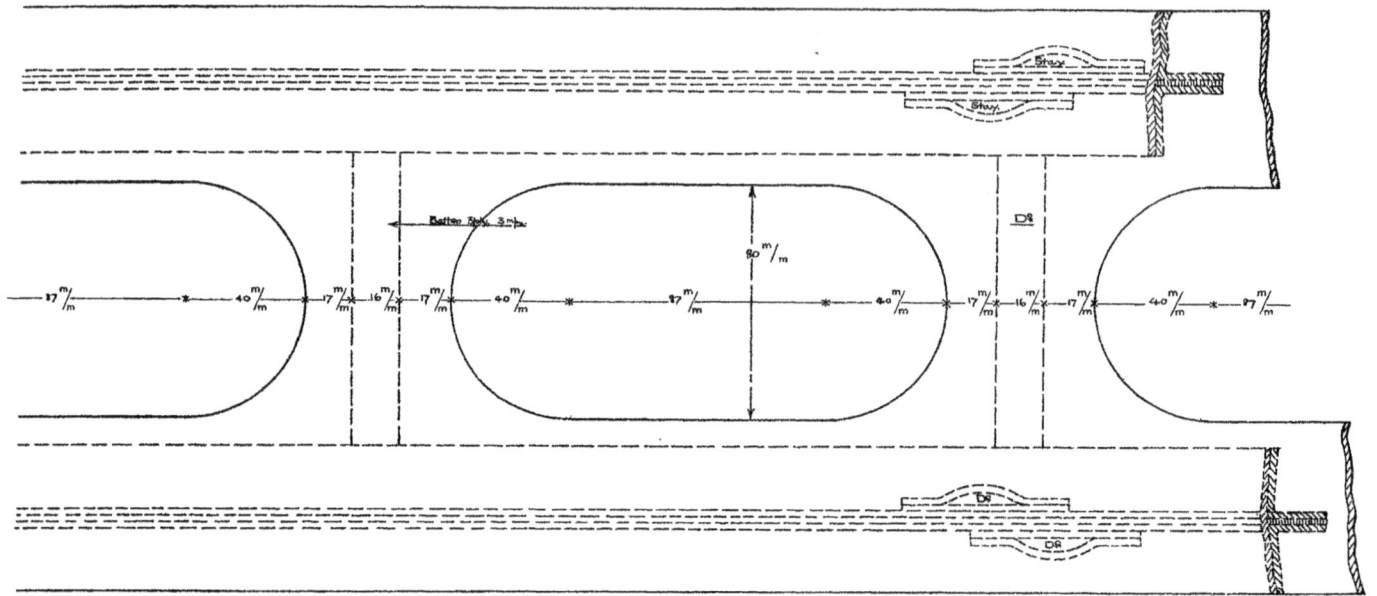

Elevation.

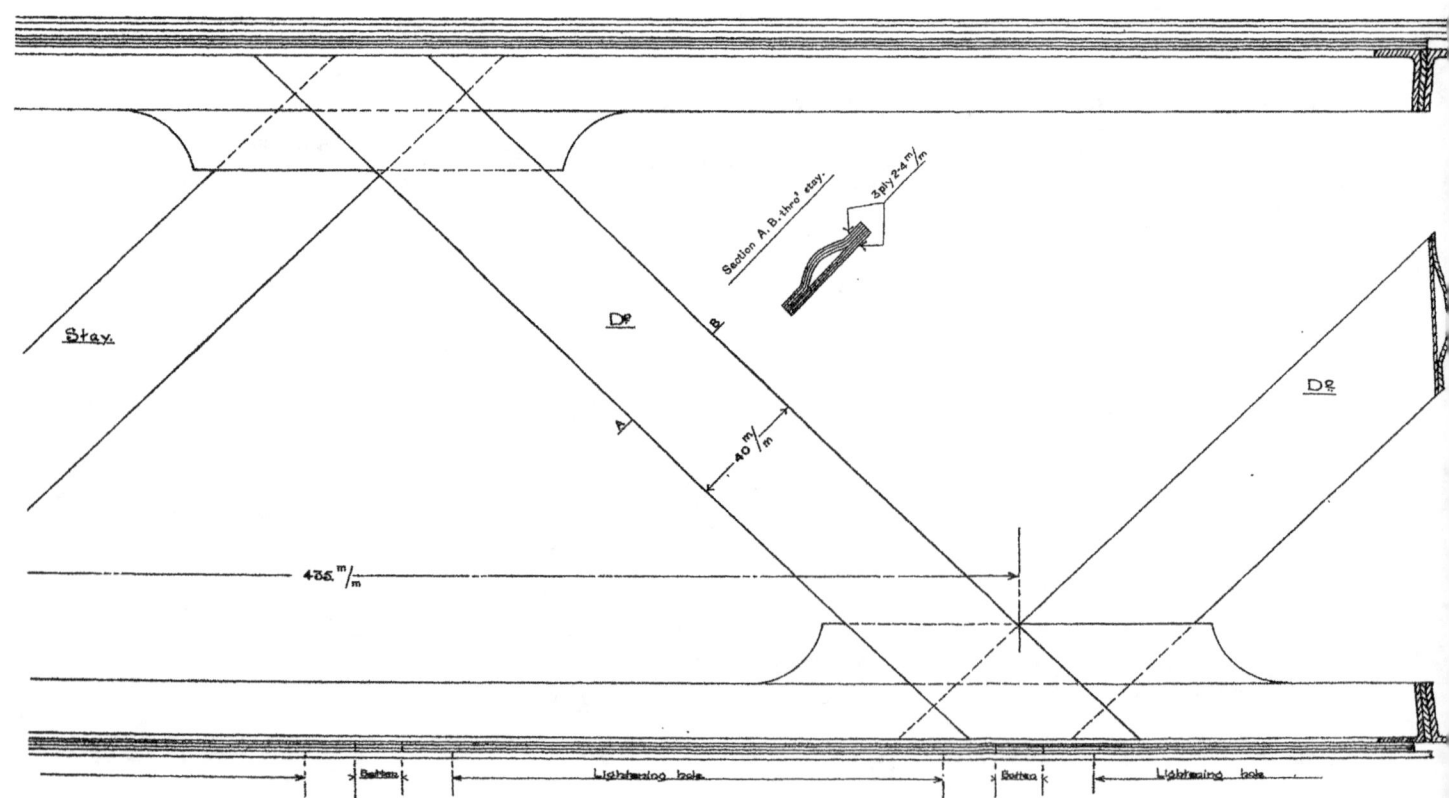

C.B. 1265.
**Plate 53**
See Chapt. III.

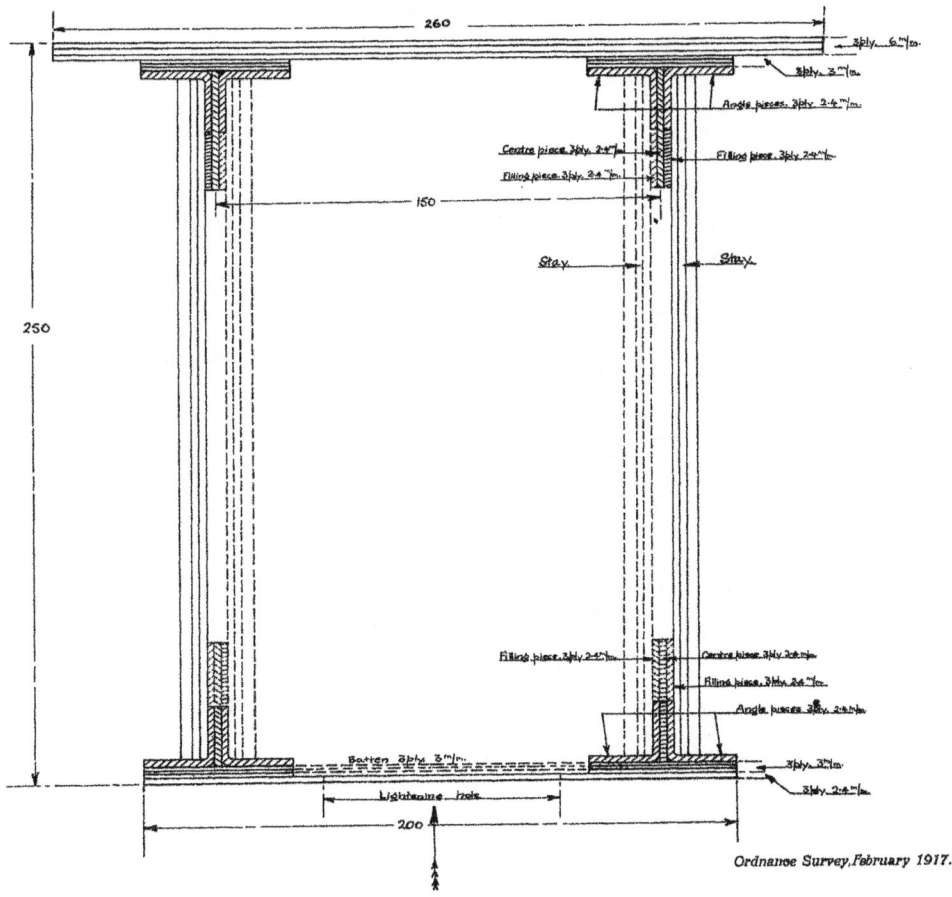

Ordnance Survey, February 1917.

C.B. 1265. **Plate 54.** See Chapt. IV.

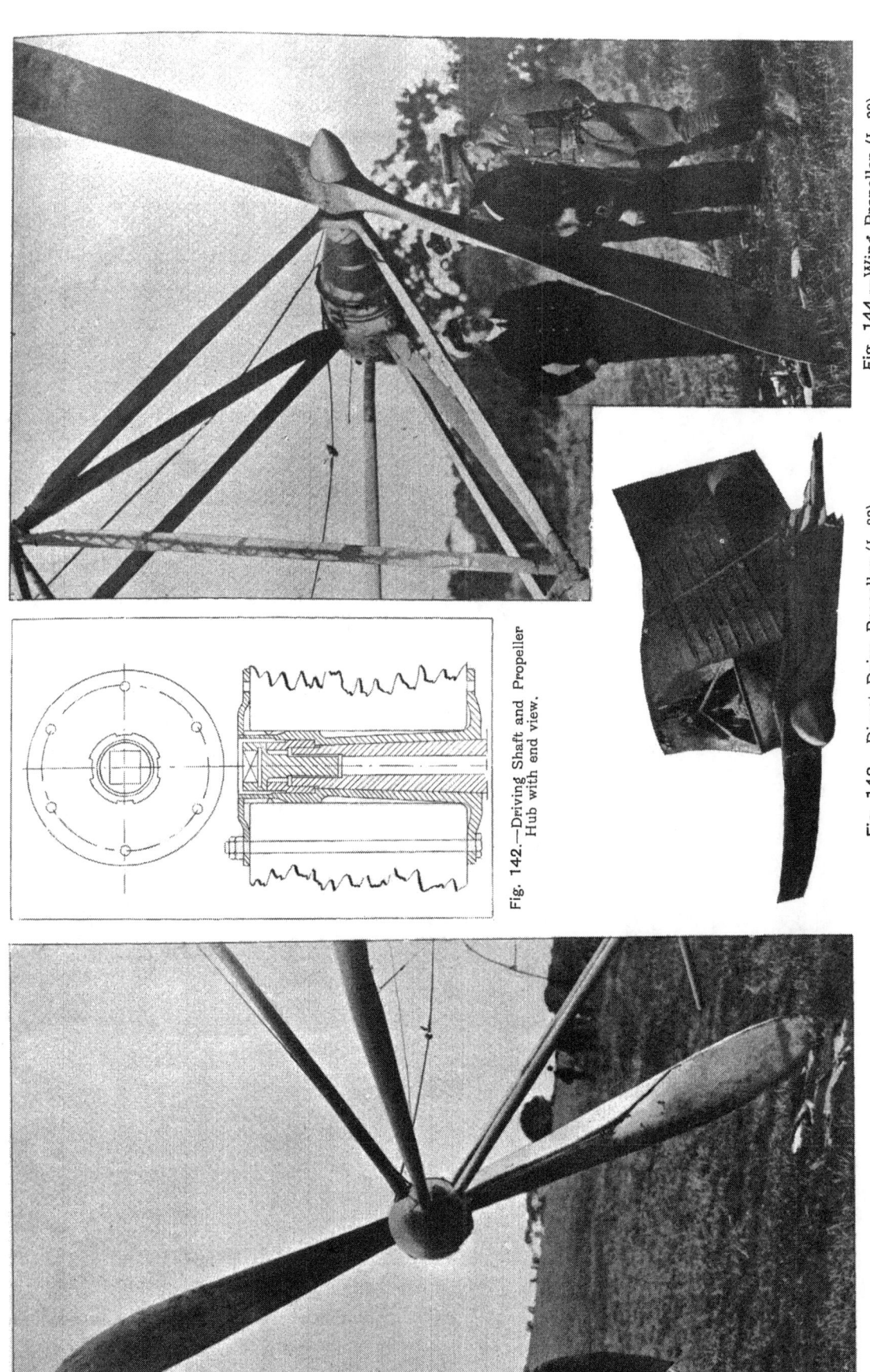

Fig. 144.—Wing Propeller (L 33).

Fig. 143.—Direct Drive Propeller (L 33).

Fig. 142.—Driving Shaft and Propeller Hub with end view.

Fig. 141.—Wing Propeller (L 33).

Ordnance Survey, February, 1917.

C.B. 1265.
**Plate 55.**
See Chapt. IV.

## VIEWS OF ENGINES—L 33.

Fig. 145.

Fig. 146.

Fig. 147.

Fig. 148.

Fig. 149.

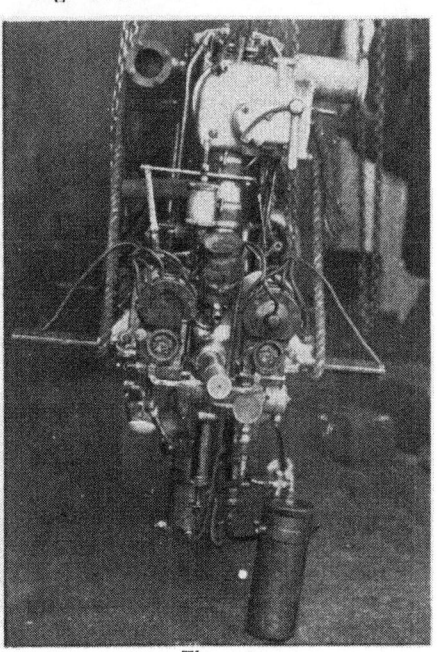

Fig. 150.

*Ordnance Survey, February, 1917.*

Fig. 151.
Top View of Cylinder (L 33).

Fig. 152.
Cylinder (L 33).

Fig. 153.
Piston (L 33).

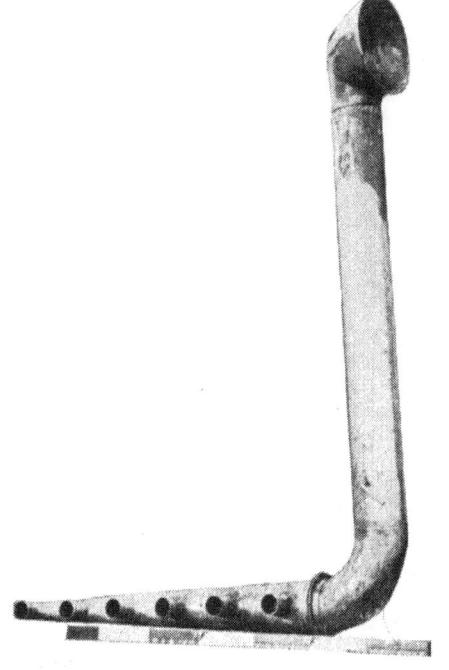

Fig. 154.
Crank Case Ventilator Outlet (L 33).

Fig. 155.
Connecting Rod (L 33).

C.B. 1265.
**Plate 57.**
See Chapt. IV.

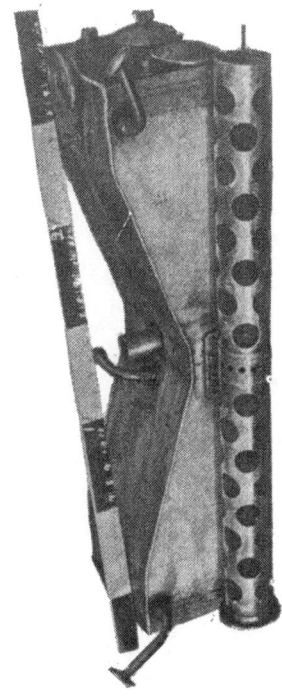

Fig. 157.
Oil Sump from under Engine (L 33).

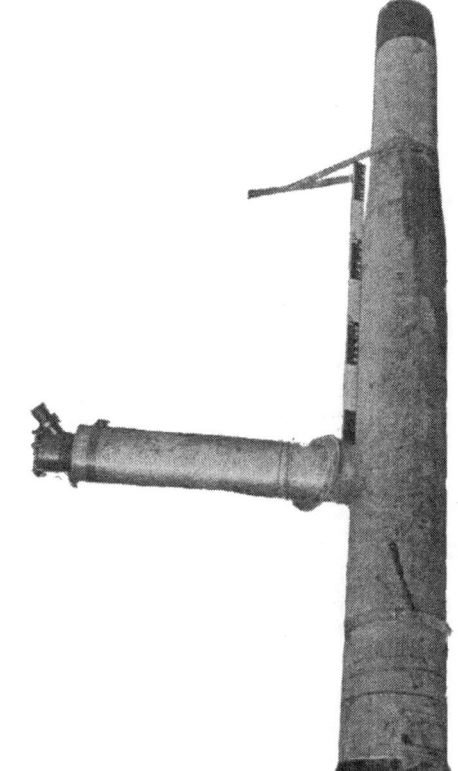

Fig. 159.
Silencer (L 33).

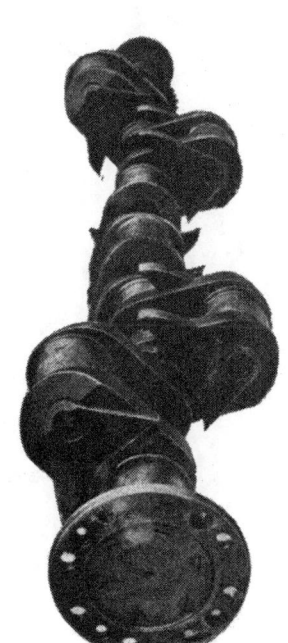

Fig. 156.
Crank Shaft (L 33) Showing Lubricators.

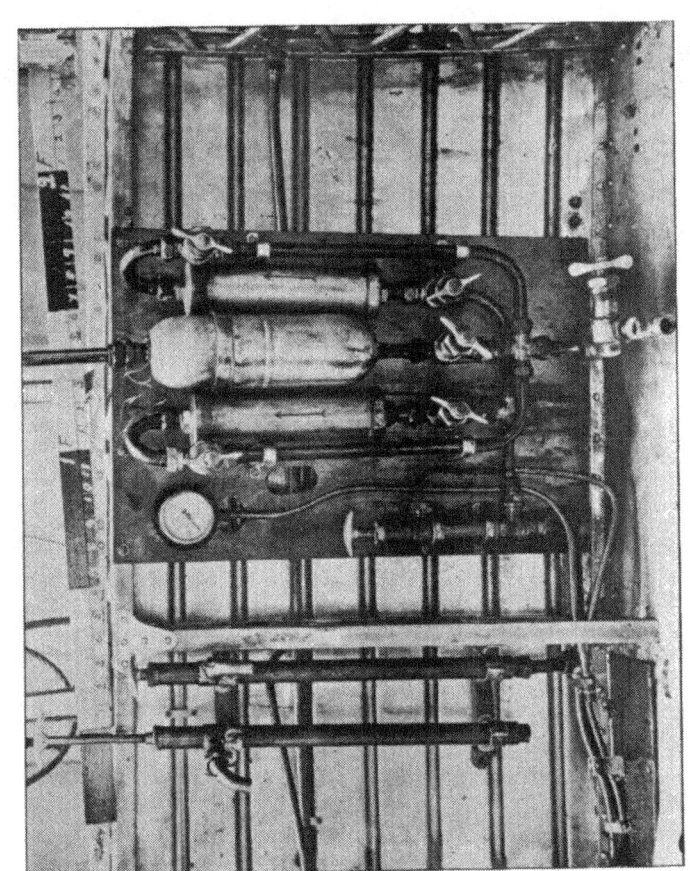

Fig. 158.
Oil Filters (L 33).

Ordnance Survey, February, 1917.

C.B. 1265
Plate 58
See Chapt. IV.

Fig. 160.—After End of Wing Propeller Box.

Fig. 161.—Fore End of Wing Propeller Box.

Fig. 162.—Wing Propeller Box.

Fig. 163.—Reverse Box.

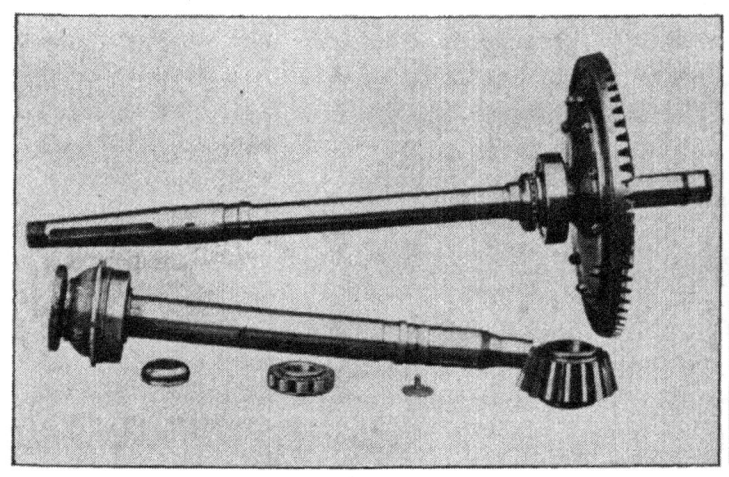

Fig. 164.—Wheel and Pinion of Wing Propeller Box.

Fig. 165.—Wheels and Pinion of Reverse Box.

Ordnance Survey, February, 1917.

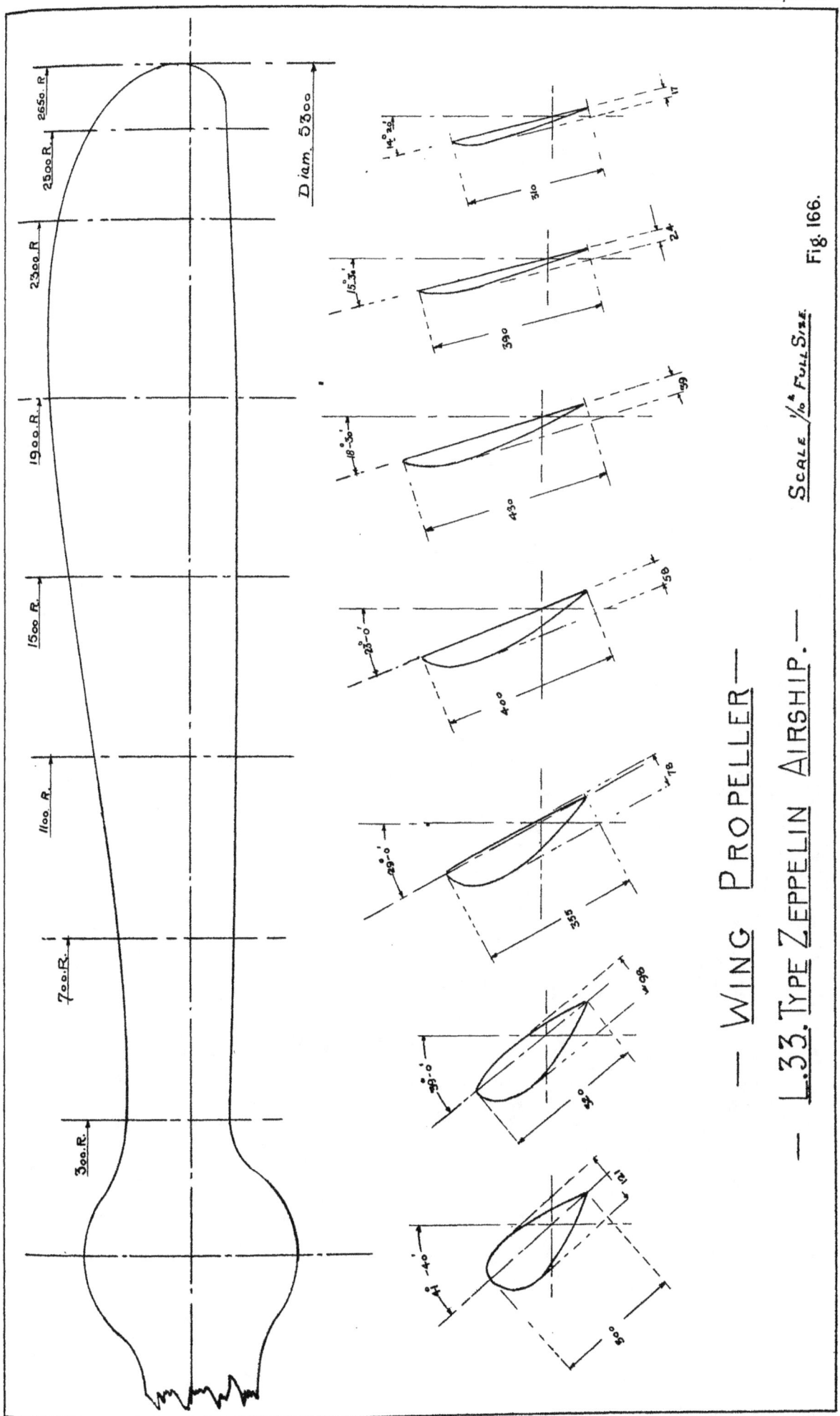

Fig. 166. — WING PROPELLER — L.33. TYPE ZEPPELIN AIRSHIP. — Scale 1/10 Full Size.

C.B. 1265.
**Plate 60.**
See Chapt. IV.

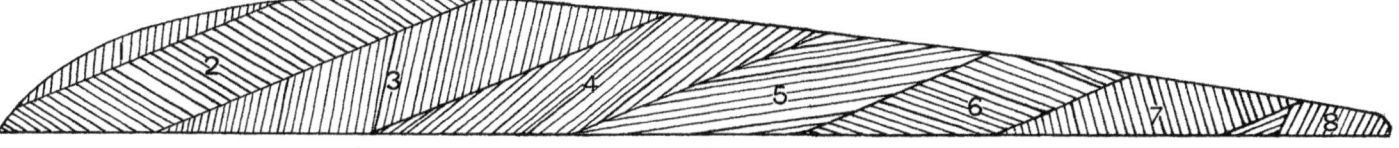

LITTLE WIGBOROUGH BLADE I. FIG. 167.

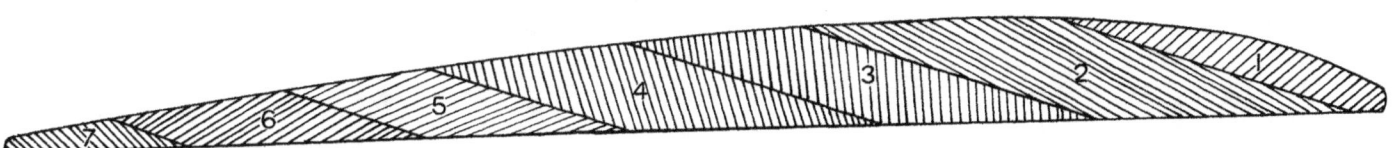

LITTLE WIGBOROUGH BLADE II. FIG. 168.

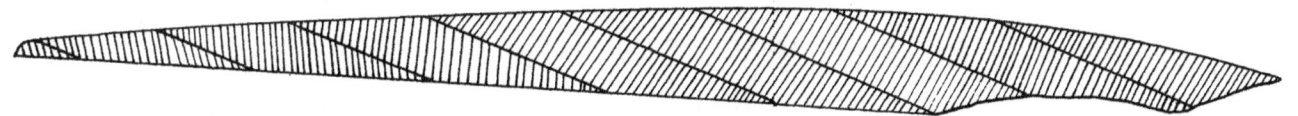

LITTLE WIGBOROUGH BLADE III. FIG. 169.

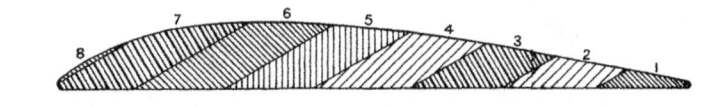

BILLERICAY III. FIG. 170.

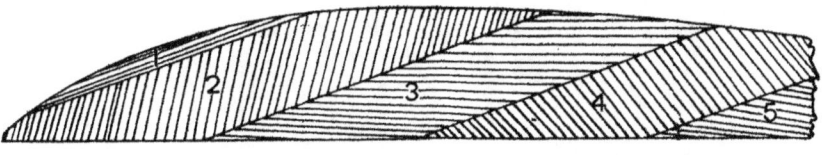

POTTERS BAR BLADE I. FIG. 171.

Ordnance Survey, February, 1917.

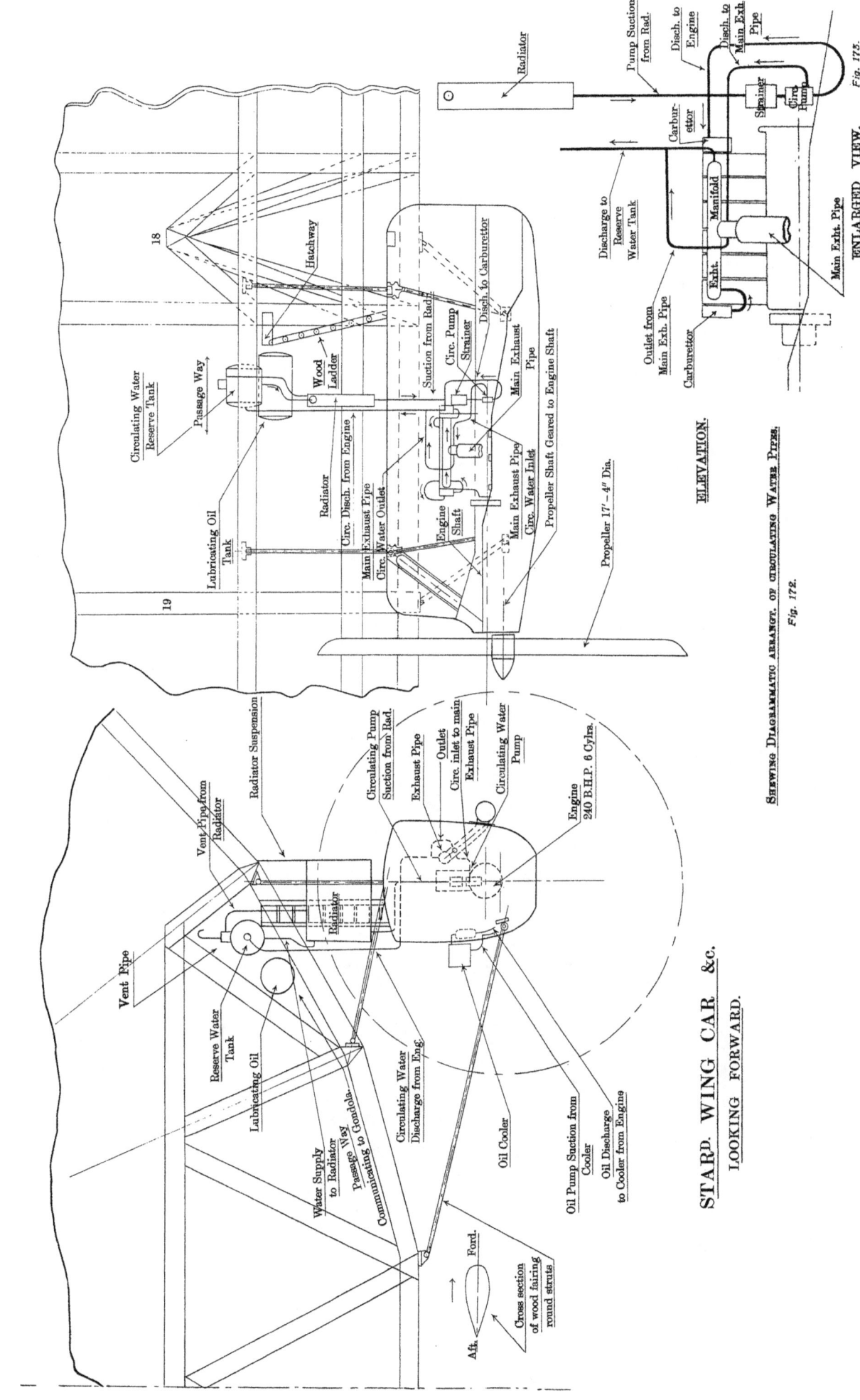

C.B. 1265.
**Plate 62.**
See Chapt. IV.

## TELEGRAPH DIAL IN STAR<sup>D</sup> WING CAR.

*Fig. 174.*

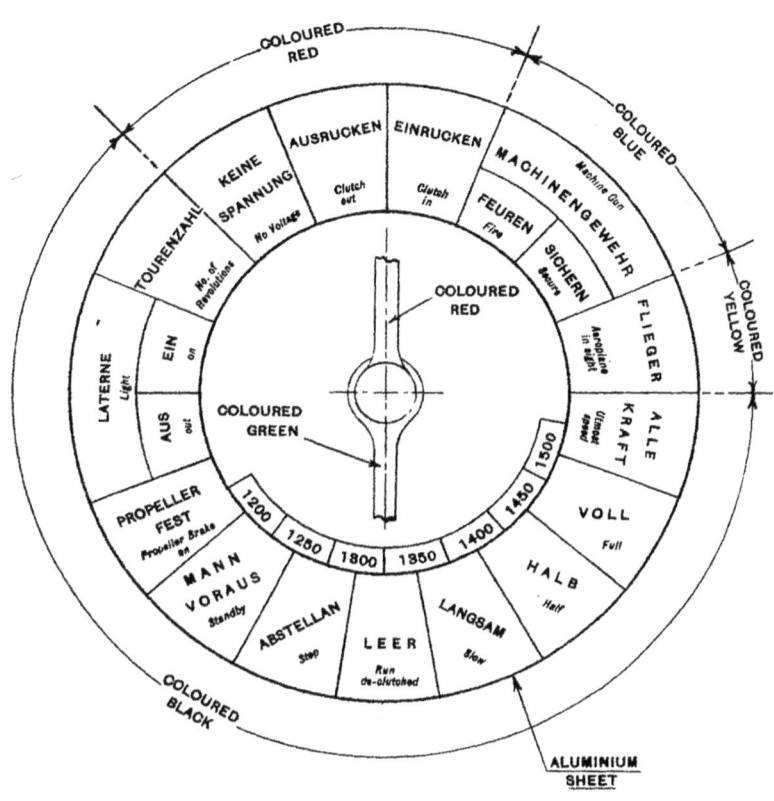

## OIL SUMP UNDER ENGINE CRANK CASE.

CAPACITY 1·18 CUB. FT. (APPROX).

*Fig. 175.*

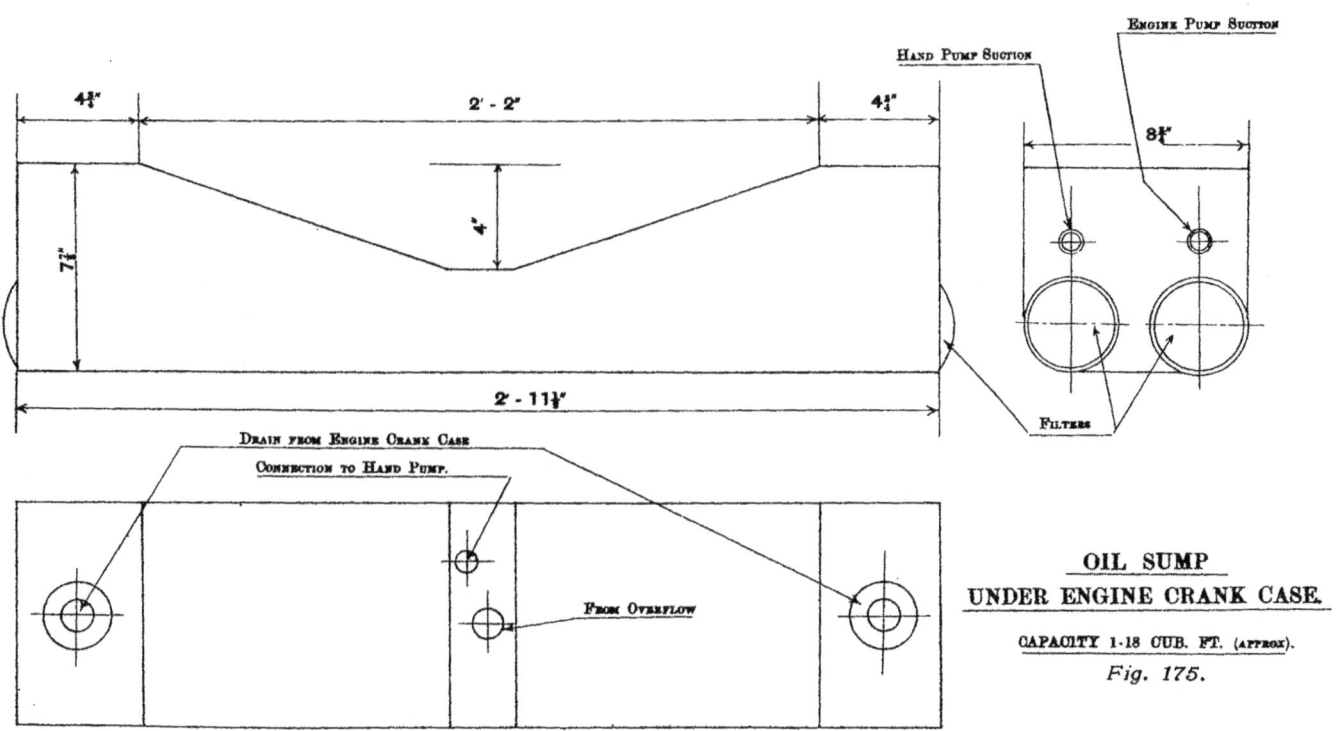

Ordnance Survey, February, 1917.

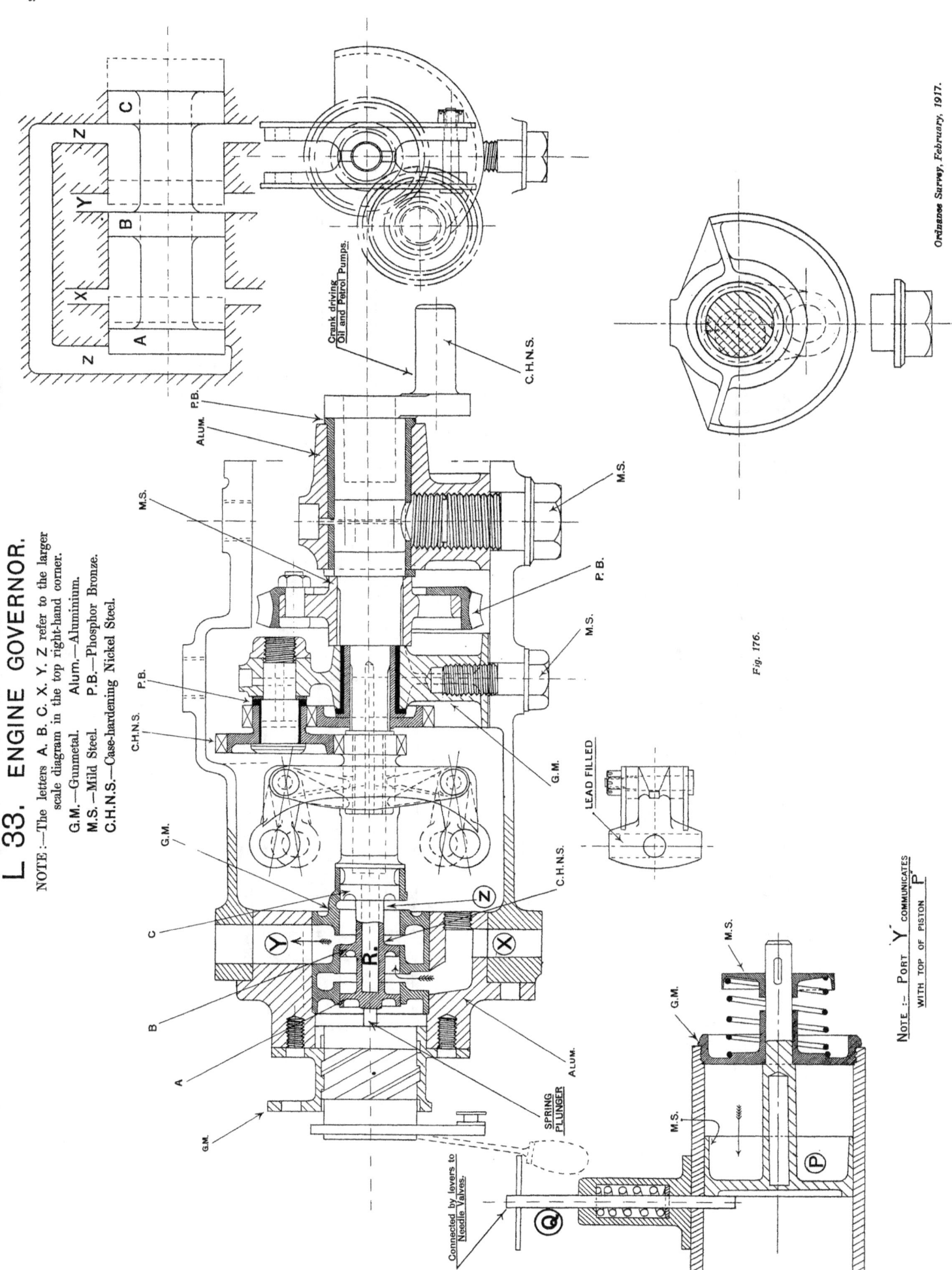

Fig. 176.

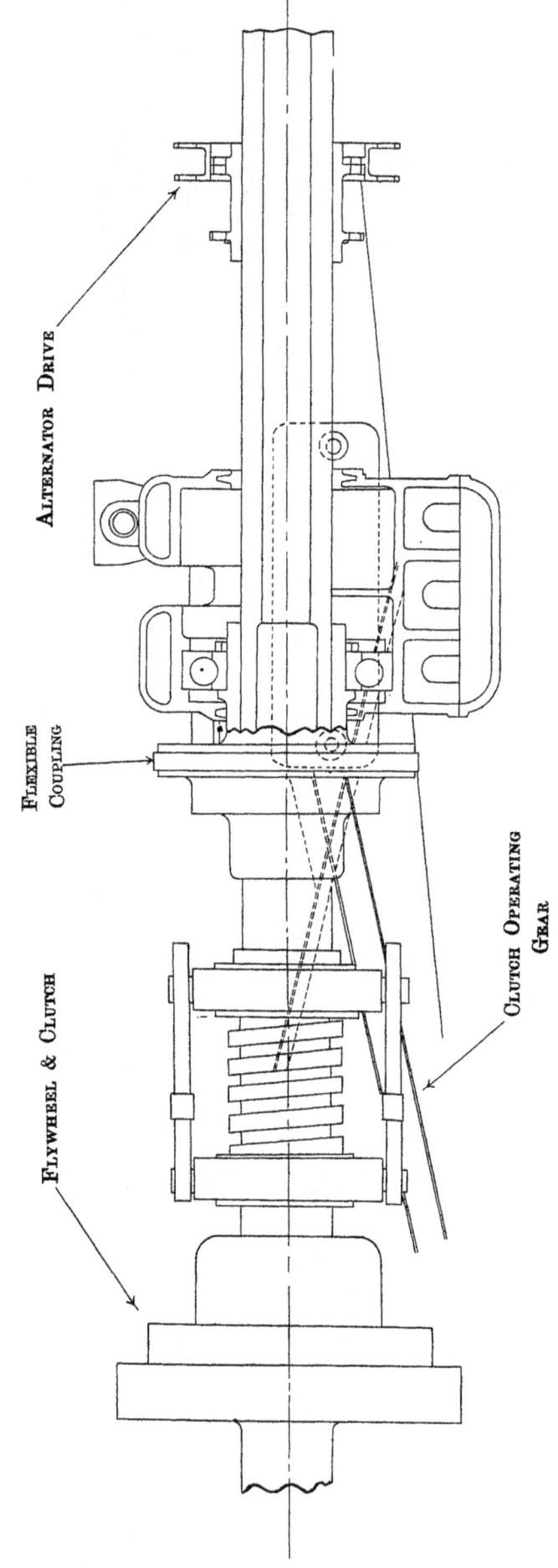

L33.—ARRANGMENT OF TRANSMISSION GEAR FOR NON-REVERSIBLE PROPELLER.

Scale — 3" = 1 Foot.

ELEVATION.

Fig. 177.

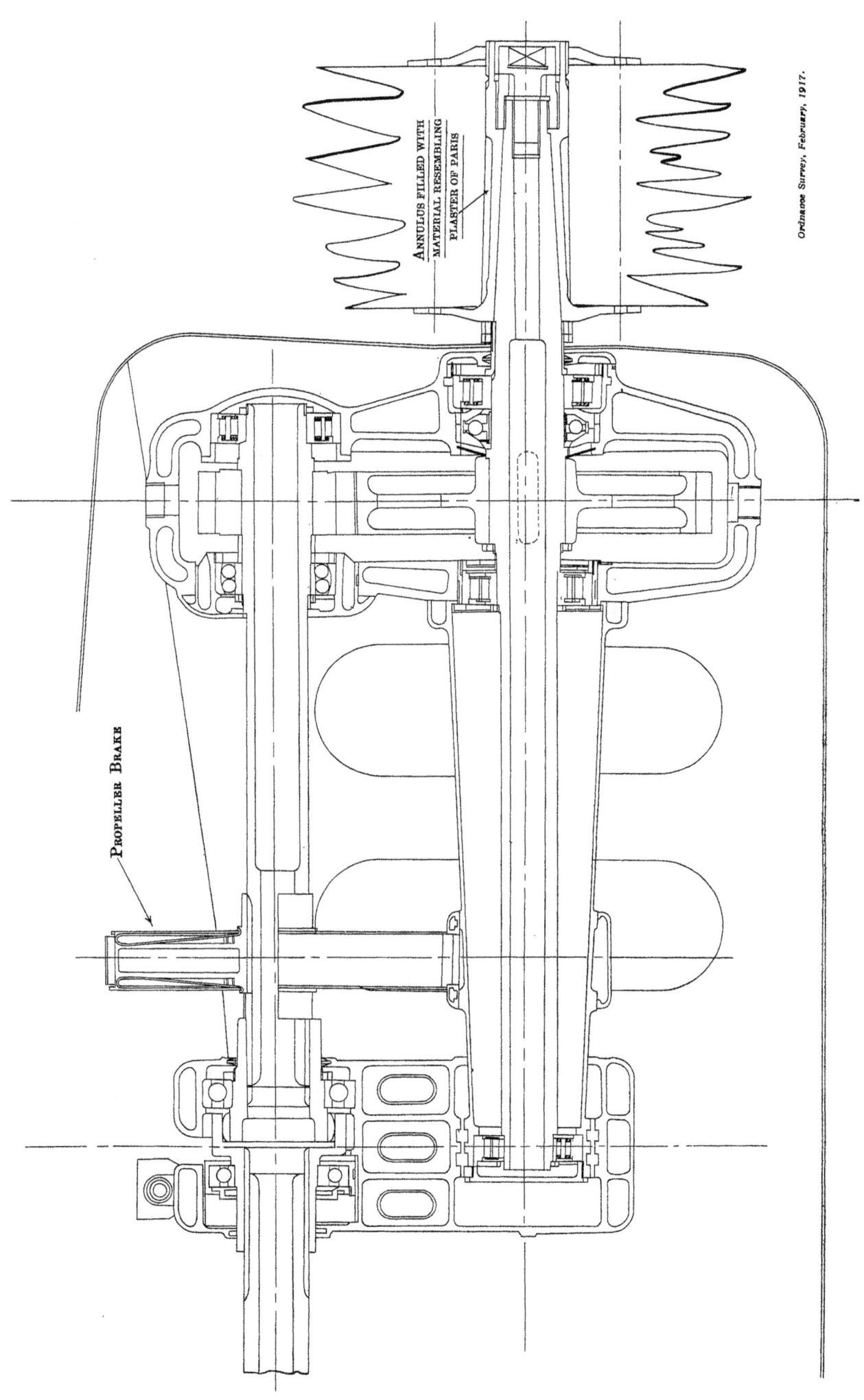

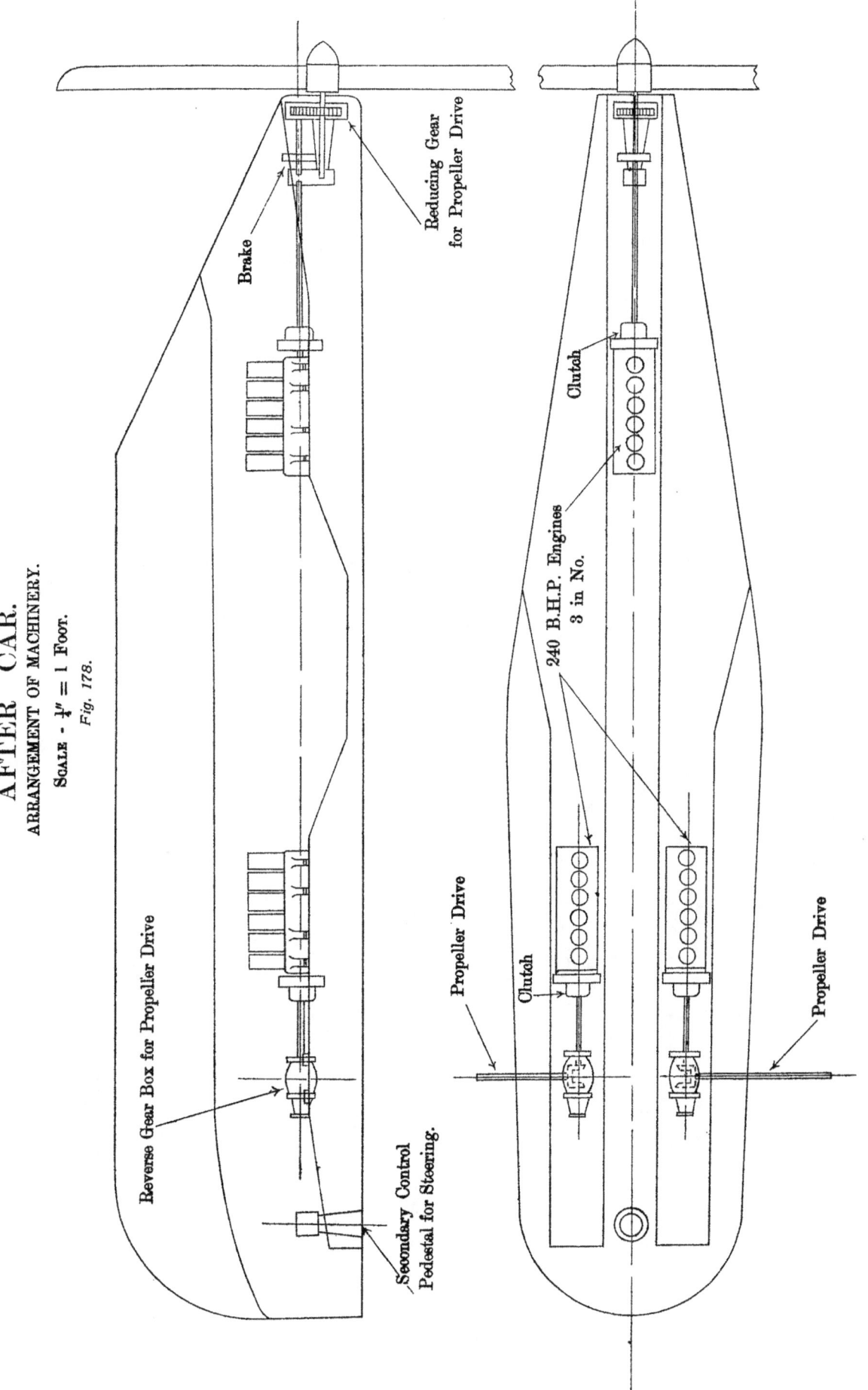

## STARBOARD WING CAR.

### Fig. 179.

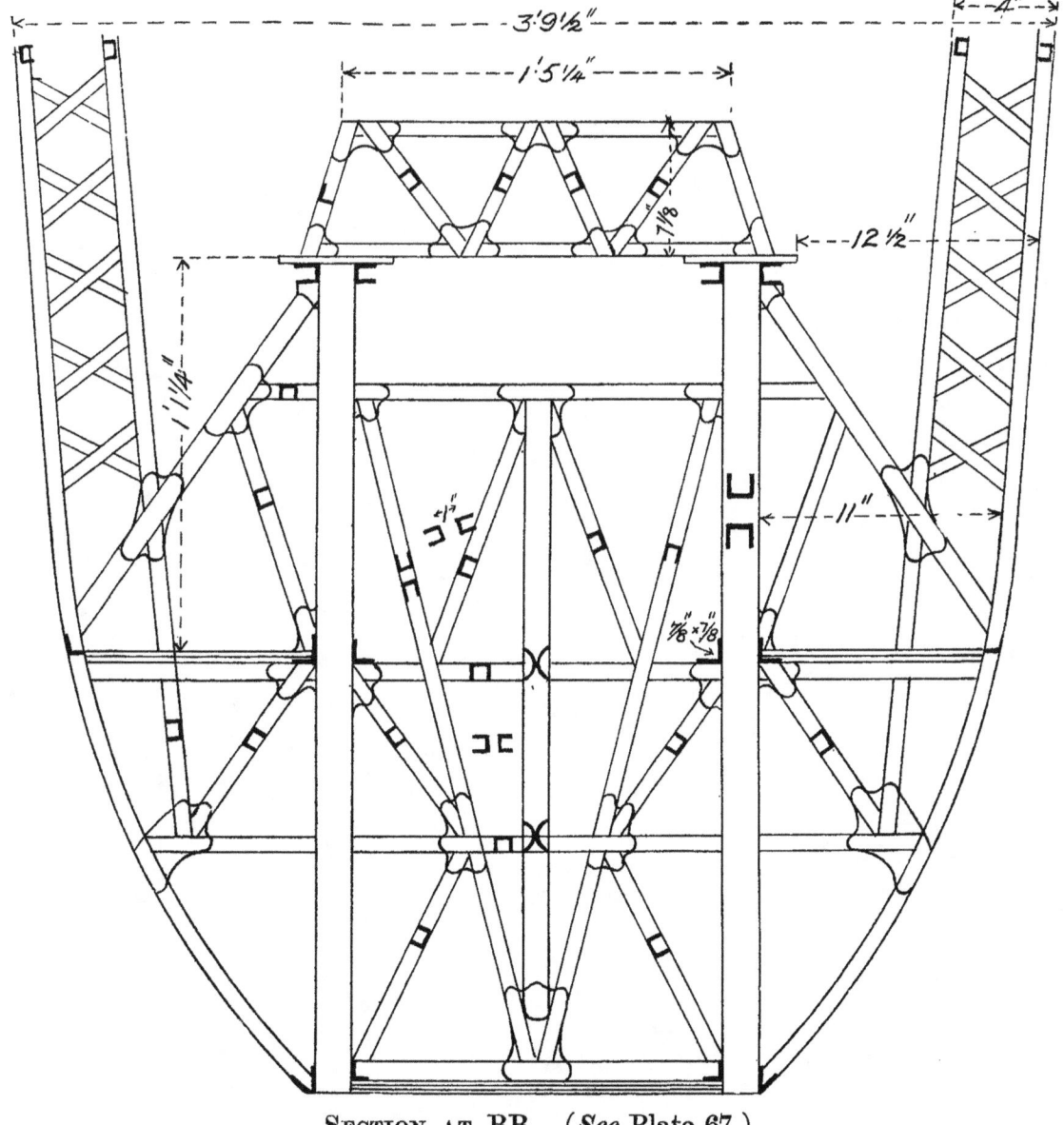

SECTION AT BB. (*See* Plate 67.)

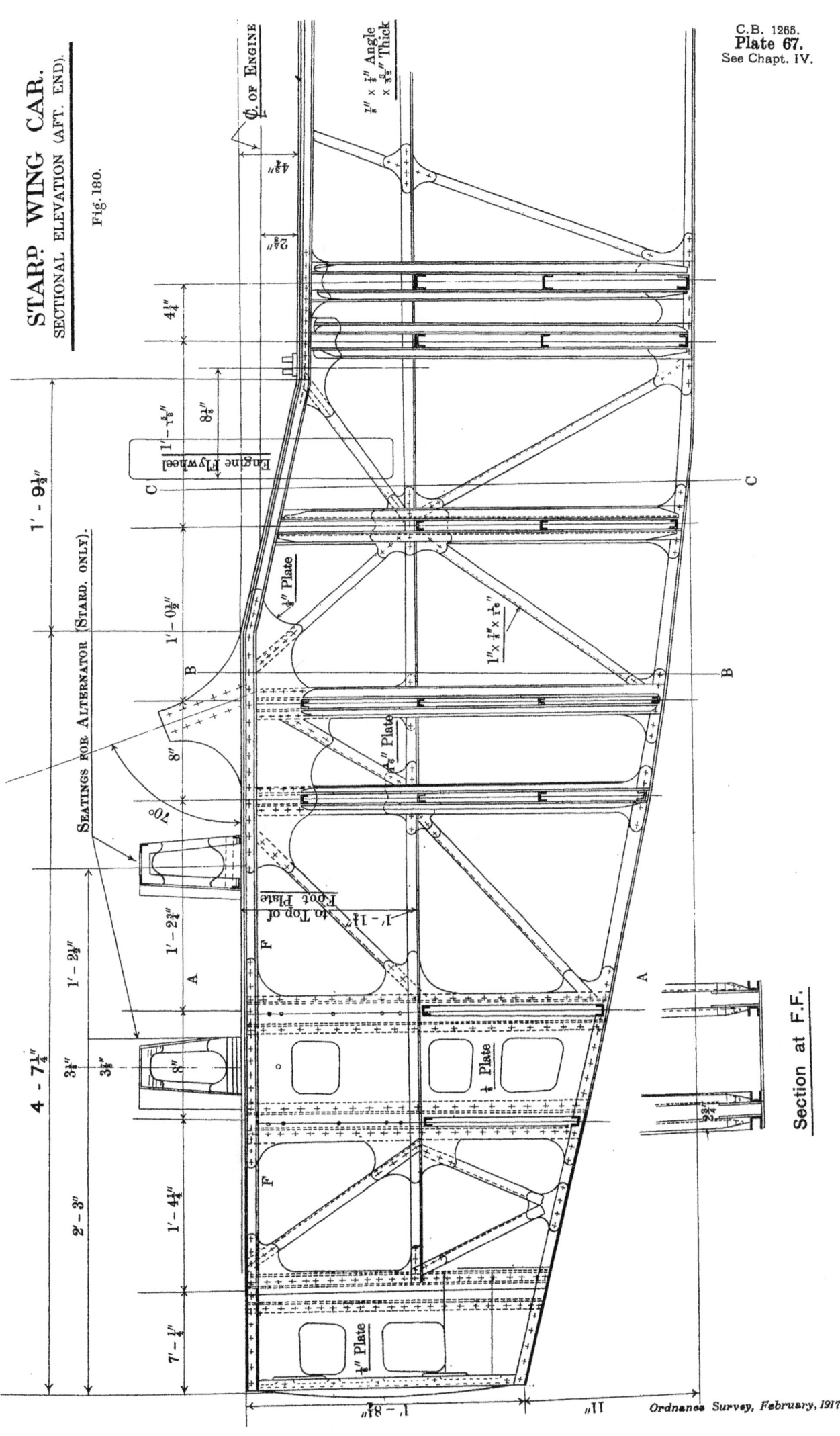

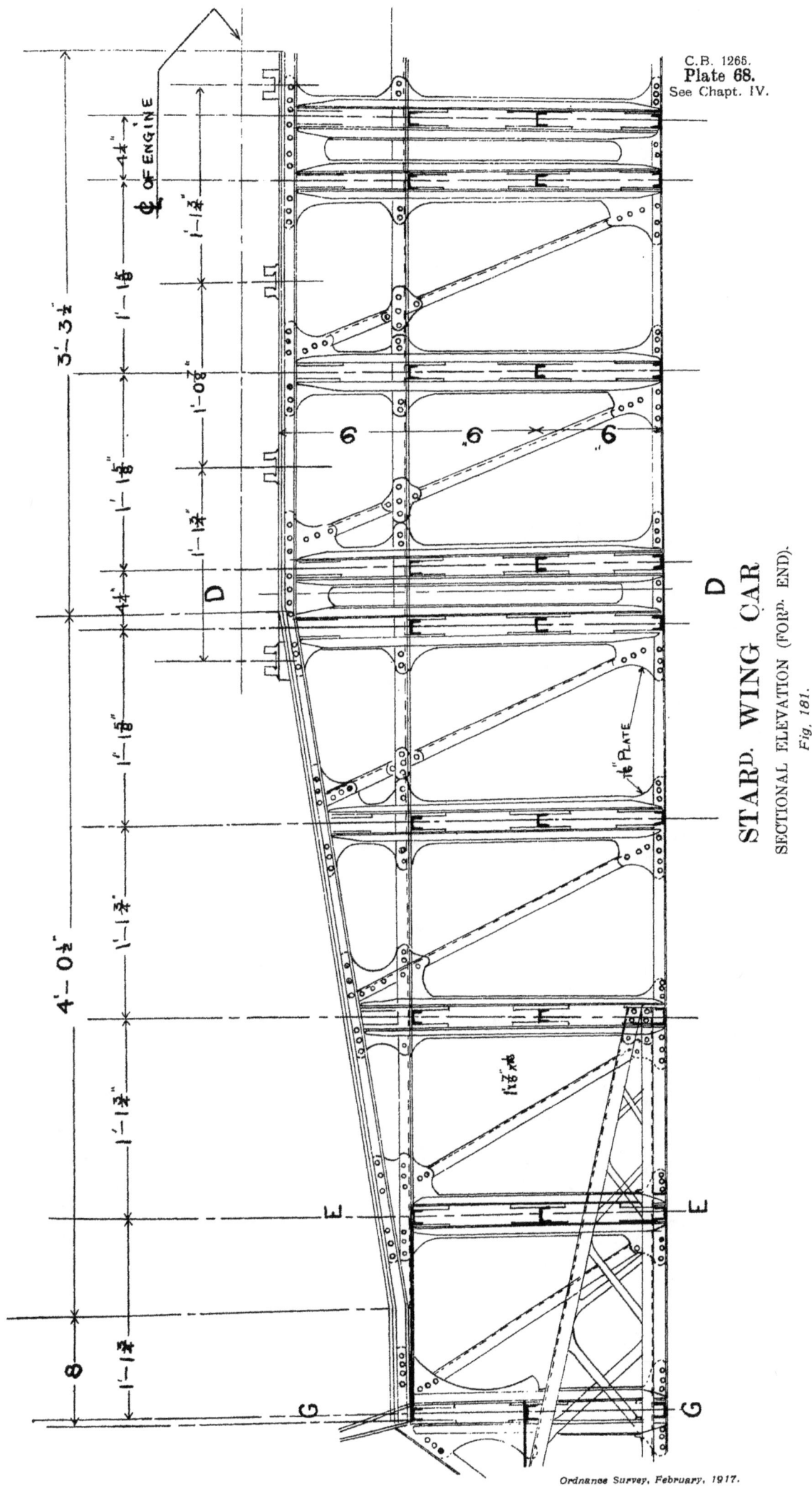

STARD. WING CAR
SECTIONAL ELEVATION (FORD. END).
Fig. 181.

# WING CARS.

Plate 69.
See Chapt. IV.

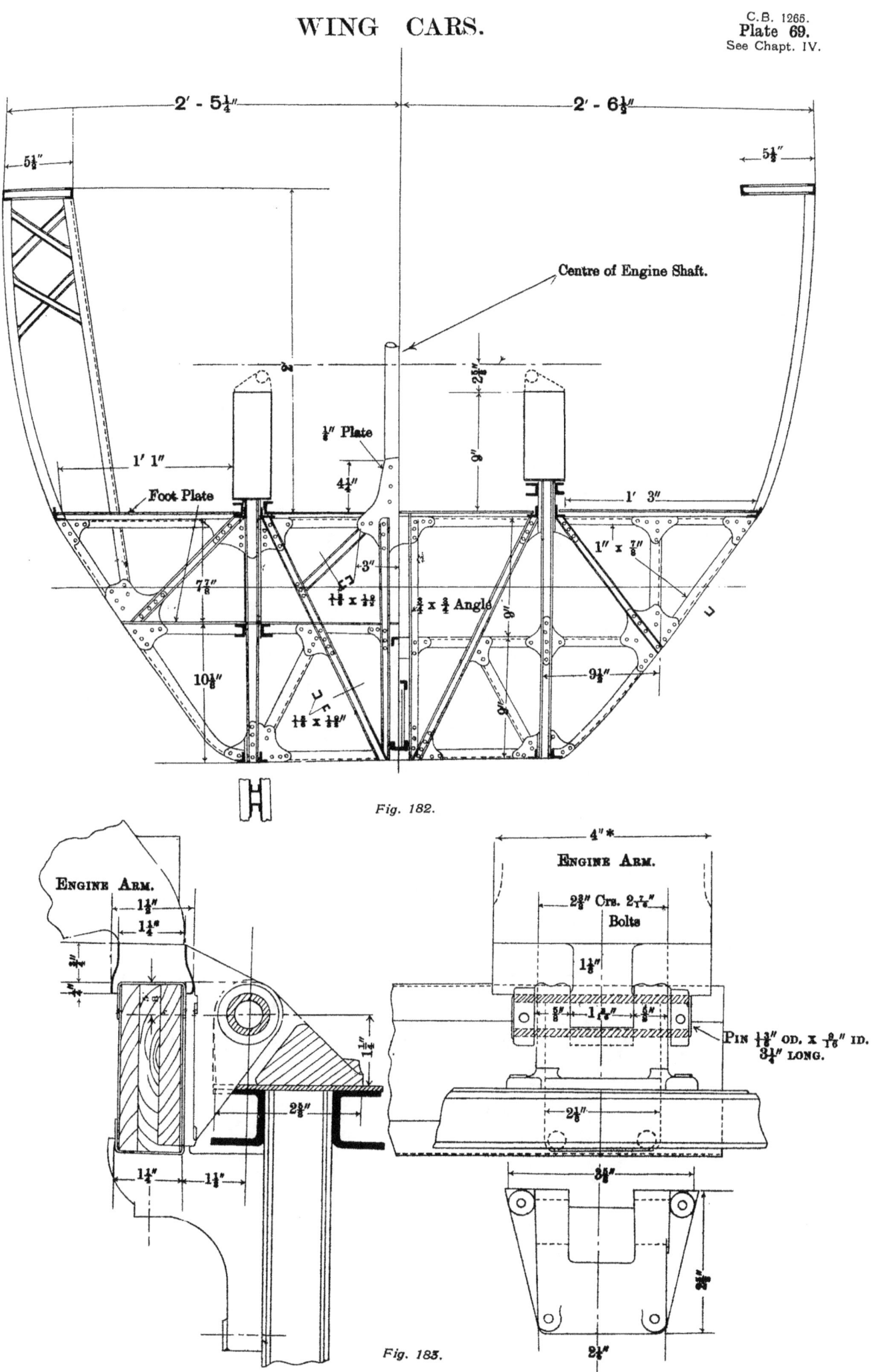

Fig. 182.

Fig. 183.

**Fore and Aft Engine Brackets.**

# PETROL TANK & SLING.

Fig. 184.

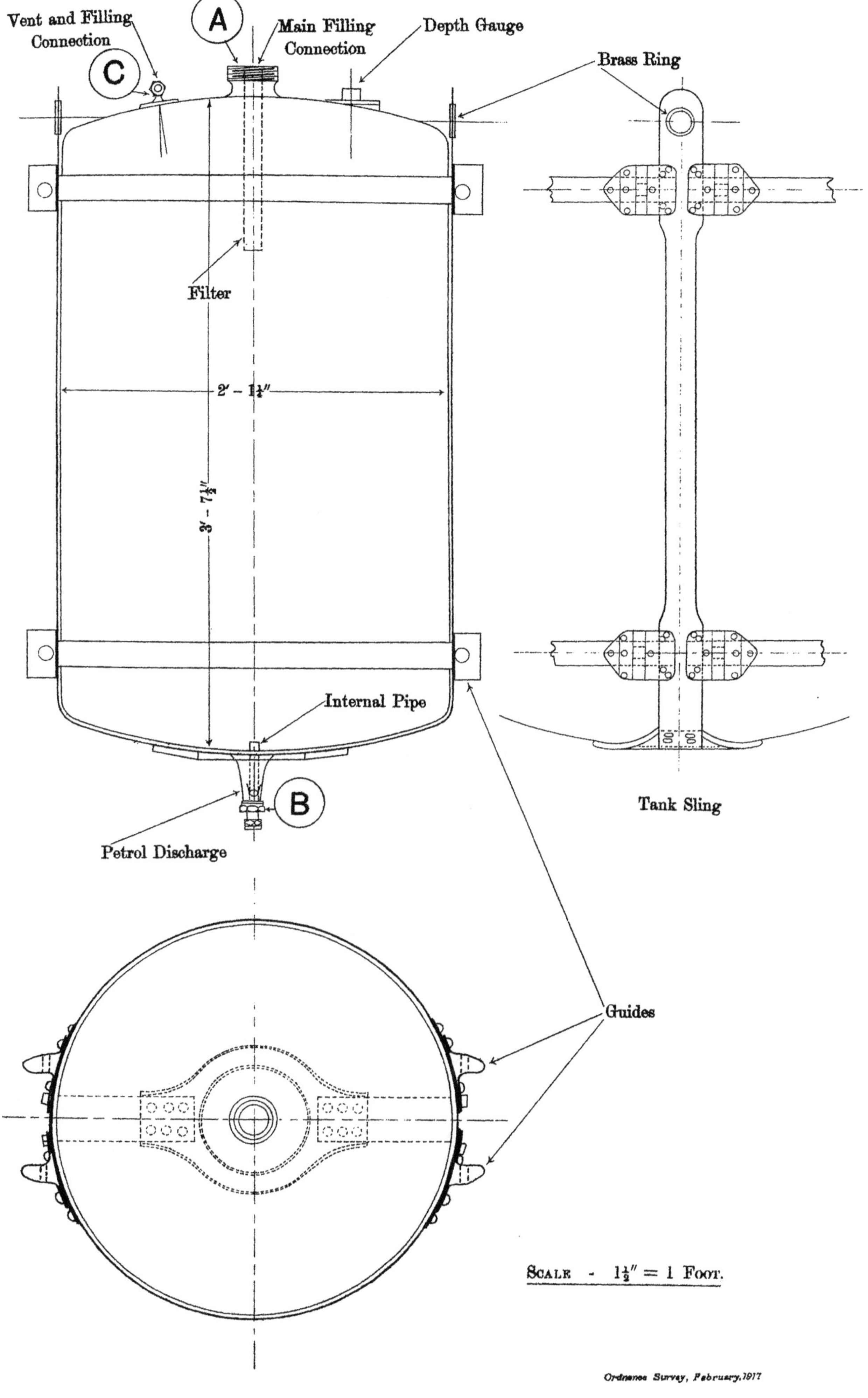

C.B. 1265.
Plate 70.
See Chapt. IV.

- A — Main Filling Connection
- C — Vent and Filling Connection
- Depth Gauge
- Brass Ring
- Filter
- 2' – 1¼"
- 3' – 7½"
- Internal Pipe
- B — Petrol Discharge
- Guides
- Tank Sling

Scale - 1½" = 1 Foot.

Ordnance Survey, February, 1917.

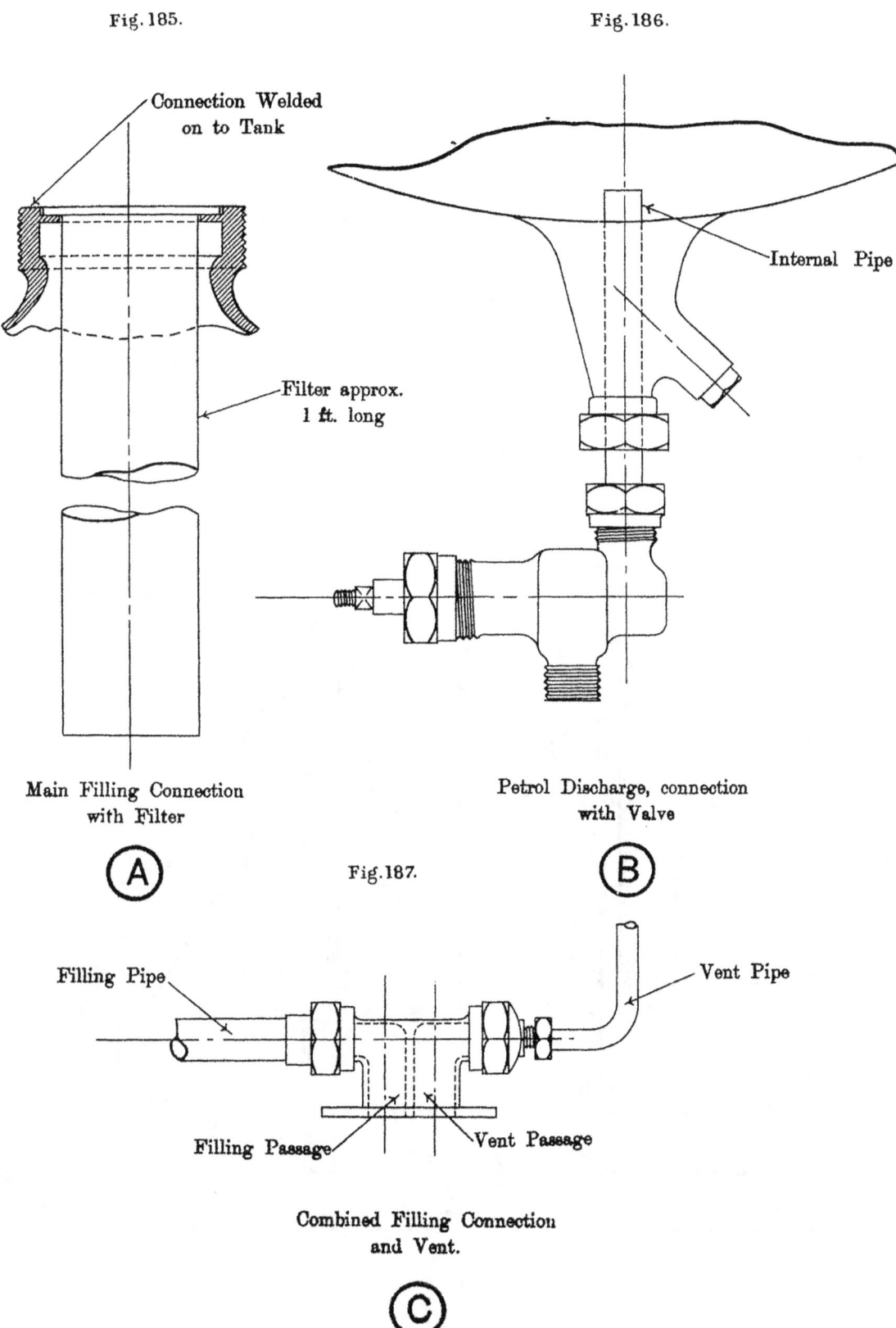

PETROL TANK FITTINGS.

SCALE - HALF SIZE.

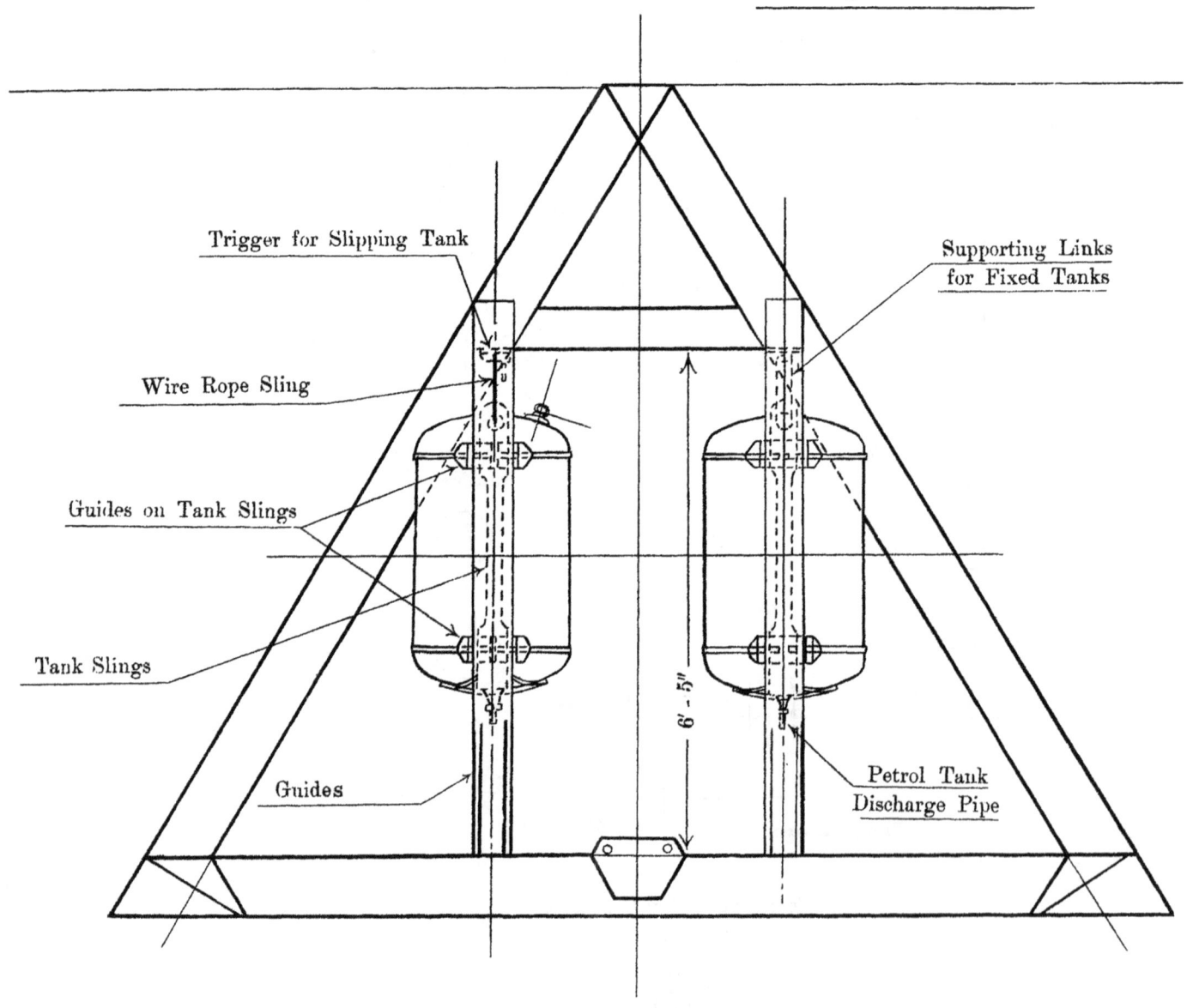

Fig. 188.

C.B. 1265.
**Plate 72.**
See Chapt. IV.

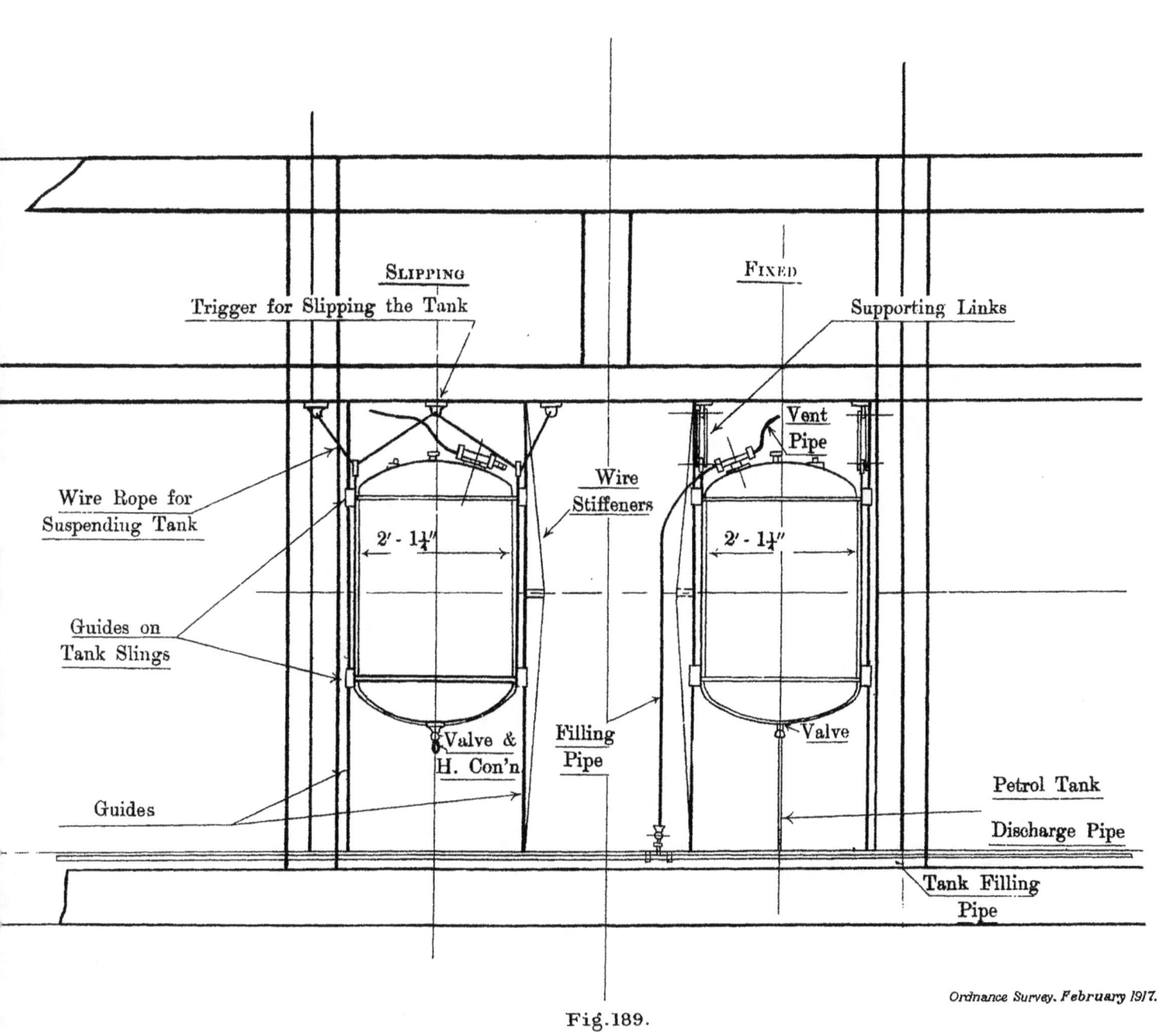

Fig. 189.

Ordnance Survey. February 1917.

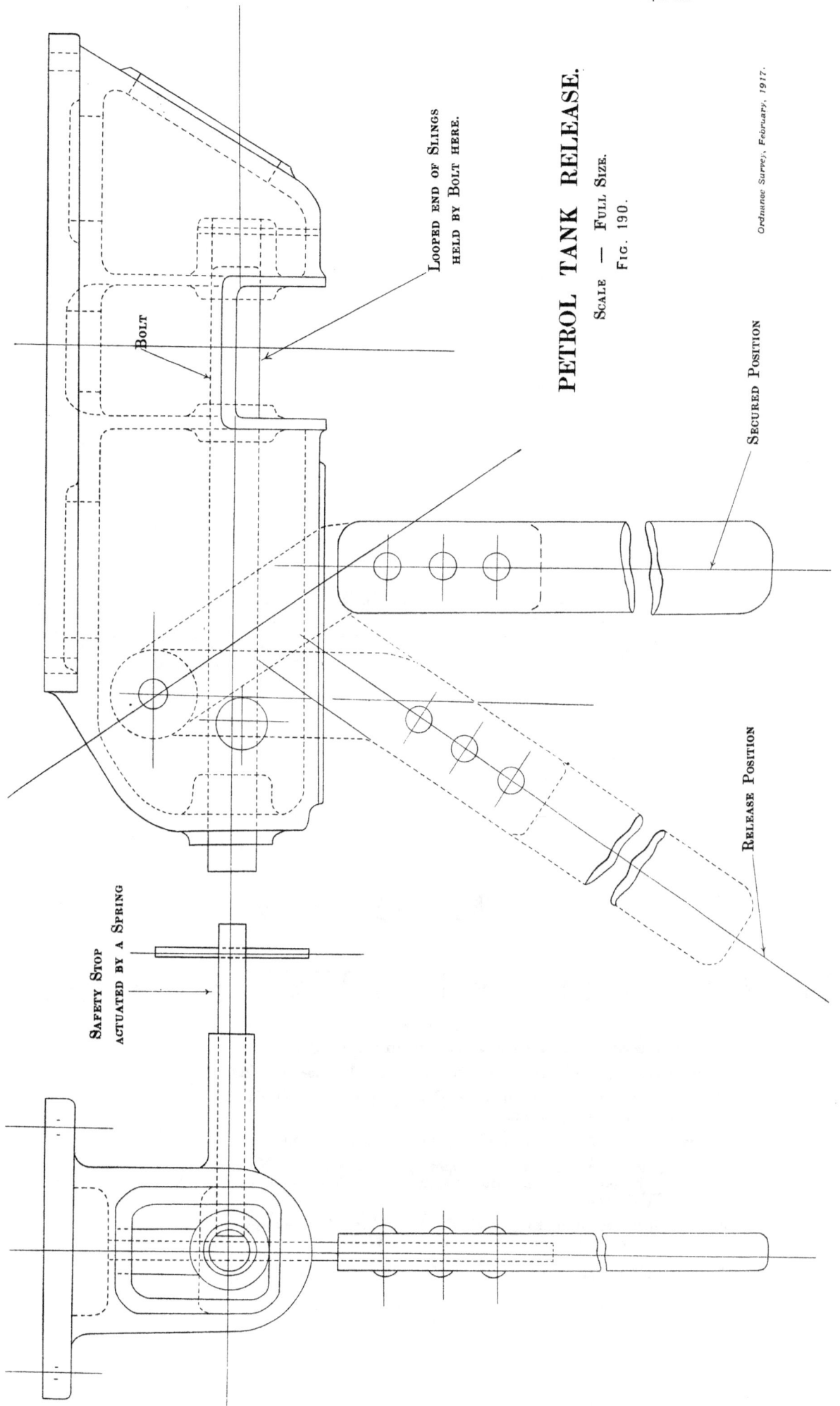

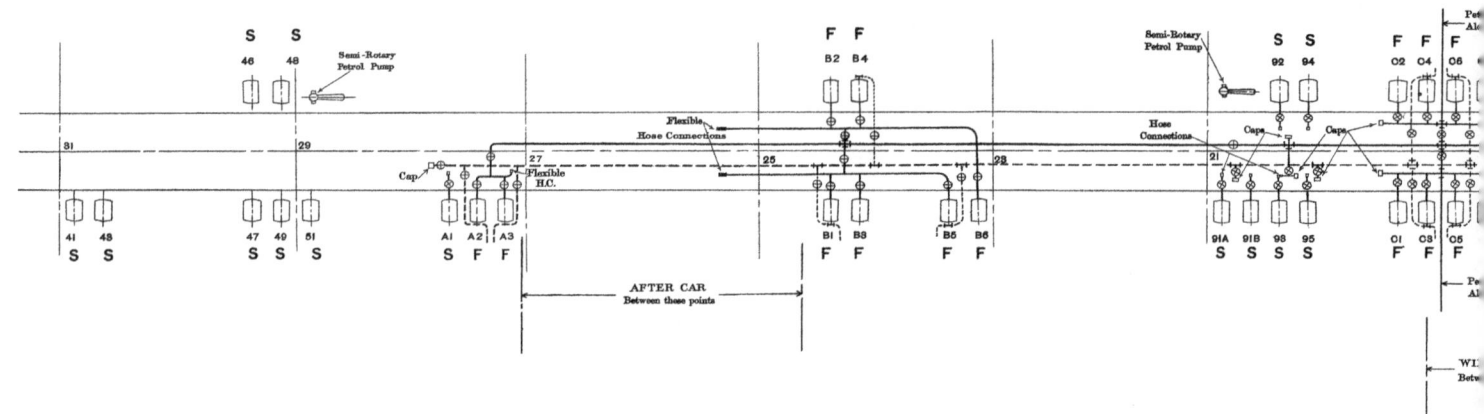

# L 33

## PETROL ARRANGEMENT (DIAGRAMMATIC)
Fig. 191.

NOTE.—The true position of tanks is as shown on plate XVI.

Chain lines indicate petrol tank filling, and full lines, running

All petrol tanks fitted with vent pipe, petrol gauge and c filling in shed (see detail sketch).

Actual number of tanks carried depends on requirements of vo

Running down pipes are of copper $\frac{3}{4}''$ ext. dia. 18 W.G. those which branch to wing and ford. Gondolas and are 18 W.G. thick.

Filling pipes are of brass $\frac{1}{2}''$ ext. dia. 24 W.G. thick except l main leads which are $\frac{3}{4}''$ ext. dia. 18 W.G. thick.

"F" signifies fixed tank. "S" signifies slipping tank.

⊕    "    stop valve.

C.B. 1265.
**Plate 74.**
See Chapt. IV.

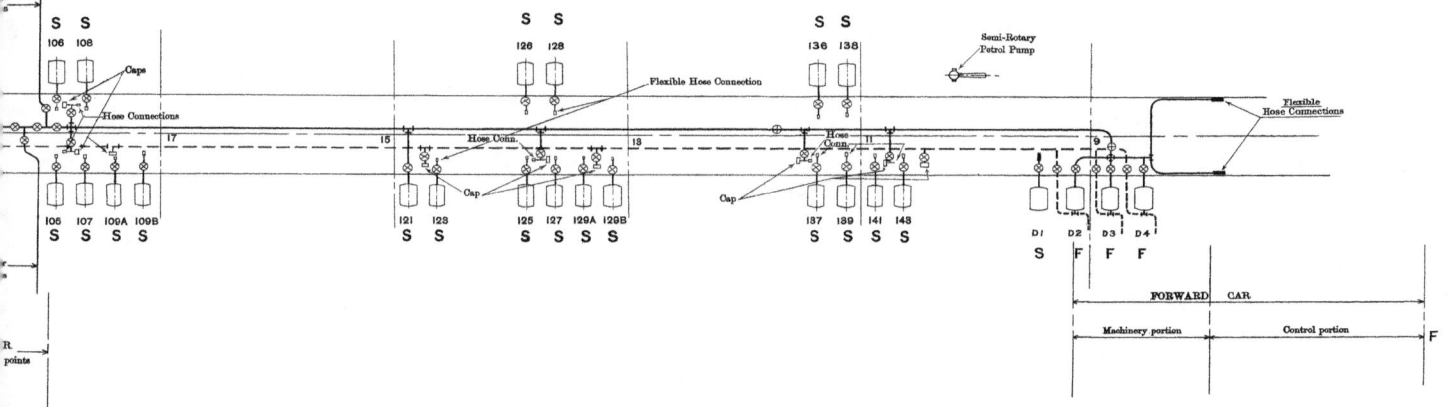

Ordnance Survey, February, 1917.

C.B. 1265.
Plate 75.
See Chapt. V.

Fig. 192.—Lower Rudder with Fabric in position. (L 32).

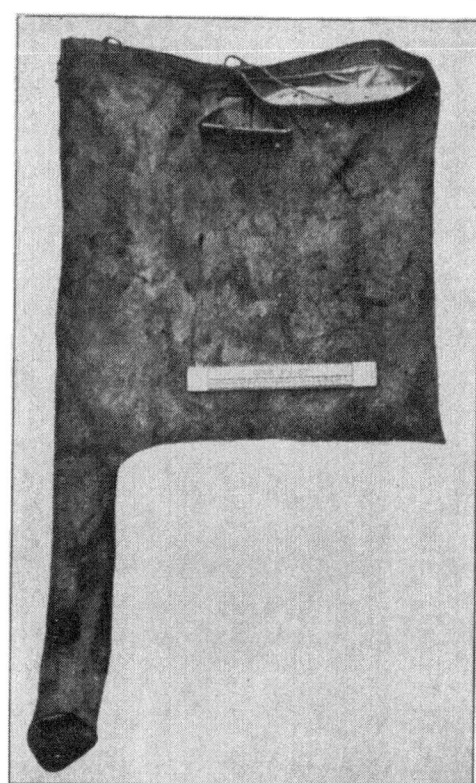

Fig. 193.—Emergency Water Ballast Bag from L 33.

Fig. 194.—Lower Rudder of L 32. Fabric turned back (on left) showing lacing along edge of Rudder.

Fig. 195.—Lower Rudder of L 32, showing (at top of picture) the lacing of cover to girder at root of sector.

*Ordnance Survey, February, 1917.*

C.B. 1265.
**Plate 76.**
See Chapt. V.

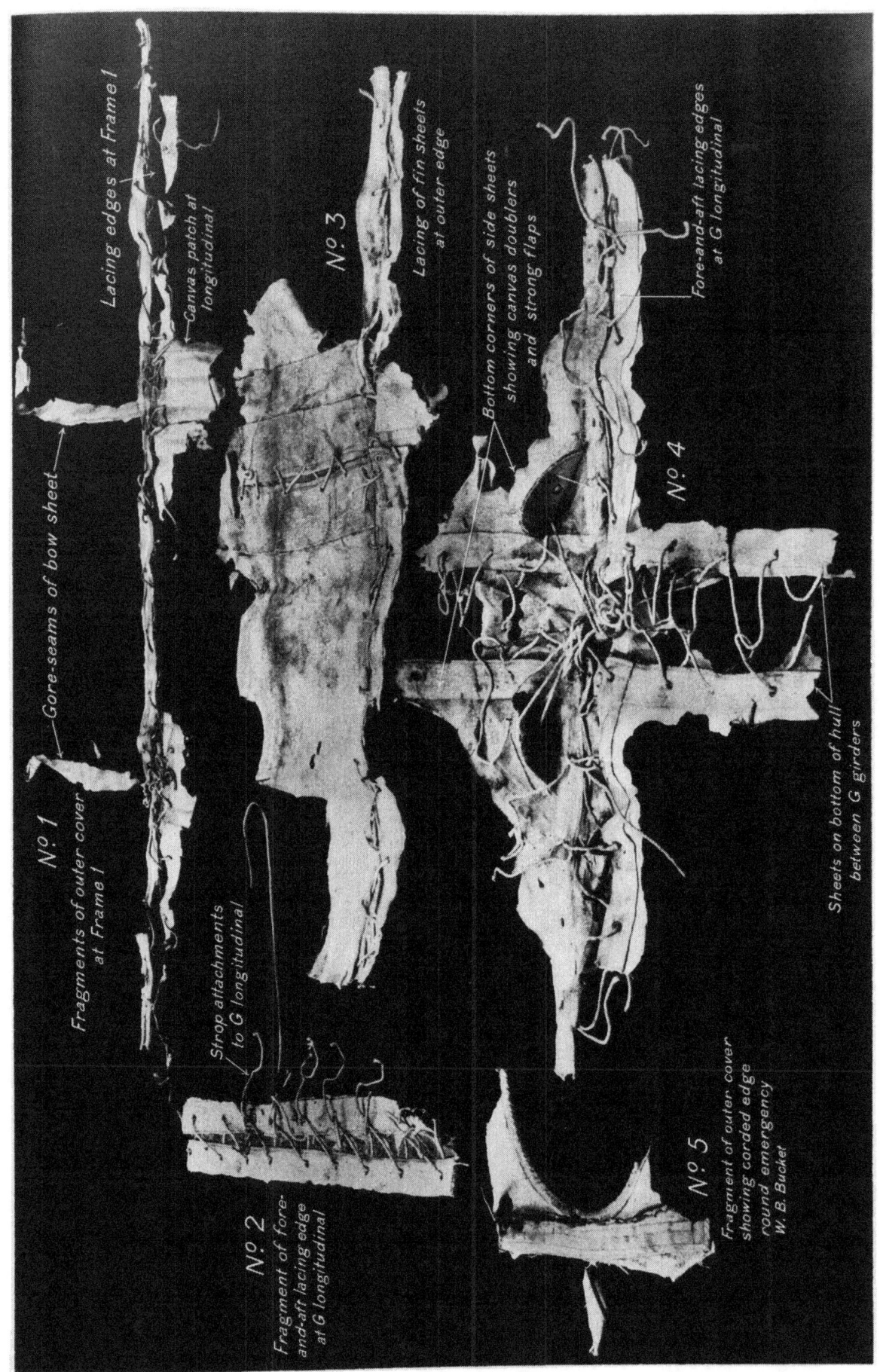

Ordnance Survey, February, 1917.

C.B. 1265.
**Plate 77.**
See Chapt. III.

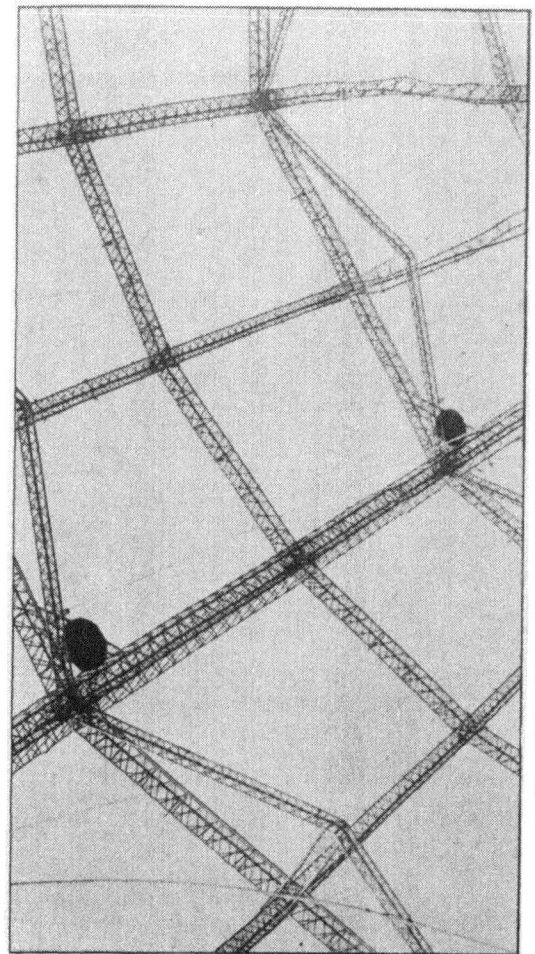

Fig. 197.—Manoeuvring valves in position alongside "A" longitudinal of L 33.

Fig. 199.—Manoeuvring valve from LZ 85.

Fig. 196.—Manoeuvring valve and structure in L 33.

Fig. 198.—Manoeuvring valve from LZ 85.

*Ordnance Survey, February, 1917.*

# ZEPPELIN "L 33."

PAD PIECE AND TURNBUCKLE AT LOWER END OF SHEAR WIRE.
Fig. 200.

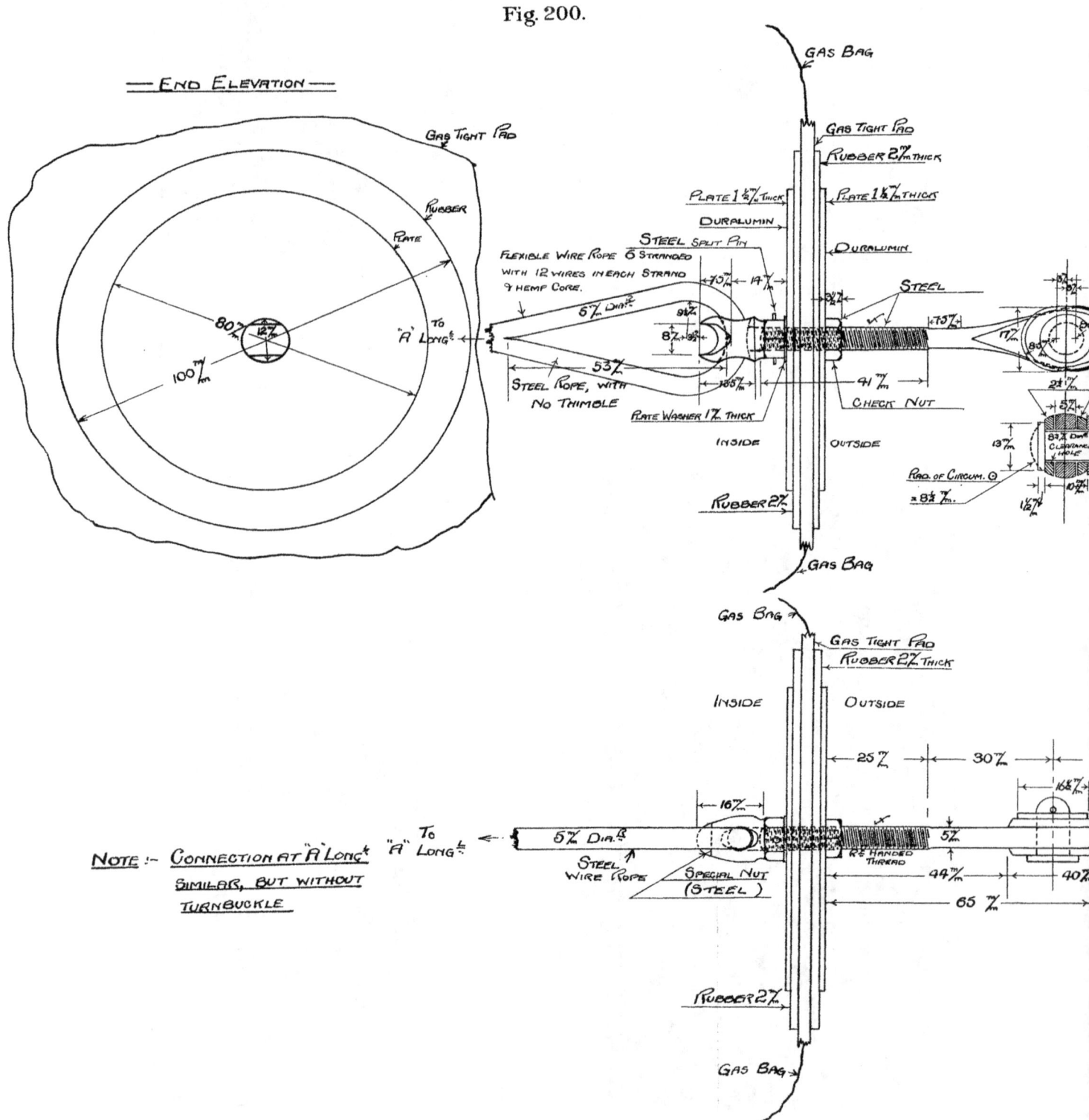

C.B. 1265.
**Plate 78.**
See Chapt. V.

— ELEVATION —

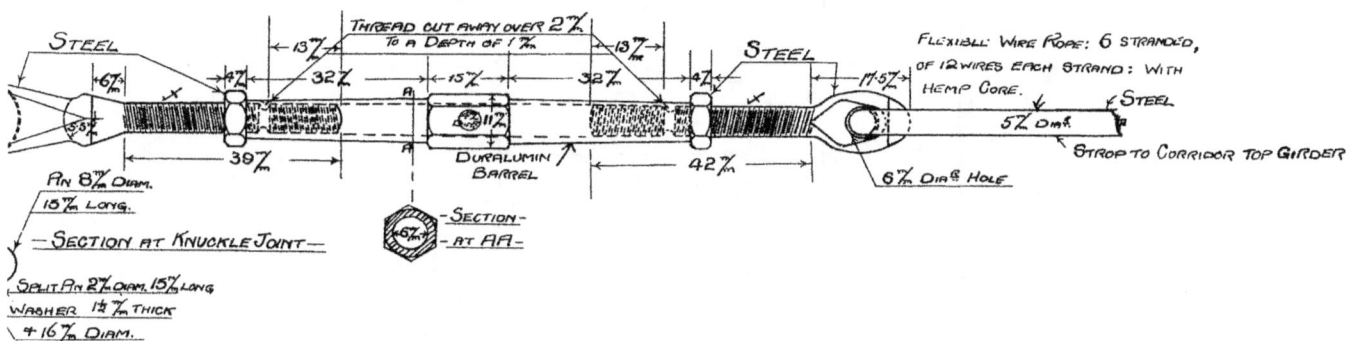

— PLAN —

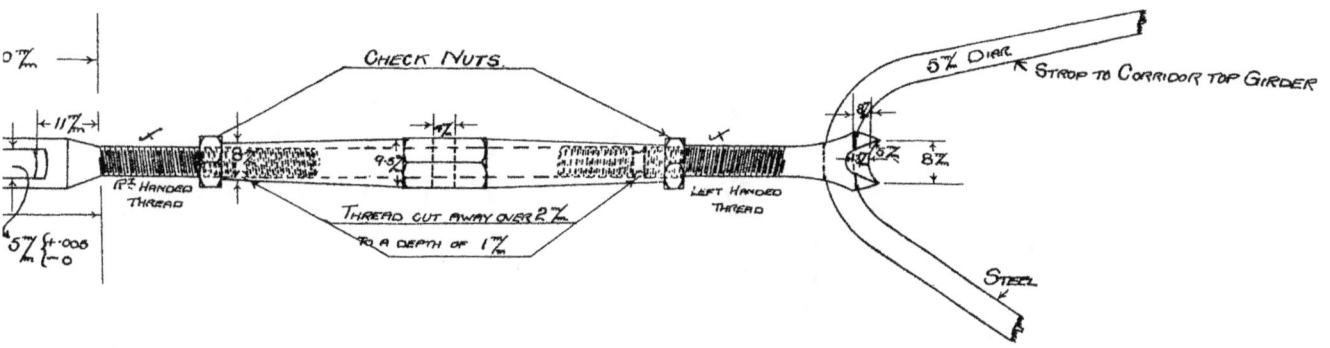

Ordnance Survey, February, 1917.

# L 33.—OUTER COVER ON PARALLEL PORTION.

*Fig. 201.*

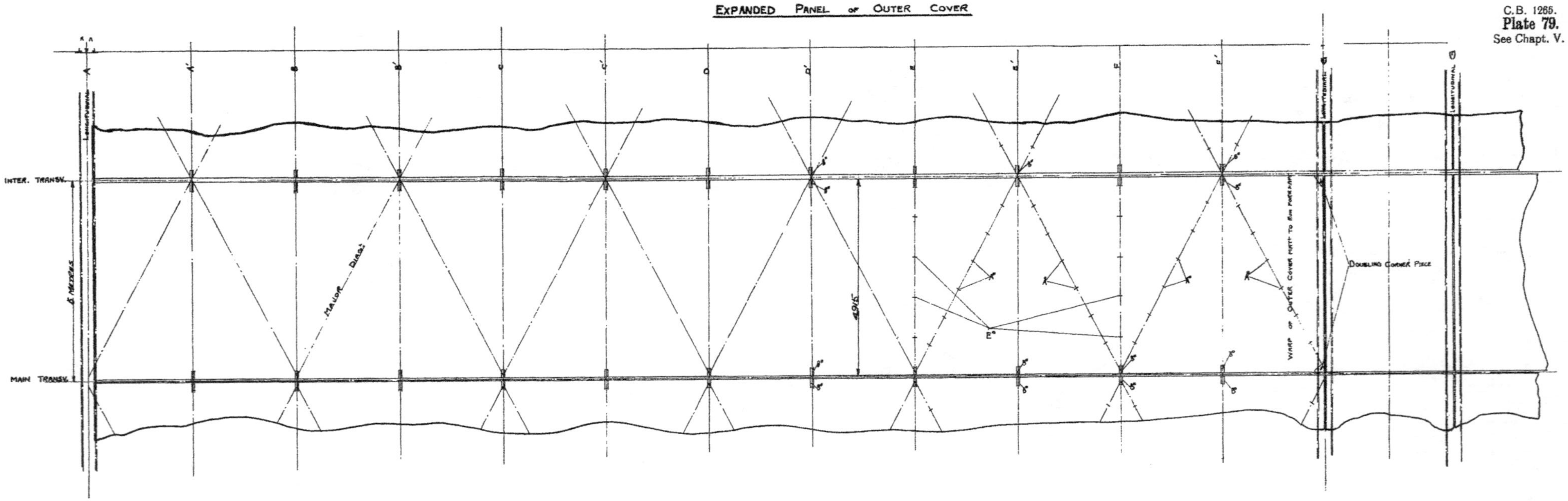

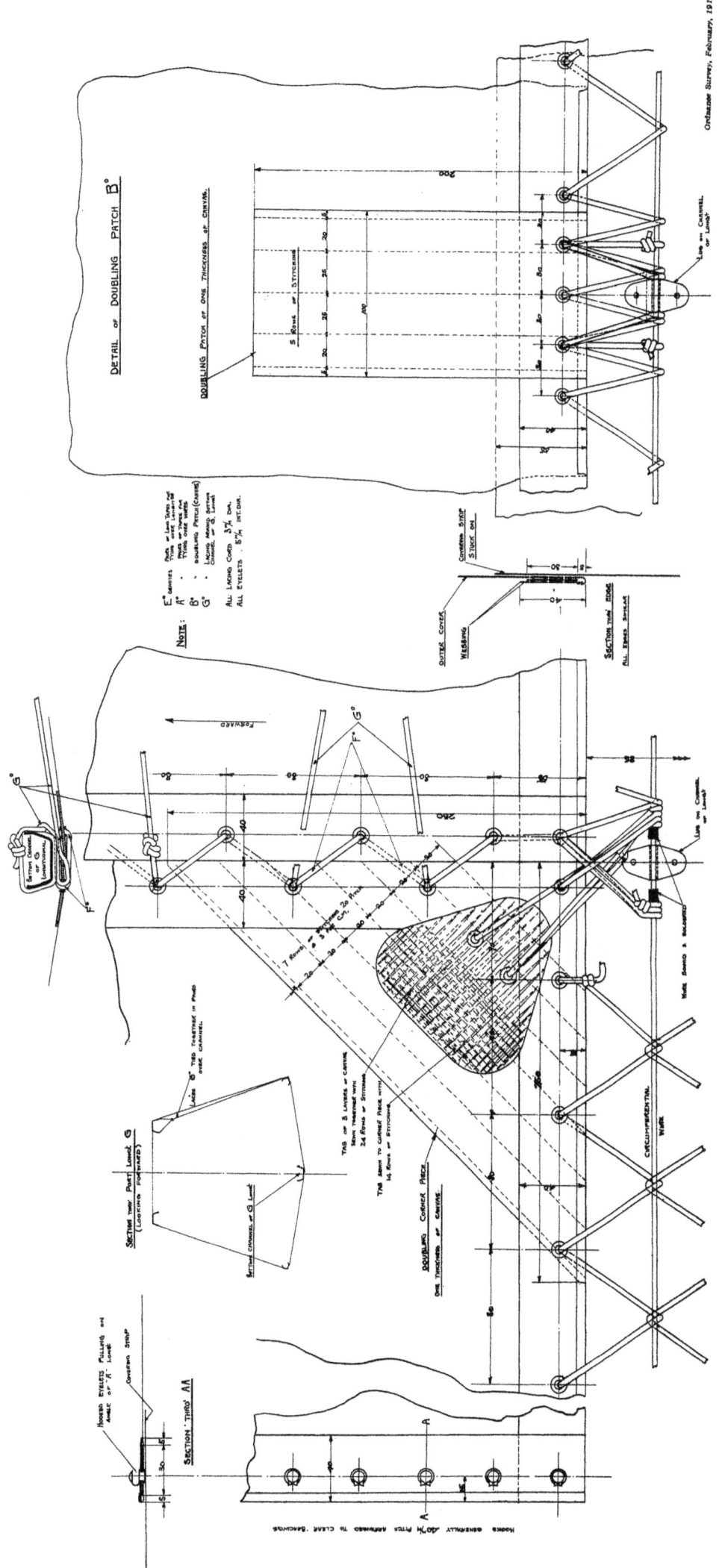

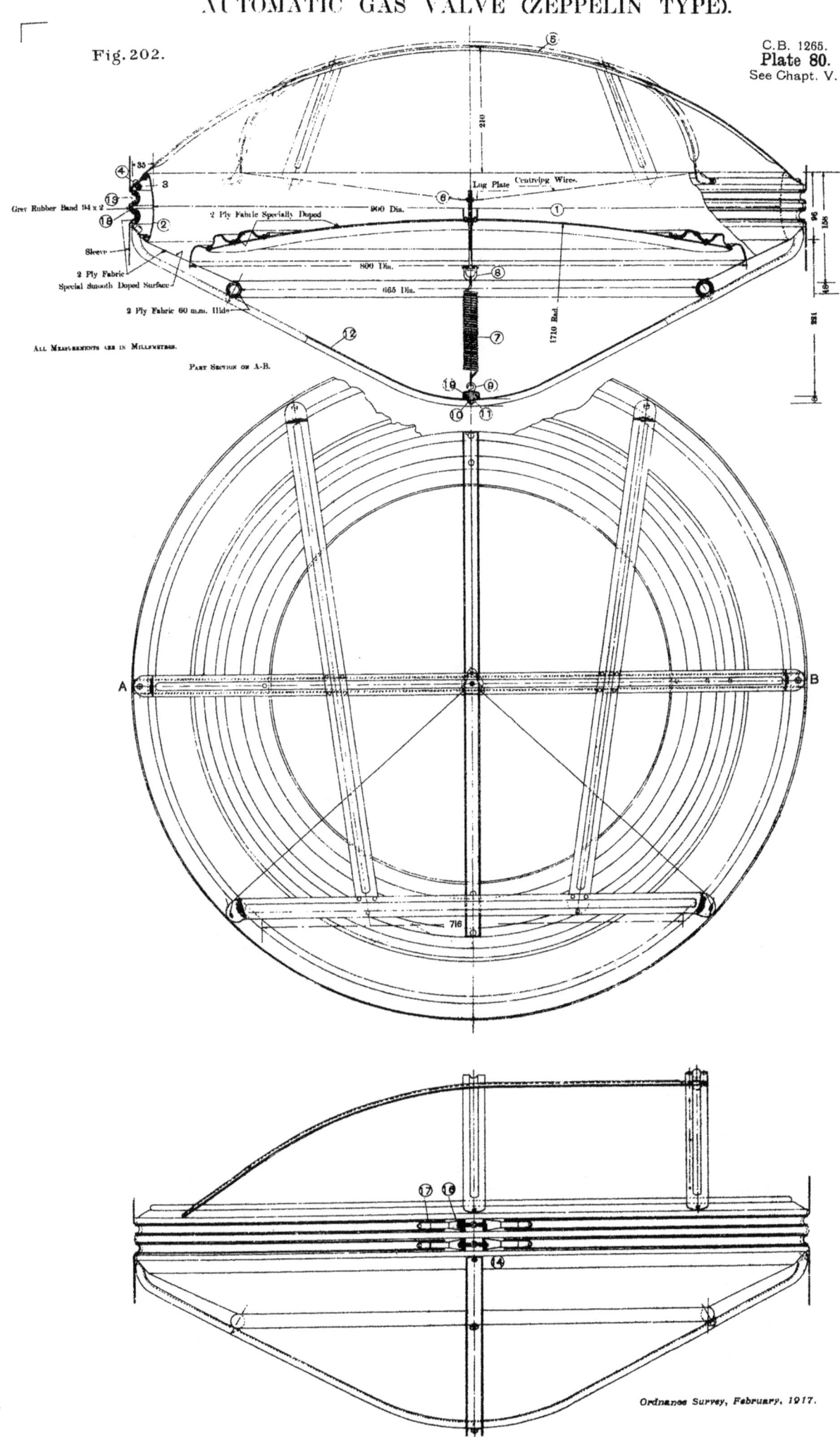

C.B. 1265.
**Plate 81.**
See Chapt. V.

# SECTIONAL ELEVATION
## OF
## GAS VALVE FOR ZEPPELIN L33.

All Measurements are in Millimeters.

Fig. 203.

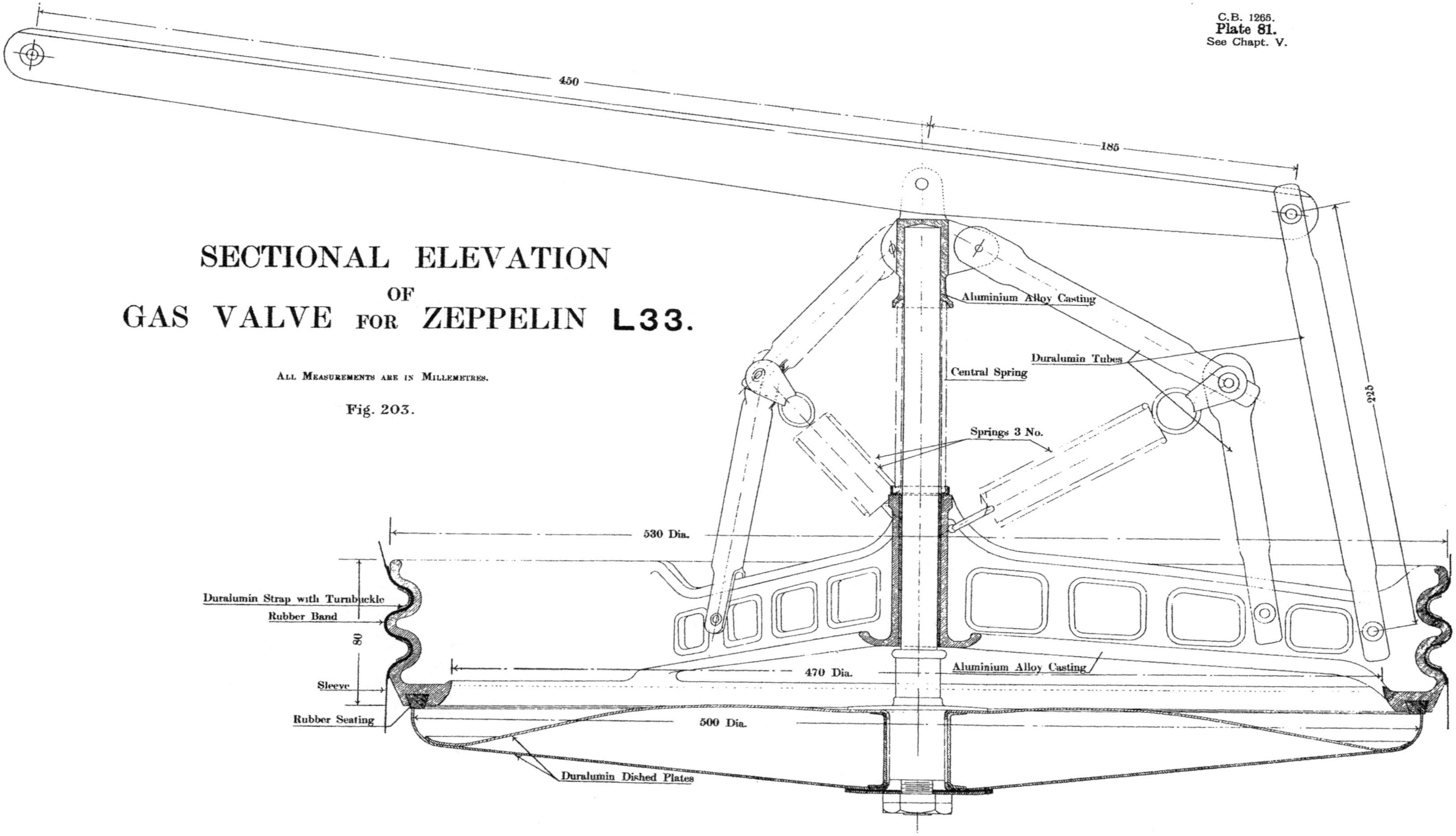

Ordnance Survey, February 1917.

Fig. 204.
Bomb Release from L 33 in "Fired" position.

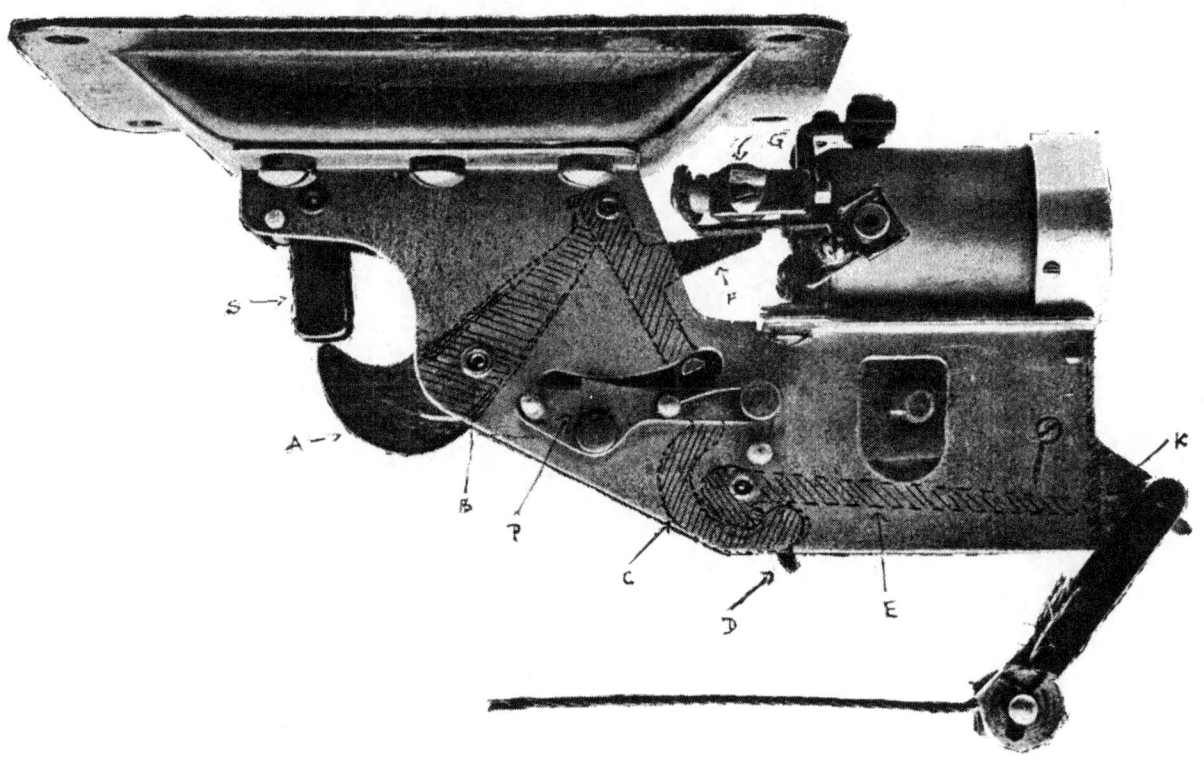

Fig. 205.
Bomb Release from L 33. Dotted Lines show positions of Tumbler, Sear, and Trigger in cocked position. Safety Lever in "Ready to Fire" position.

C.B. 1265.
**Plate 83.**
End of Chapt. VI.

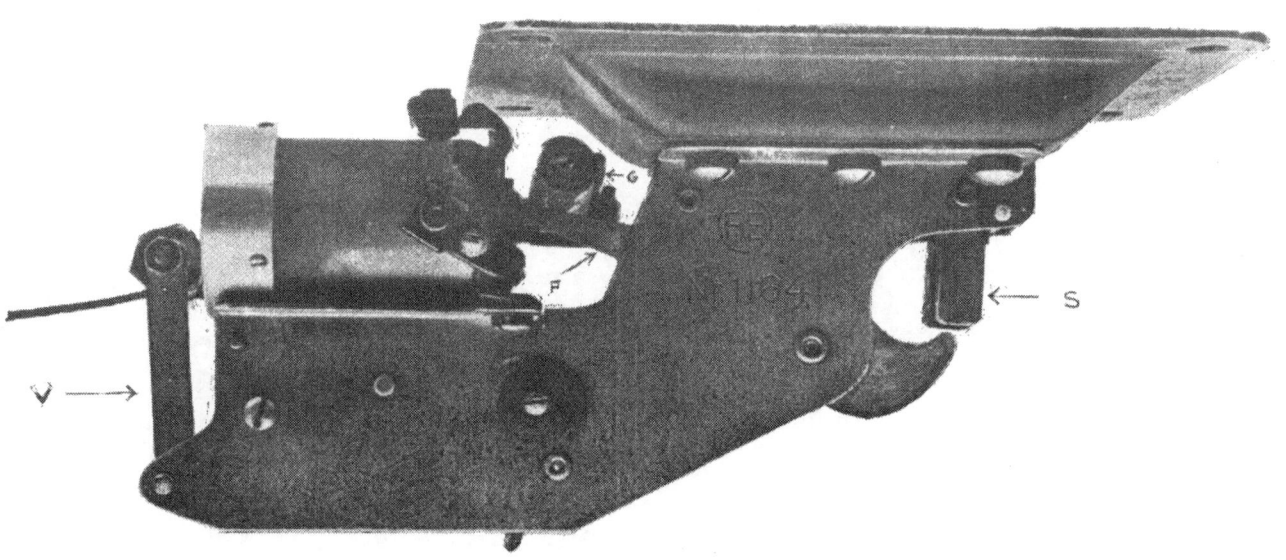

Fig. 206.
Bomb Release from L 33.   Cocked position.   Safety Lever at "Safe."

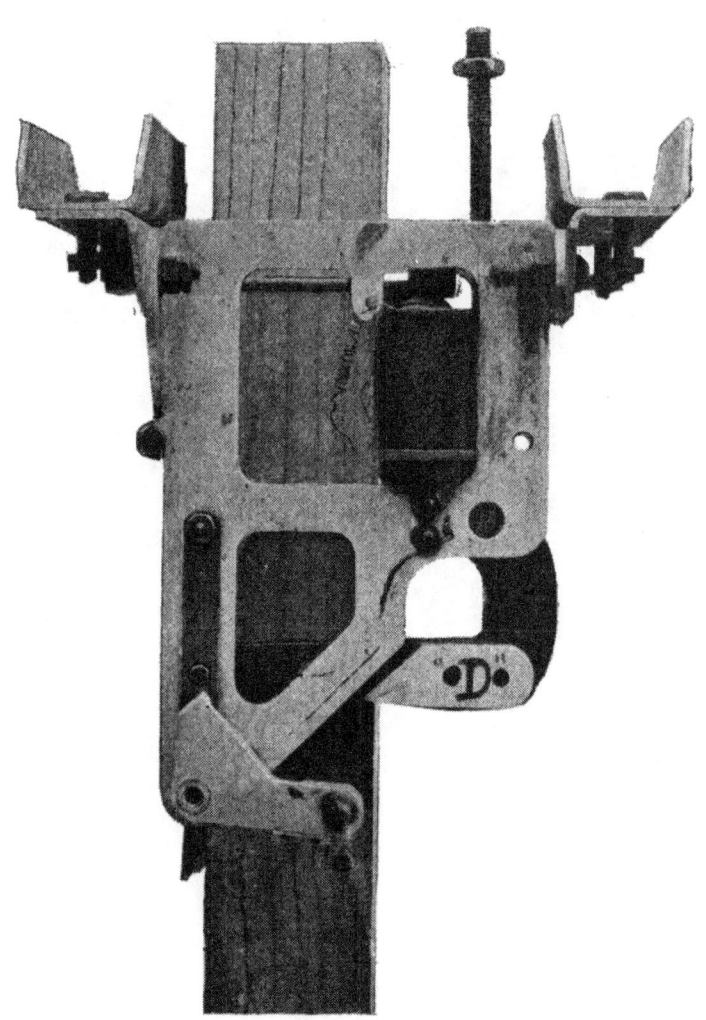

Fig. 207.
Electro Magnetic Bomb Release from LZ 85.

*Ordnance Survey, February, 1917*

C.B. 1265.
**Plate 84.**
See Chapt. VI.

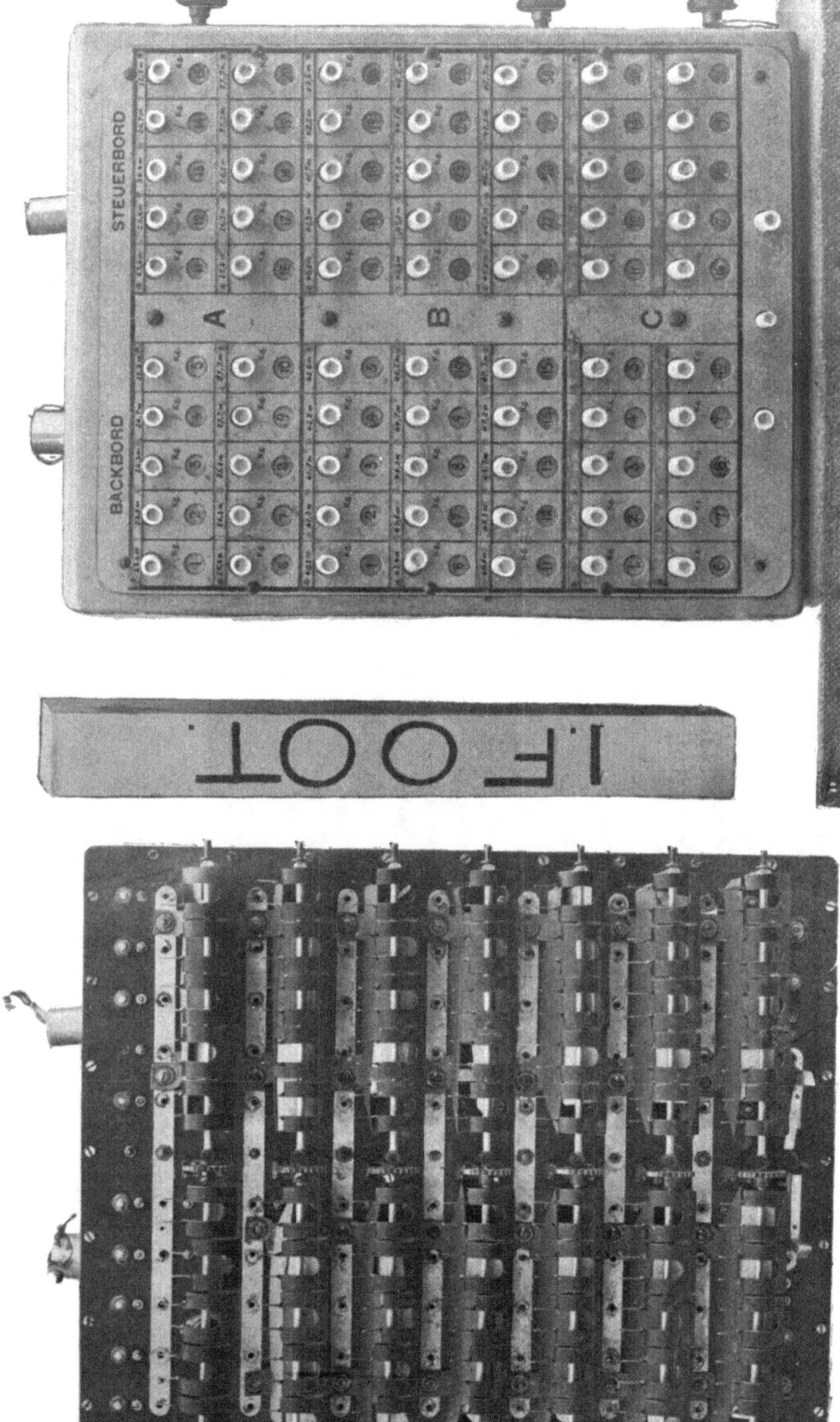

Fig. 209.
Distribution Board of LZ 85. Front View with cover on.

Fig. 208.
Distribution Board of LZ 85. Front View with cover removed.

Ordnance Survey, February, 1917.

Fig. 210.
Main Switchboard from LZ 85 with two Electro Magnetic Switches.
Back view with cover removed.

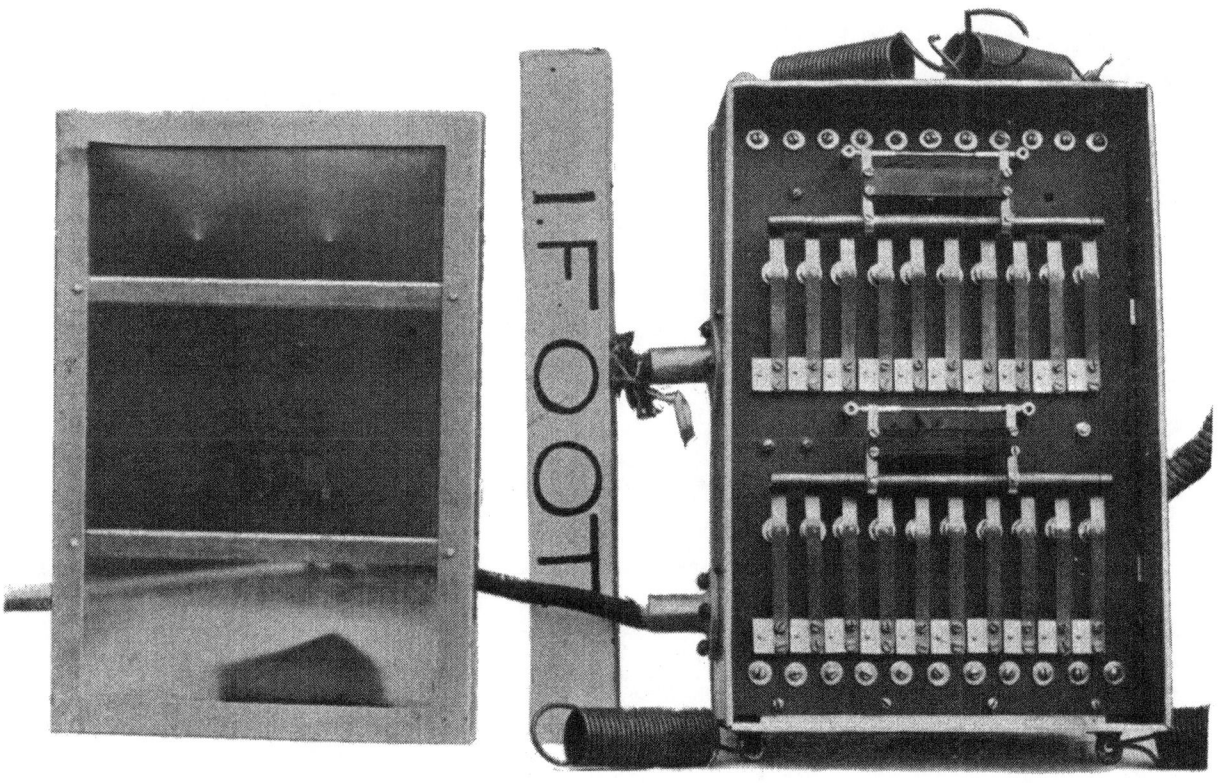

Fig. 211.
Main Switchboard from LZ 85 with two Electro Magnetic Switches.
Front view with cover removed.

C.B. 1265.
**Plate 86.**
See Chapt. VI.

Fig. 213.
Back view of Distribution Board of LZ 85
(see Fig. 209).

Fig. 212.
Main Switchboard of LZ 85 with three Electro Magnetic Switches.
Front view with cover removed.

Ordnance Survey, February, 1917.

C.B. 1265.
**Plate 87.**
See Chapt. VI.

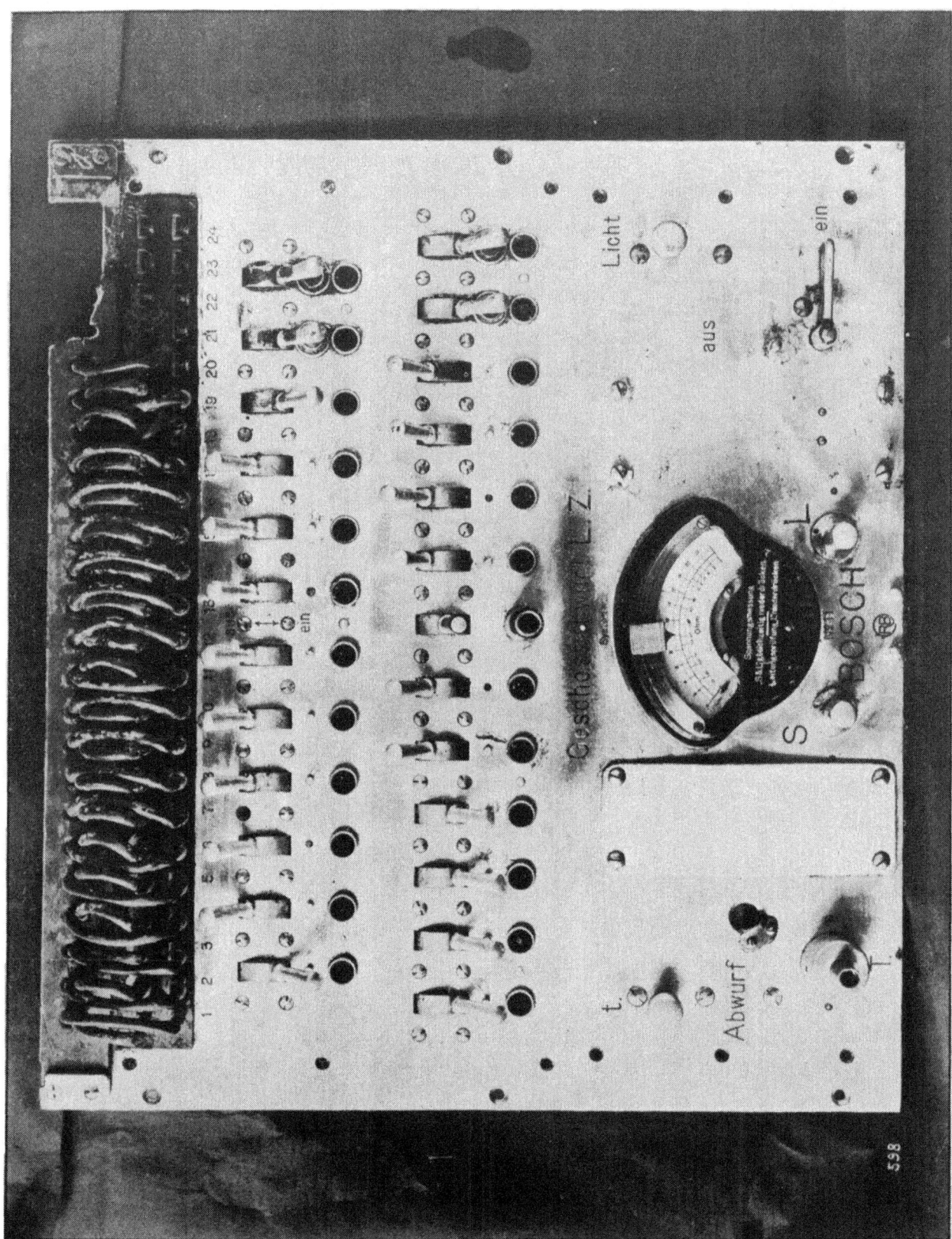

Fig. 214.—Bomb Release Switchboard from LZ 77 and L 33.

Ordnance Survey, February 1917.

C.B. 1265.
**Plate 88.**
See Chapt. VI.

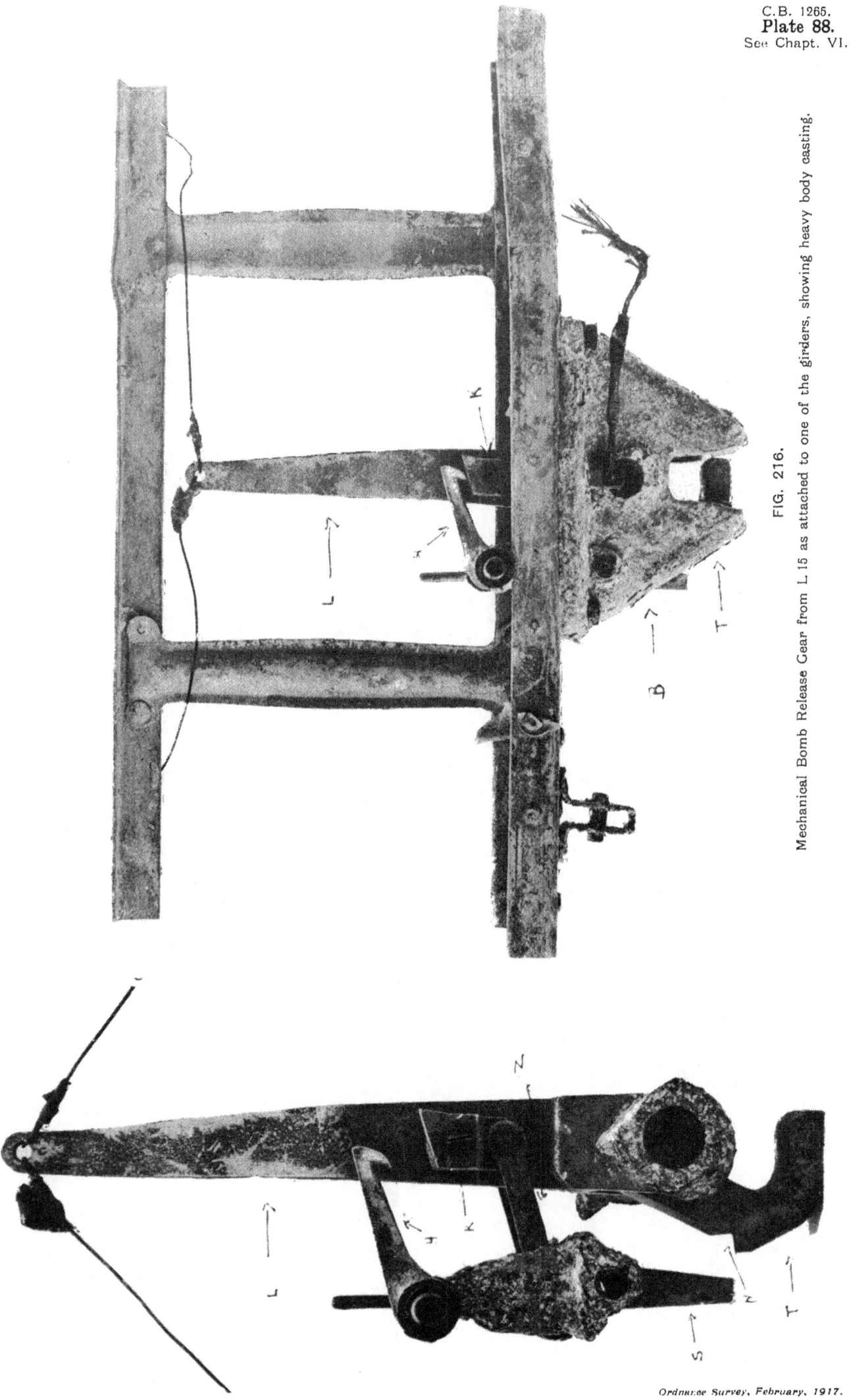

FIG. 216.

Mechanical Bomb Release Gear from L 15 as attached to one of the girders, showing heavy body casting.

FIG. 215.

Mechanical Bomb Release Gear from L 15 with body casing removed, showing Lever, Tumbler, Bent, and Sear.

Ordnance Survey, February, 1917.

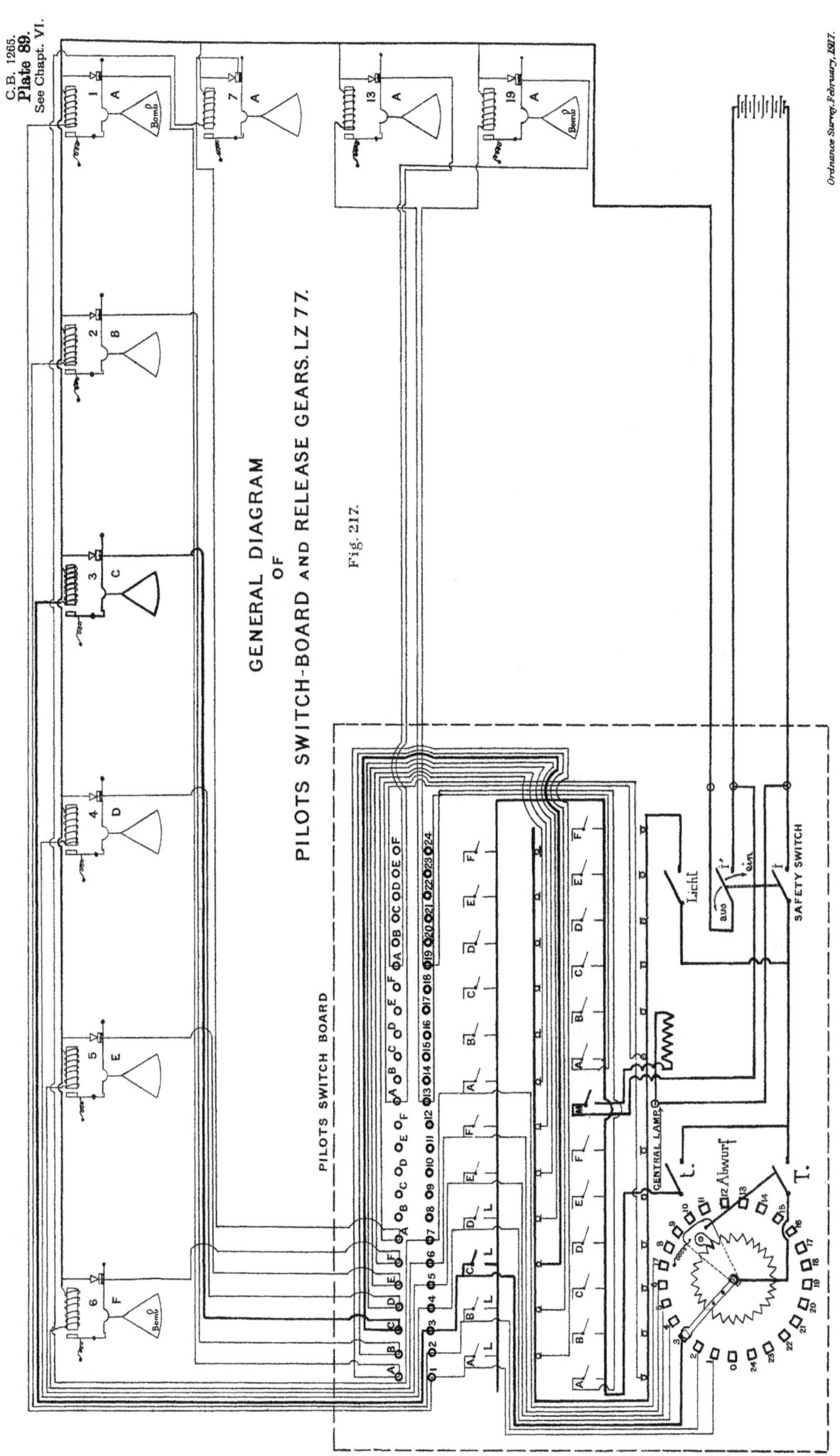

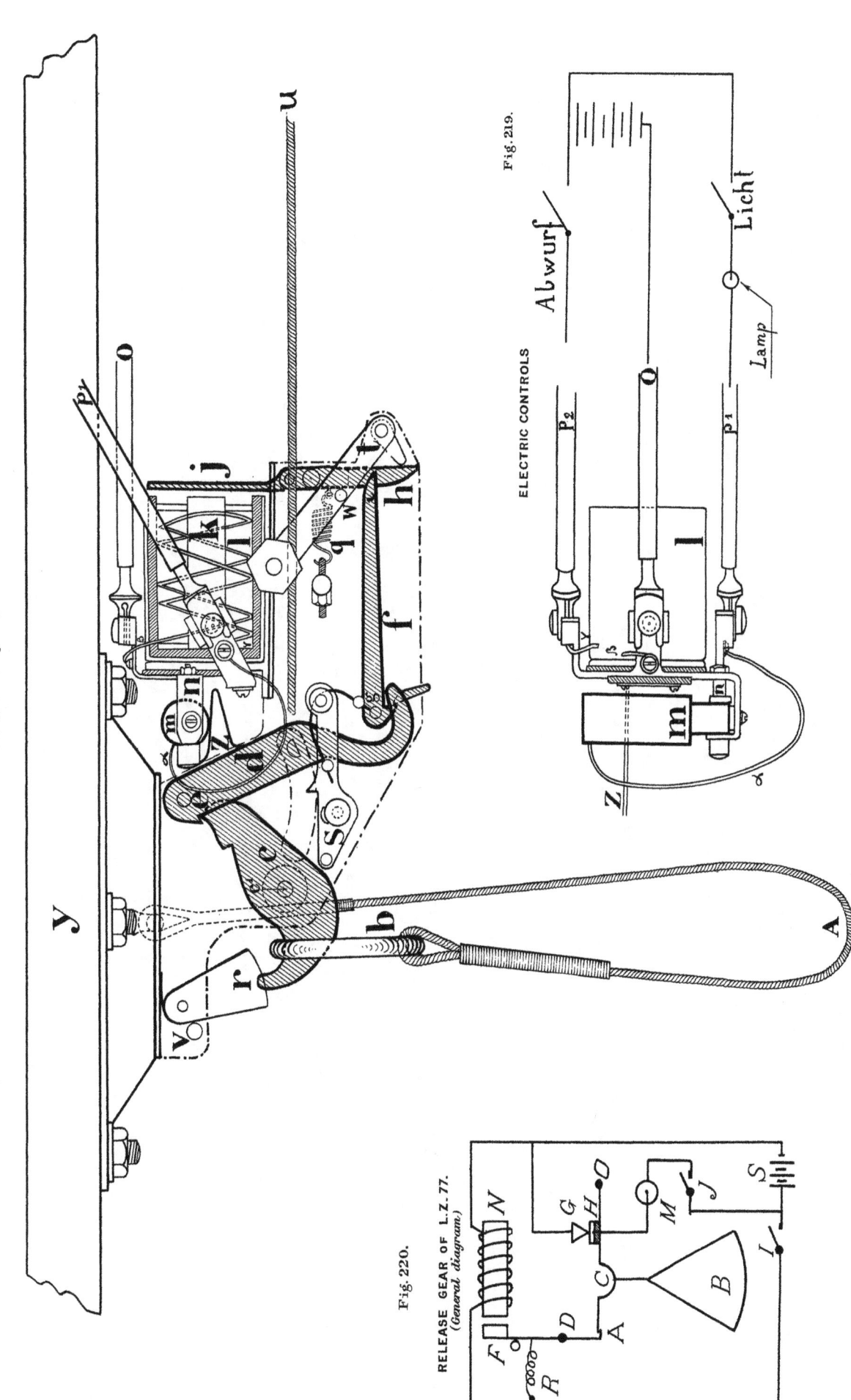

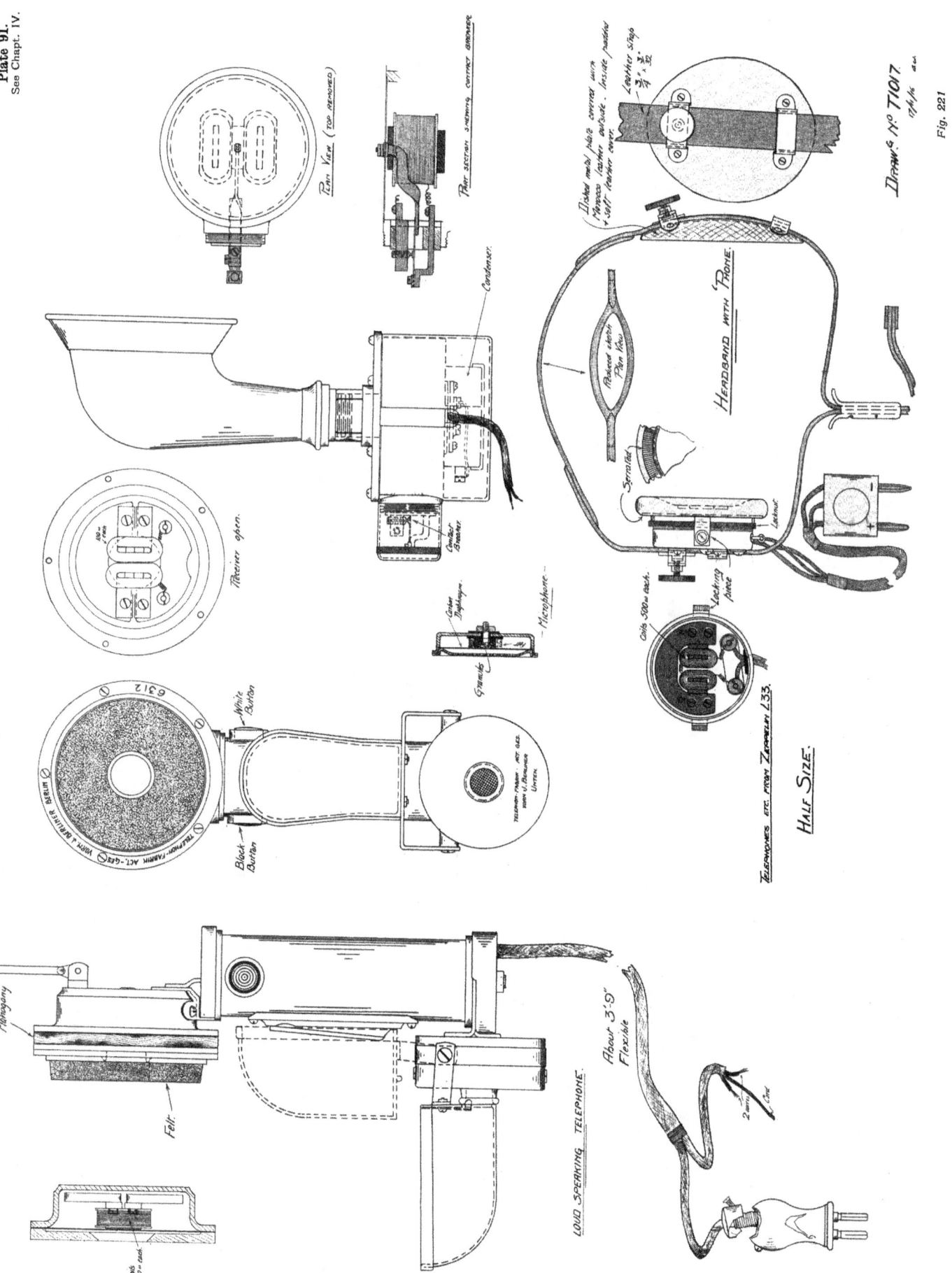

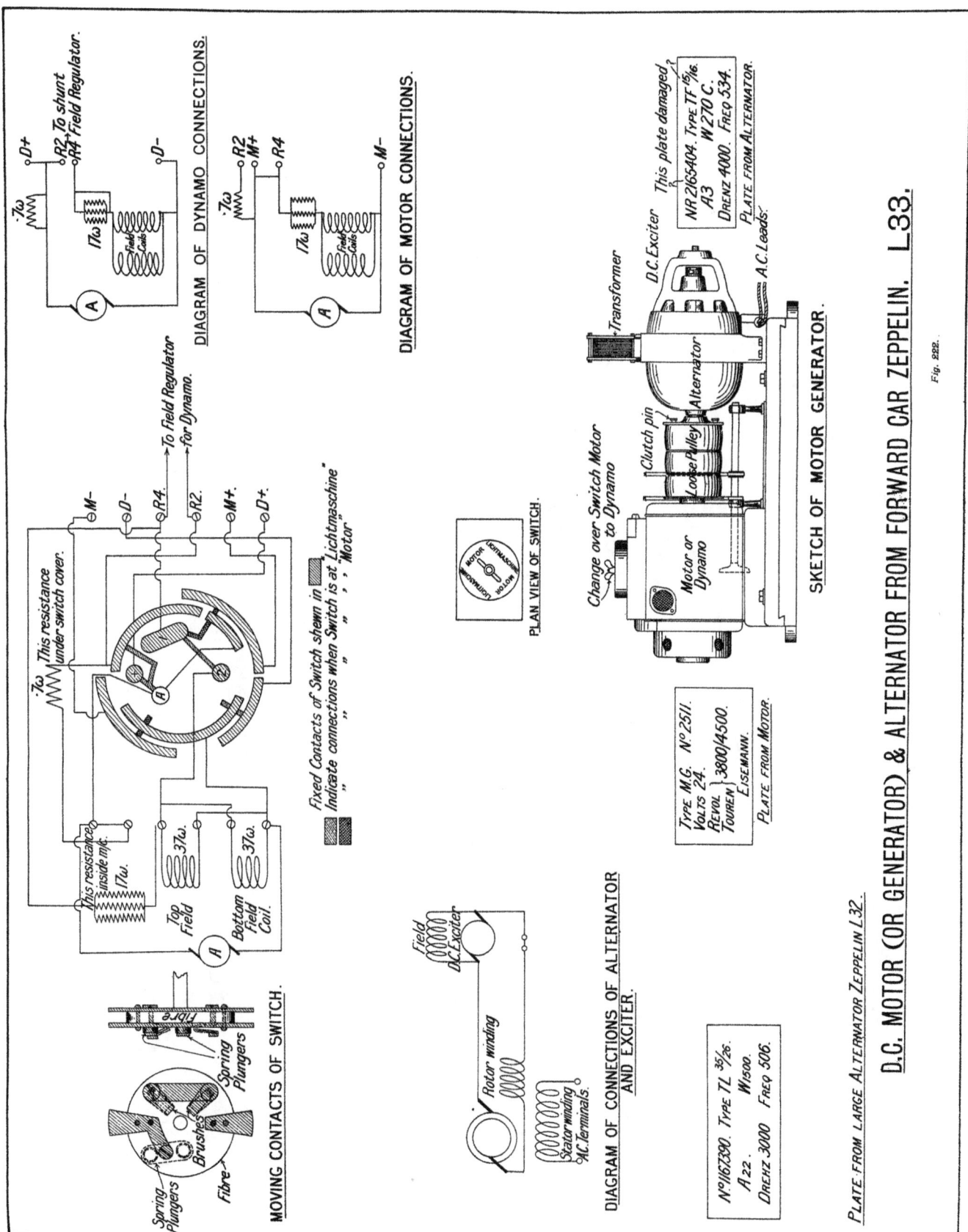

Fig. 222.

D.C. MOTOR (OR GENERATOR) & ALTERNATOR FROM FORWARD CAR ZEPPELIN. L33.

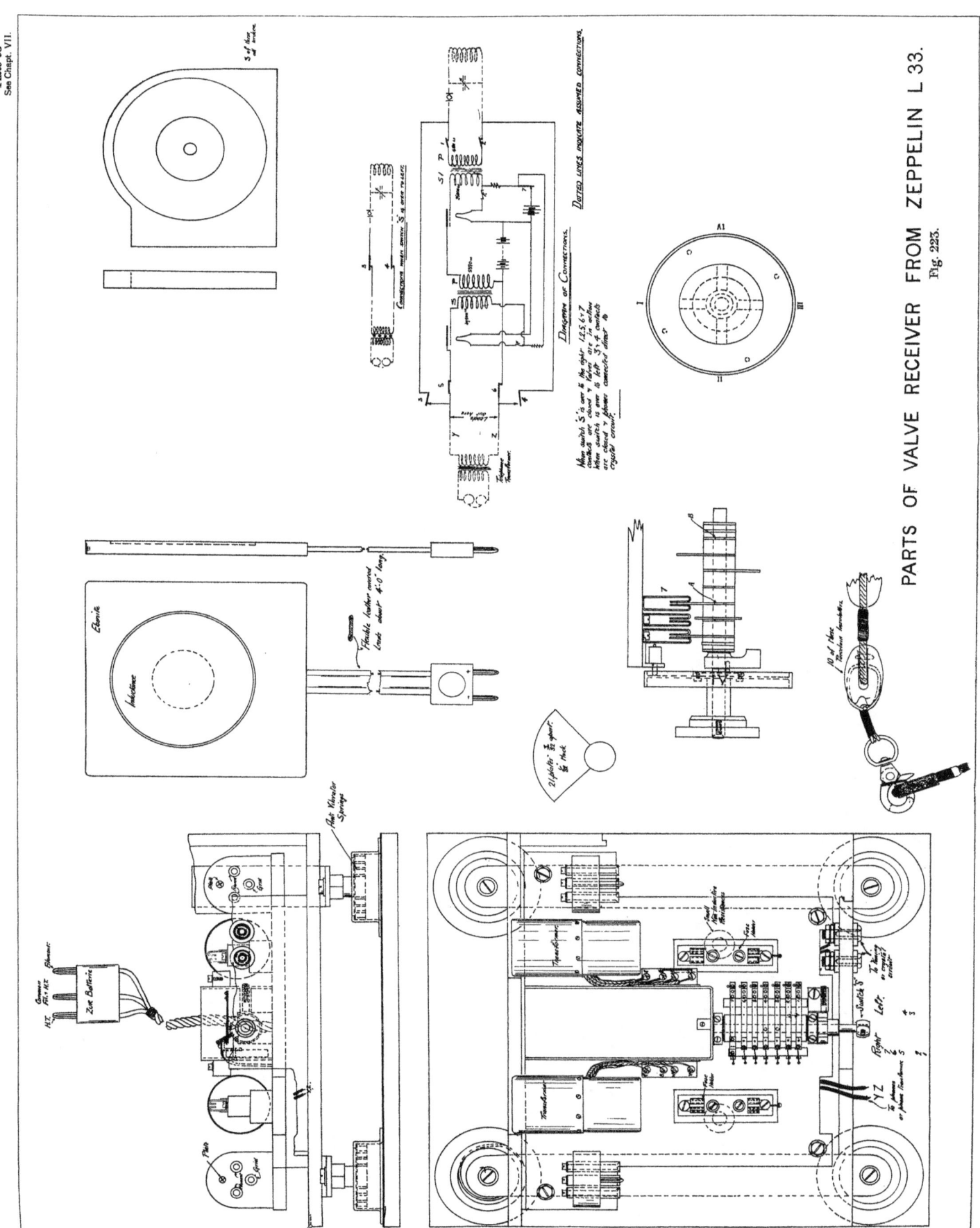

PARTS OF VALVE RECEIVER FROM ZEPPELIN L 33.
Fig. 223.

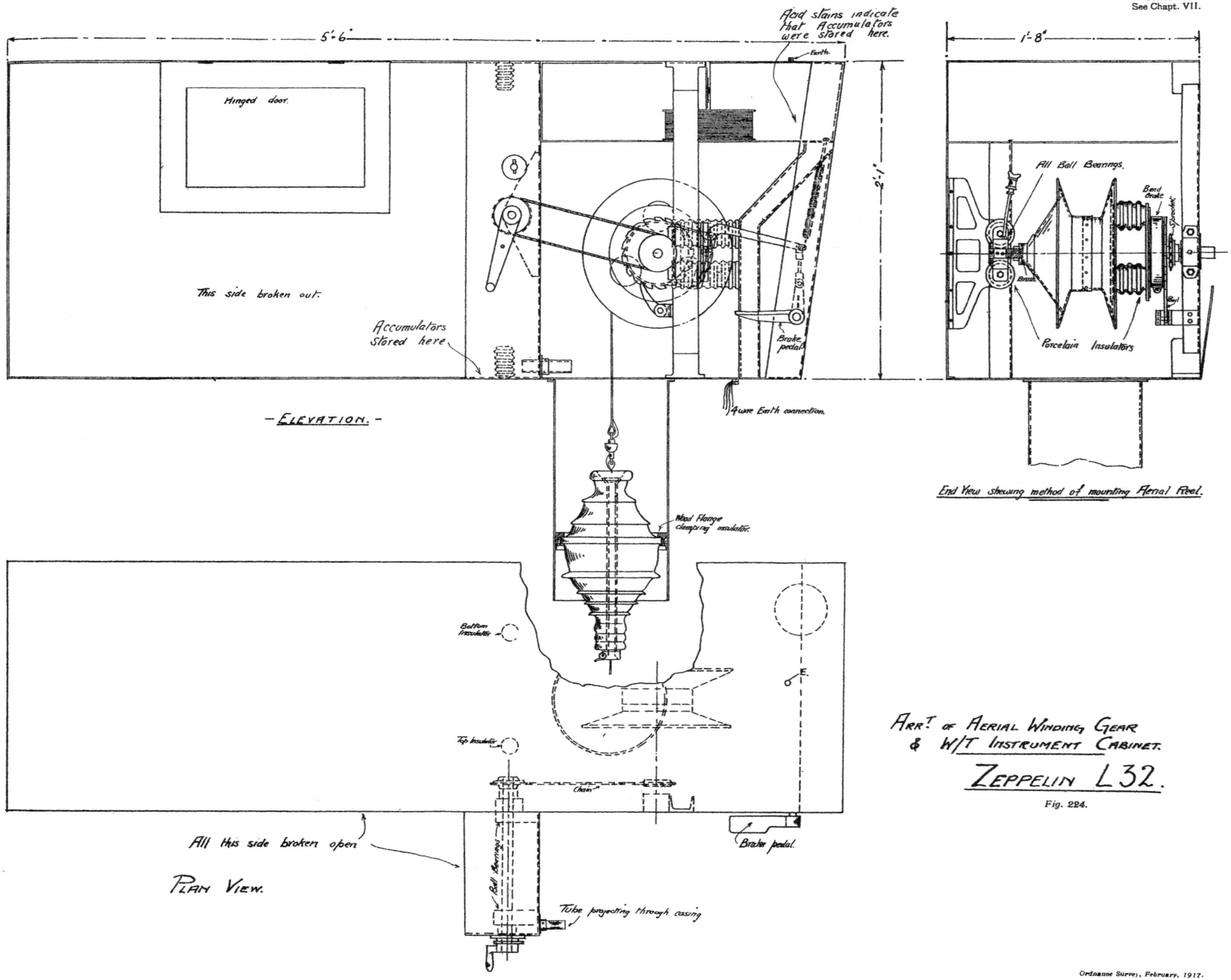


ARR.T OF AERIAL WINDING GEAR & W/T INSTRUMENT CABINET.

ZEPPELIN L32.

Fig. 224.

Ordnance Survey, February, 1917.

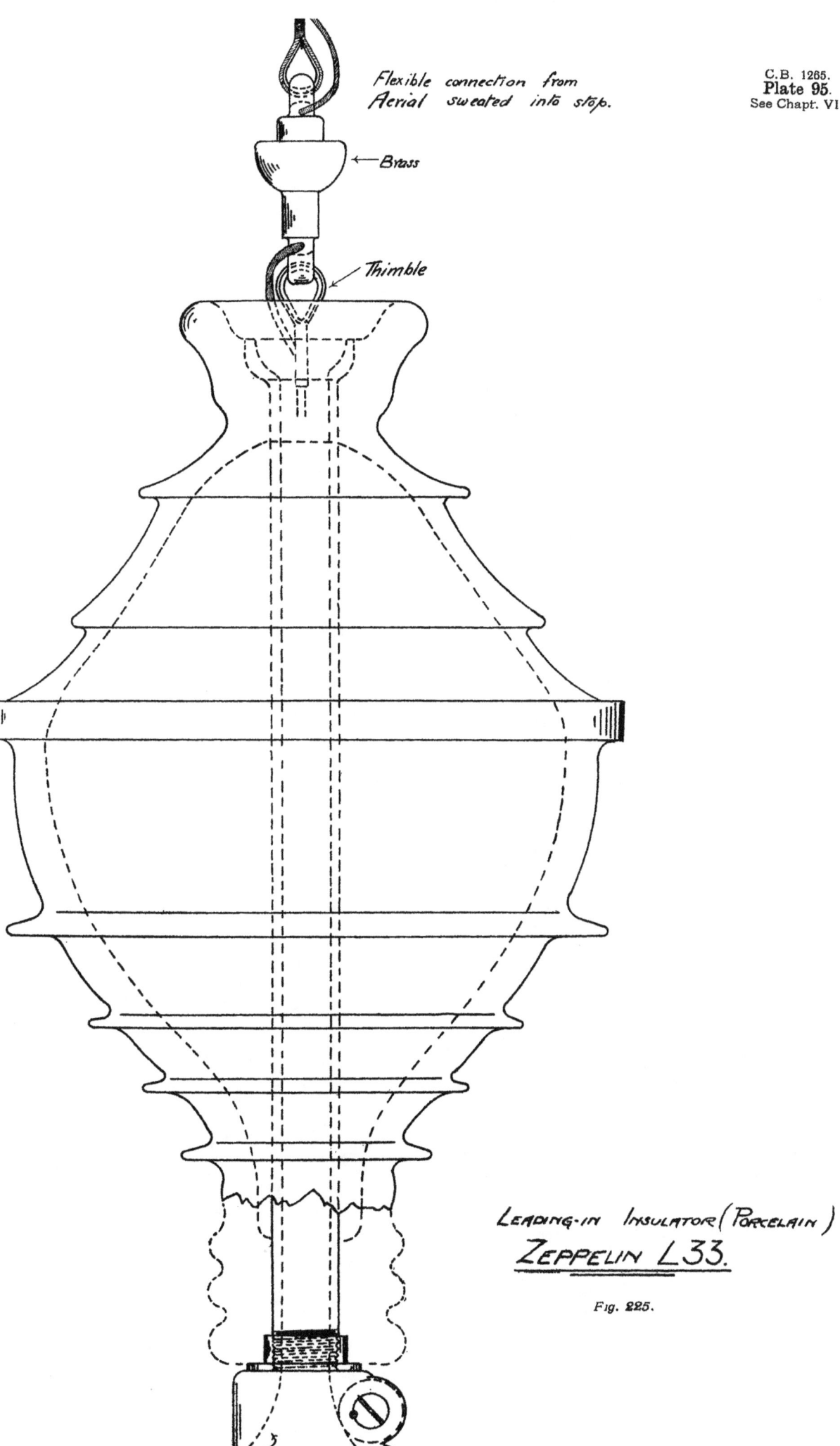

Fig. 225.

Leading-in Insulator (Porcelain) Zeppelin L33.

C.B. 1265.
Plate 96.
See Chapt. VIII.

FIG. 226. SCALE ¼.

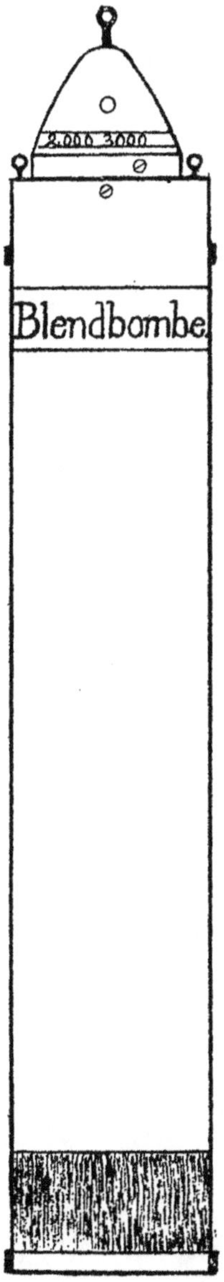

*Illuminating Flare.*

FIG. 227. SCALE ⅛.

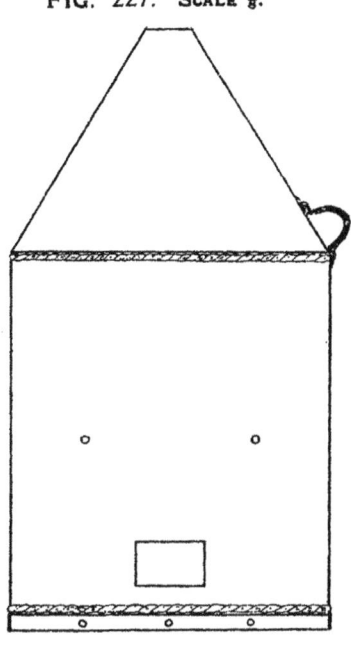

*Water Flare.*

FIG. 229. SCALE ½.

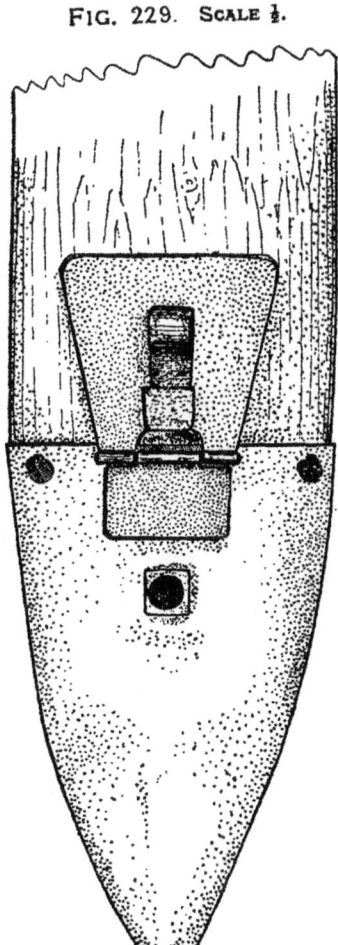

*Detail of Water Flare shown in Fig. 228.*

FIG. 228. SCALE ¼.

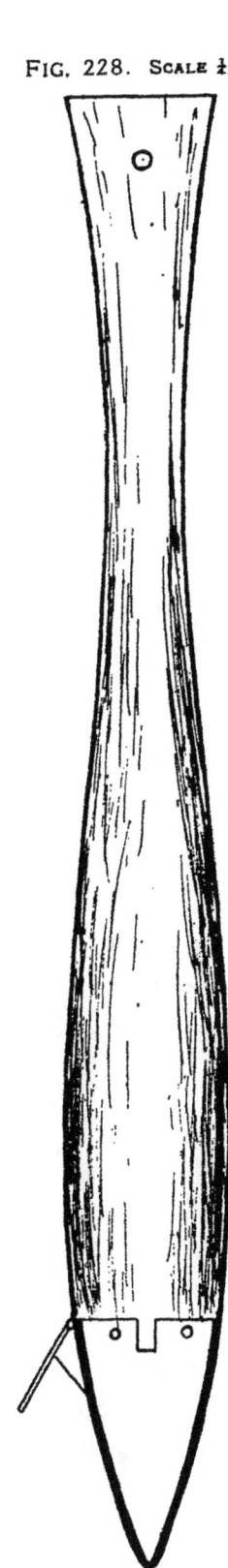

*Water Flare.*

Ordnance Survey, February, 1917.

C.B. 1265.
**Plate 97.**
See Chapt. VIII.

Fig. 232.
Tripod Gun Mounting from top of platform of L 33.

Figs. 230 and 231.
Maxim Nordenfeldt G.A. pattern 8 m.m. bore Machine Gun from L 33.

Ordnance Survey, February, 1917.

C.B. 1265.
**Plate 98.**
See Chapt. IX.

Fig. 234.—Centre position of Sight from L 31.

Fig. 235.—Ground Speed Stop-Watch from LZ 85.

Fig. 233.—Centre position of Sight from L 31 with Curve Drum removed.

*Ordnance Survey, February, 1917.*

C.B. 1265.
**Plate 99.**
See Chapt. IX.

Fig. 236.—Front View of Machine Gun Sight.

Fig. 237.—Rear View of Machine Gun Sight.

*Ordnance Survey, February, 1917.*

C.B. 1265. **Plate 100.**
See Chapt. X.

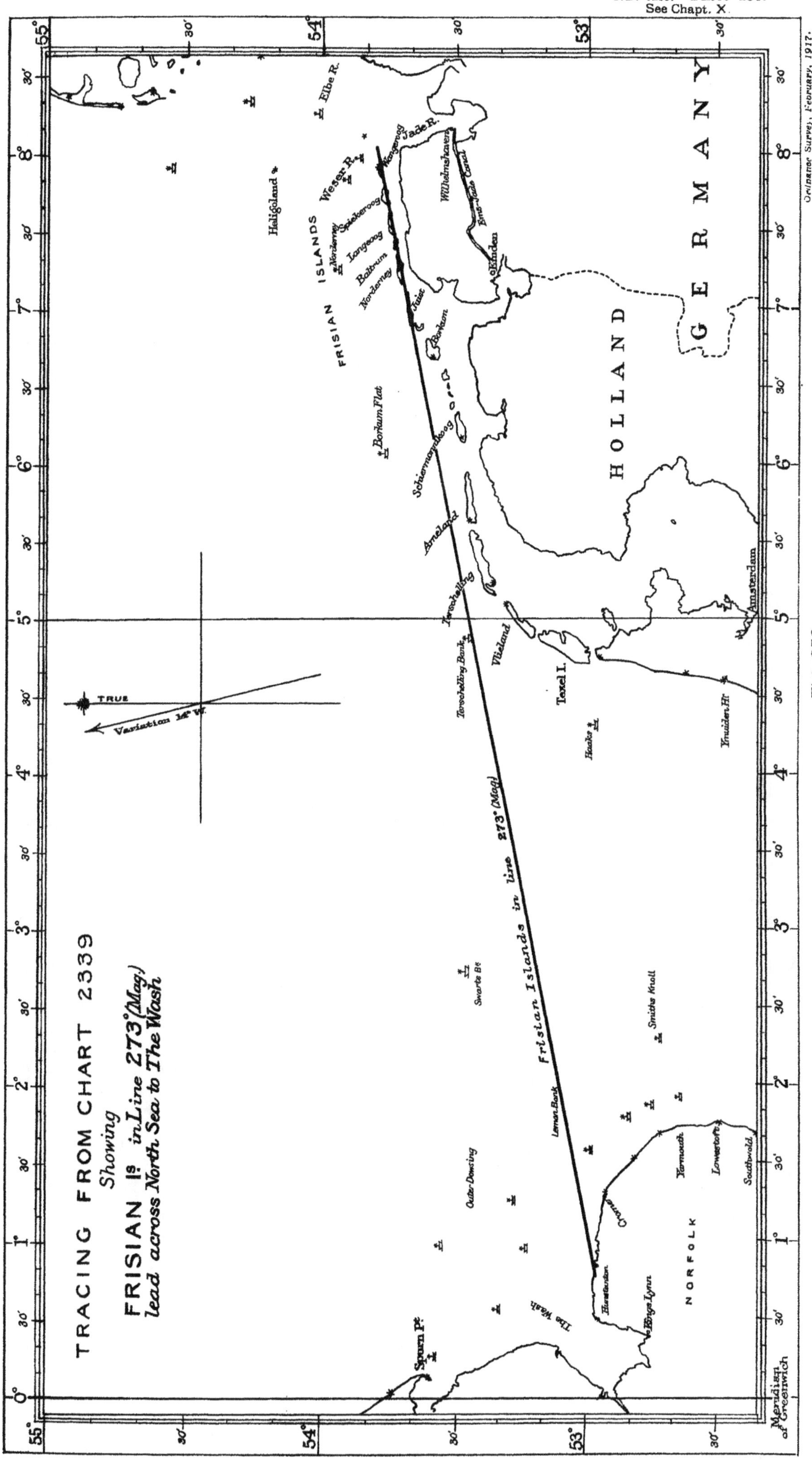

TRACING FROM CHART 2339
*Showing*
**FRISIAN I$^s$ in Line 273°(Mag.)**
*lead across North Sea to The Wash*

Fig. 258.

C.B. 1265.
**Plate 101.**
See Chapt. XI.

Fig. 239.—Aneroid from L 33.

Fig. 240.—Fire Extinguisher from L 15.

Fig. 241.—Statascope from L 32.

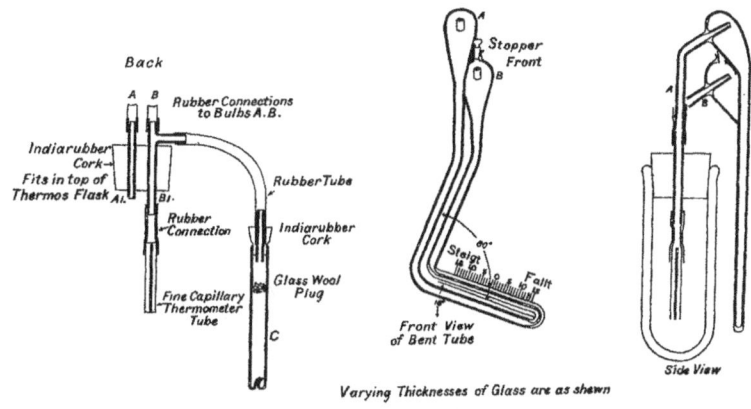

Fig. 242.—Diagram of Details of Statascope from L 15.

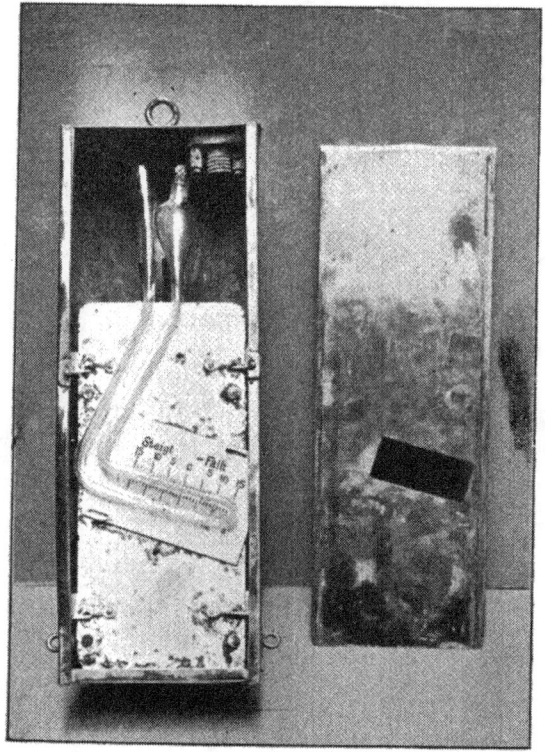

Fig 243.—Statascope from L 15 with cover removed.

Fig. 244.—Statascope from L 15 in metal casing.

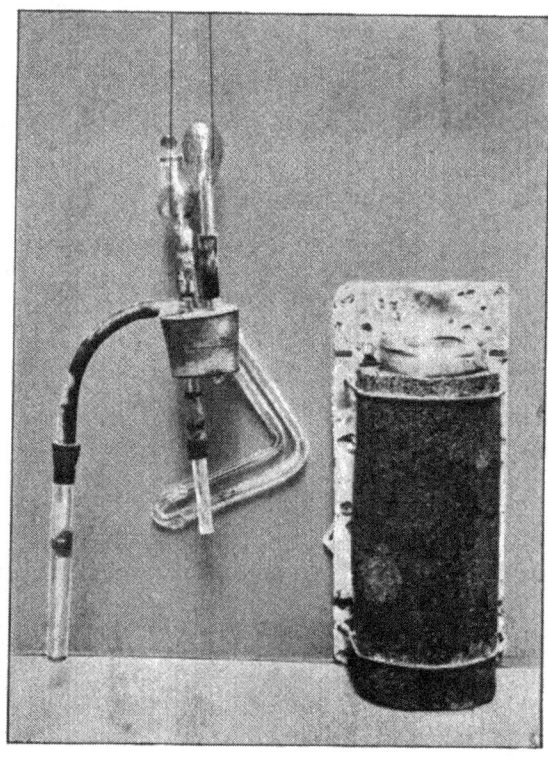

Fig. 245.—Statascope from L 15 with casing removed and parts separated.

*Ordnance Survey, February, 1917.*

C.B. 1265.
**Plate 102.**
See Chapt. XI.

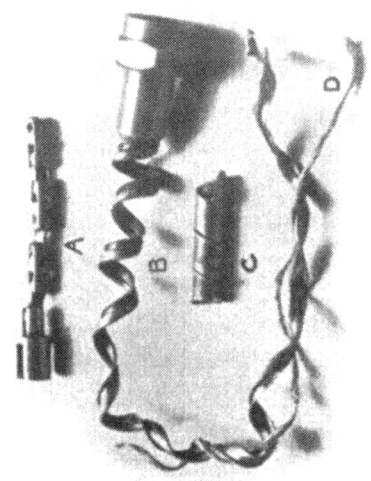

Fig. 248.—Engine Revolution Indicator. Flexible Shaft and Metallic Tube.

Fig. 252.—Petrol Gauge.

Fig. 247.—Engine Revolution Indicator with face removed.

Fig. 251.—Electric Thermometer from L 15.

Fig. 246.—Parts of Engine Revolution Indicator.

Fig. 250.—Lifebelt. Back View.

Fig. 249.—Lifebelt from a Zeppelin. Front View.

*Ordnance Survey, February, 1917.*

# Fig. 253. Diagrammatic Sketch of Drift Indicator.

C.B. 1265.
Plate 103.
See Chapt. XI.

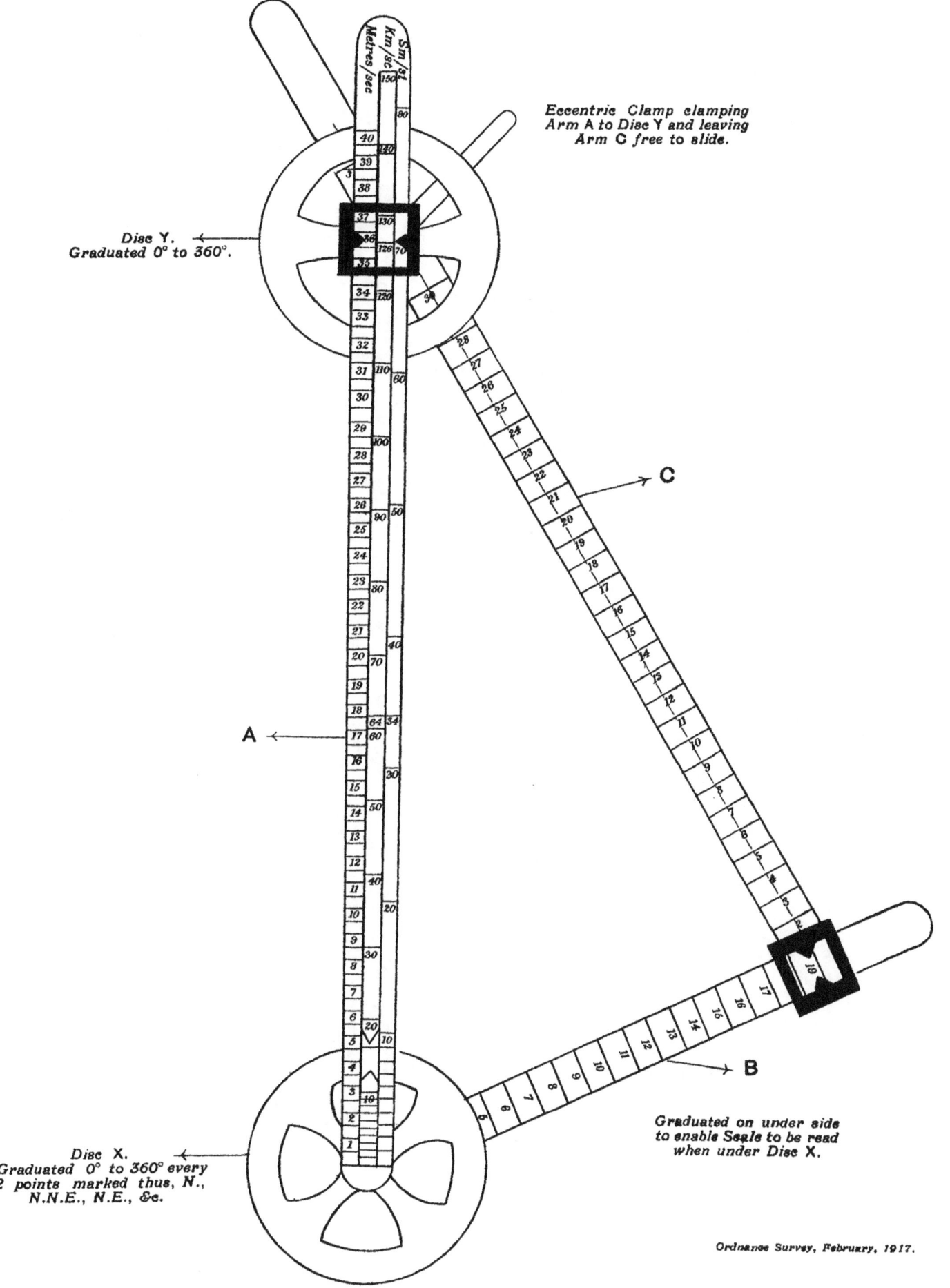

Ordnance Survey, February, 1917.

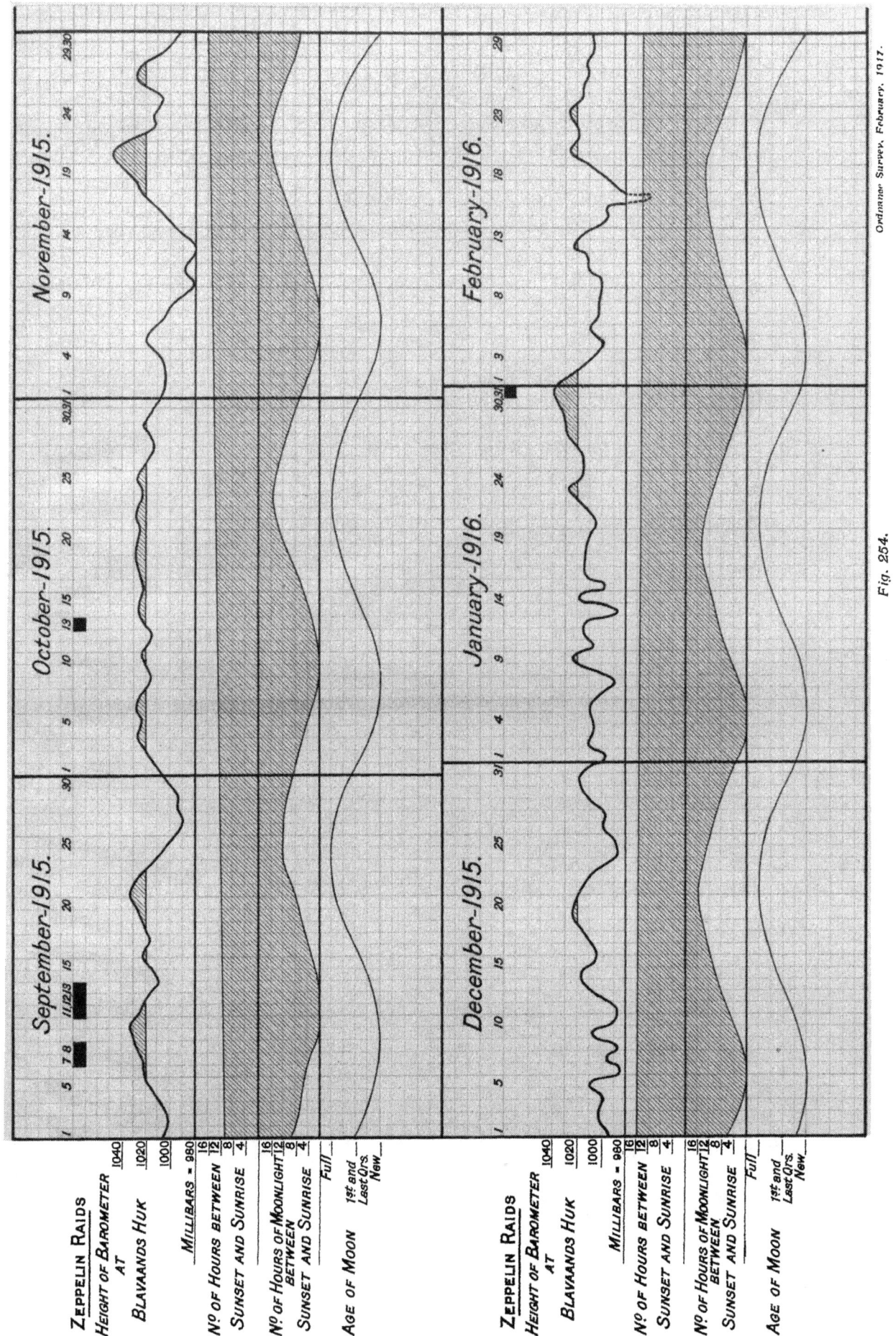

Fig. 254.

C.B. 1265.
Plate 105.
See Chapt. XII.

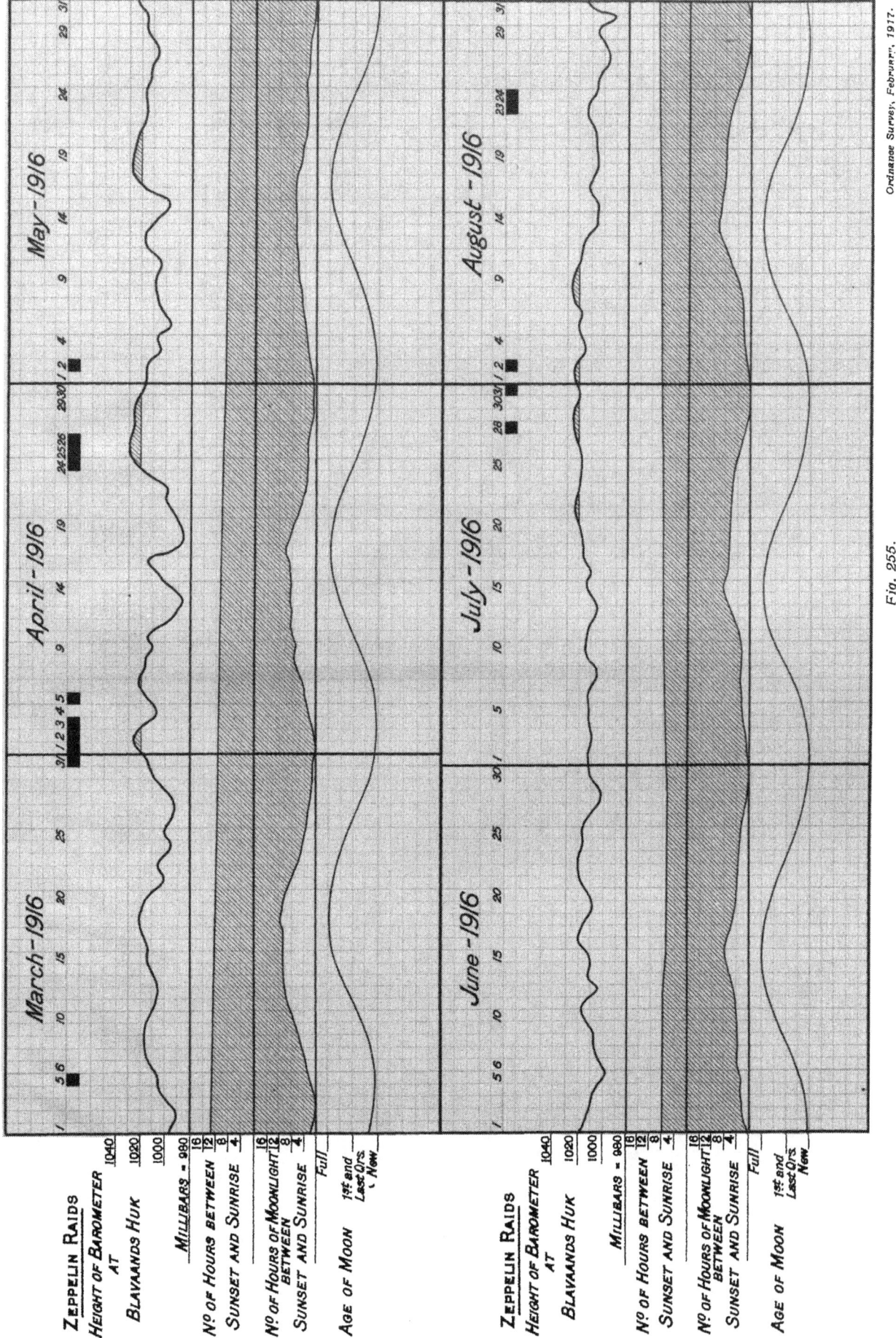

Fig. 255.

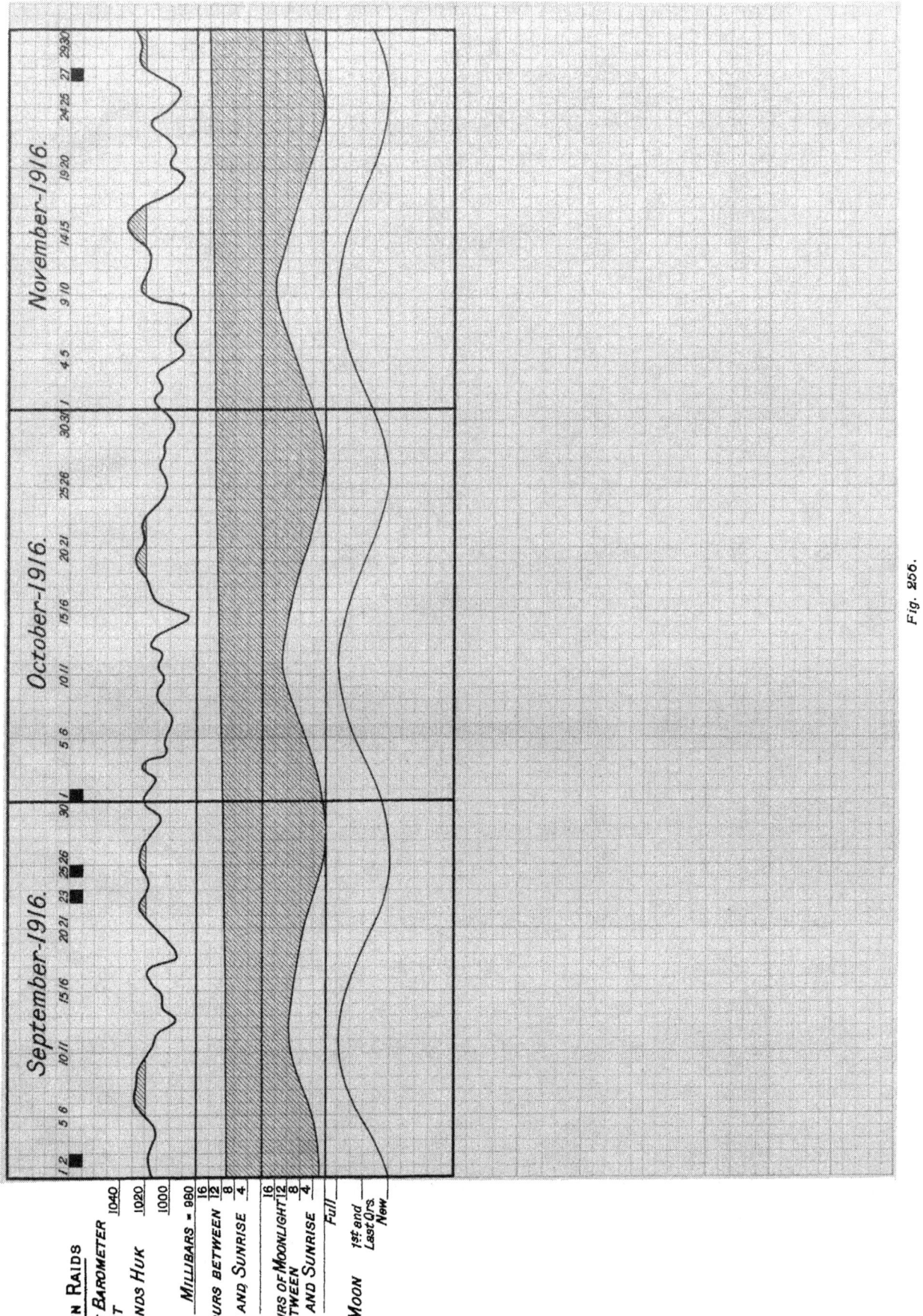

Fig. 256.

www.ingramcontent.com/pod-product-compliance
Lightning Source LLC
Chambersburg PA
CBHW081435300426
44108CB00016BA/2374